Springer Series in Computational Mathematics

50

More information about this series at http://www.springer.com/series/797

Francisco-Javier Sayas

Retarded Potentials and Time Domain Boundary Integral Equations

A Road Map

Springer

Francisco-Javier Sayas
Department of Mathematical Sciences
University of Delaware
Newark, DE, USA

ISSN 0179-3632 ISSN 2198-3712 (electronic)
Springer Series in Computational Mathematics
ISBN 978-3-319-26643-5 ISBN 978-3-319-26645-9 (eBook)
DOI 10.1007/978-3-319-26645-9

Library of Congress Control Number: 2015955727

Mathematics Subject Classification (2010): 65M38, 65M12, 65R20, 65J08, 45A05, 45B05, 35L20, 35D30, 35Q61

Printed on acid-free paper

This Springer imprint is published by Springer Nature
The registered company is Springer International Publishing AG Switzerland

Dedicated to my parents
(Elena y Antonio, mamá y papá)
for their unconditional support
of my crazy American adventure

Preface

In these notes I try to show the way through the relatively difficult theory of retarded layer potentials and integral operators for the acoustic wave equation in two and three dimensions. I will also introduce convolution quadrature techniques for the time discretization of potentials and integral equations, giving the reader a taste of their challenging but exciting theory and their huge and partly unexplored potentialities. Part of the aim of these notes has been to set a clear path to learn the mathematical techniques needed to understand time domain boundary integral equations. This is part of a joint effort with Antonio Laliena (Universidad de Zaragoza, Spain), Lehel Banjai's research group (formerly at the Max-Planck-Institut in Leipzig, Germany, and at the time of finishing these notes at Heriot-Watt University, Edinburgh, UK), and my own group at the University of Delaware. This monograph is intended as a learning tool, and that is why the tone will be somewhat colloquial. Apart from some more narrative sections (those with less mathematical rigor), everything else will be duly divided into paragraphs—in the LaTeX sense of the word—(plus propositions and their proofs) so that at each moment we know where we are. A more computational approach to convolution quadrature for wave propagation can be found in the notes written in collaboration with Matthew Hassell [54].

An informal table of contents. Here is a guide of what each of the ten chapters contains:

1. We start with an informal presentation of the retarded layer potentials. We then derive the corresponding boundary integral calculus based on a few elements (the potentials and a uniqueness theorem) and its use for scattering problems. This chapter is, again, informal.
2. We introduce the basic tools for vector-valued distributions and their Laplace transforms. (Not all the proofs will be given here, but all steps will be duly sketched.) We next give the distributional form of the problem of scattering by an obstacle. We study the Laplace transform of the single layer potential and operator and prove estimates depending on the Laplace transform parameter for all of them.

3. We start this chapter by giving formulas for (strong) inversion of the Laplace transform and the differentiation theorem and then use them to delimit a precise class of symbols (Laplace transforms) and their time domain distributional counterparts. Convolution operators with this class of distributions are the setting for the remainder of the theory. We show how layer potentials, boundary integral operators, and their inverses (whenever they exist) are in this class. A rigorous proof of Kirchhoff's formula (the integral representation theorem for causal acoustic waves) gives the necessary justification for the Calderón-type calculus we had introduced in the first chapter. We finally have a look at how causality, finite speed of propagation, and some kind of coercivity are hidden in the Laplace transform of the potentials and operators.
4. Of the two classes of convolution quadrature methods, we introduce here the one that is based on multistep methods. We present an almost finished portrait of the theory of these methods applied to the class of convolution operators and equations that was introduced in the previous chapter and detail the kind of results that are derived in the case of scattering by a sound-soft obstacle.
5. We go back to the single layer retarded potential and go as far as we can with the Laplace domain techniques to prove estimates for the full discretization (Galerkin in space, convolution quadrature in time) of the model equation that solves the scattering problem by a sound-soft obstacle using an indirect formulation. Once we have finished with the single layer operator, we will repeat the process for the double layer potential representation of the scattering by sound-hard obstacles.
6. This chapter is a simplified introduction to a class of abstract differential equations of the second order in Hilbert spaces. The hypotheses are much reduced with respect to what is common in the Hille-Yosida theory, but they will be those that we will meet later on. All the results of this chapter are proved in Appendix B using quite rudimentary arguments of separation of space-time variables, arguments which are related to the discrete spectrum of a given unbounded operator.
7. The techniques of Chapter 6 are now used to prove again all the estimates for the single layer retarded potential and operator (as well as general Galerkin semidiscretization-in-space for the associated equation) using time domain tools. We will develop a streamlined way of proving the time domain results, by working on a cut-off domain and identifying the resulting solution with the beginning of the evolution of the potential solution.
8. We now repeat all the time domain arguments on the double layer potential and its use for an indirect formulation of the scattering problem by a sound-hard obstacle. As the reader will easily realize when reading this chapter, the arguments end up being very similar in each particular situation, and we will only have to take care of whatever is different in each concrete problem.
9. We next mix the time domain theory with convolution quadrature and, case by case, prove new estimates for the fully discrete methods for one model problem. This chapter shows how classical techniques for the numerical analysis of low-order time discretizations can be easily extended to the much more complicated situation we are dealing with.

10. In this chapter we collect the ideas of Chapters 7 and 8, looking for common patterns that allow us to easily guess what kind of bounds we will obtain in new situations. We also show easy (not to say straightforward) extensions to screen problems and to linear elasticity.

There are five appendices:

- Appendix A contains some Laplace domain arguments for the Maxwell transient single layer potential
- Appendix B contains the proofs of the results on evolution equations that were presented in Chapter 6.
- Appendix C presents some algorithms for the implementation of convolution quadrature.
- Appendix D contains a precise but very terse introduction of the Sobolev space background material needed for this monograph.
- Appendix E shows some numerical illustrations of two-dimensional scattering problems.

Although the theory of time domain boundary integral equations is far from finished (as its full potential in applications is only partially exploited), let me drop here some names of some of the originators of the current excitement in the area. First of all, the focus of these notes lies in the realm of integral equations for wave propagation, although boundary integral equations are also used for parabolic problems. While over ten years old, Costabel's encyclopedia article [33] contains an excellent introduction to the use of integral equations for evolutionary partial differential equations.

What follows is a highly non-exhaustive list, so please, nobody take offense if their name does not appear here.

- The theory of time domain boundary integral equations (at least, the theory that we numerical analysts use) stems from two papers by Alain Bamberger and Tuong Ha-Duong [9, 10] in 1986. Many other papers were published [7, 23, 49], and even more theses were written (unfortunately much of this material was left unpublished and is now very difficult to locate) in the buoyant French numerical analysis school. The names of Jean-Claude Nédélec (at the Polytechnique) and Alain Bachelot (at the University of Bordeaux) are attached to quite a lot of these doctoral dissertations. Much of this is reported and referenced in the survey paper [50]. Touffic Abboud and Isabelle Terrasse can be held responsible for the practical development of these methods, evolving in their (to the best of my knowledge) only commercial implementation. Interest in research aspects of this approach seems to be back: Abboud and Terrasse together with Patrick Joly (INRIA Rocquencourt, France) and Jerónimo Rodríguez (Santiago de Compostela, Spain) have recently developed one of the few sets of integral transparent boundary conditions [1].
- Convolution quadrature originated as a completely independent tool for approximation of convolutions. It came to age very much at the same time as time domain integral equations, with two articles by Christian Lubich [64, 65] in

1988. This technique was first devised for problems with parabolic structure (exemplified in the operators having Laplace transforms defined on a sector instead of a half-plane). A second family of convolution quadrature methods, based on Runge-Kutta methods, originated in the joint work of Lubich with Alexander Ostermann [67]. Almost at the same time, Lubich applied his ideas to problems with hyperbolic structure, including the single layer potential for the three-dimensional equation [66]. This was only natural, since CQ is based in Laplace transform methods and the theory of Bamberger and Ha-Duong is based on exactly the same principle. The theory of CQ based on RK schemes applied to hyperbolic problems was left unfinished and was only recently completed by Lubich in collaboration with Lehel Banjai and Jens Markus Melenk [18, 19].

- Not being as popular as their frequency domain cousins, time domain boundary integral equations have known a rich development in the engineering community. Galerkin methods (the original ones developed in the French school) are a particular case of marching-on-in-time (MoT) methods for time domain integral equations. The group of Eric Michielssen at the University of Michigan and a large array of researchers in European universities (with a strong group in Ghent) have developed applications to electromagnetism and searched the limits of the known world in computational time domain integral equations [6, 24, 32, 72, 74, 86, 87].
- The convolution quadrature point of view was initially not very well tended by the mathematical boundary integral community, but there was a strong development in the field of applications to elastodynamics, much of it led by the group of Martin Schanz [21, 57, 63, 69, 75, 76] at the Graz University of Technology (Austria). An early account of this development can be found in the monograph [82]. A more recent survey can be found in [22]. Applications to electromagnetism have been developed by the group of Daniel Weile and Peter Monk at the University of Delaware [28, 30, 31, 62, 89], while researchers in all corners of the world have been developing CQ-BEM [29, 46–48].
- Some papers by Stefan Sauter (University of Zurich) with different collaborators [51, 52] re-sparked the interest of numerical analysts in time domain integral equations, specially with a focus on convolution quadrature techniques. Galerkin methods with smooth basis functions have also attracted the interest of the Zurich group [79, 88]. Lehel Banjai and his group are making rapid progress in this direction [8, 11, 12, 14–17, 20]. Myself, working with my then graduate student Antonio Laliena, proved that the Laplace domain contained much more information than we had expected and that convolution quadrature techniques combined perfectly with space Galerkin discretization in many nontrivial situations of scattering of acoustic and elastic waves with penetrable obstacles, including nonhomogeneous obstacles where numerical modeling is carried out with the finite element method [60]. We were happy to find a quite general (and I want to say innovative) approach that has since been applied to electromagnetism or more complicated elastodynamic problems. Contributions of my group in the area of CQ appear in recent papers [13, 20, 43, 56, 73, 81].

- Penny Davies and Dougald Duncan in the UK have explored alternatives for MoT schemes with different shapes of basis functions or collocation in time [36–40]. The full Galerkin approach is also being jointly developed by the groups of Ernst Stephan at the University of Hanover (Germany) and Matthias Maischak at Brunel University (England) with a current focus on acoustics. Several researchers in Italy (among them, Alessandra Aimi and Mauro Diligenti at Parma) are also readdressing the full Galerkin method for the acoustic equations [2–5].

Acknowledgments The first draft of these notes (covering roughly Chapters 1 to 5) was written as support for a five-hour lecture series *Retarded boundary integral equations and applications*, delivered in the context of the closure workshop of the special semester on *Theoretical and numerical aspects of inverse problems and scattering theory*. The workshop took place in La Coruña, Spain, on July 4–8, 2011. The organizers of the event are warmly thanked for having thought of me for this occasion. The current version has been expanded using recent results on time domain analysis.

As already mentioned, a great deal of what appears in this document is the result of continuous collaboration with Antonio Laliena, Víctor Domínguez, and Lehel Banjai. Some of my current students (Matthew Hassell, Tonatiuh Sánchez-Vizuet, and Tianyu Qiu) have had a careful look at several chapters of these notes, working out all the exercises and checking proofs. Tainyu Qiu helped me with the appendix on the Maxwell equations. Another one of my students, Allan Hungria, has done an excellent job in proofreading and creating pristine LaTeX TikZ figures. My current research is partially supported by the NSF (DMS 1216356). Three different meetings at the Oberwolfach Mathematical Institute, and in particular the talks I gave there, helped me in the search of a systematic approach to the development of this theory. I am deeply grateful to the organizers of those workshops (Martin Costabel and Ernst Stephan in the first one; Ralf Hiptmair, Roland Hoppe, Patrick Joly, and Ulrich Langer for the last two) for giving me the chance to enjoy the wonders of working, thinking, and discussing in the middle of the Black Forest.

Newark, DE, USA
June 2015

Francisco-Javier Sayas

Contents

Chapter 1
The retarded layer potentials

In this chapter we are going to introduce the basic concepts of time domain acoustic layer potentials and how they can be used to represent the solutions of scattering problems. *All notions introduced in this chapter will be given at an intuitive level and with basically no formalization. The reader will find a precise sketch of the theory in the next chapters.* For the sake of clarity, let me remark here:

- Sections 1.1 through 1.3 deal with three-dimensional waves.
- Just by looking at the mathematical expressions therein, it will be clear that Sections 1.4 through 1.6 are dimension-independent.
- Section 1.7 revisits the particular three-dimensional case, using the specific formulas for the Huygens' single layer potential.
- Finally, Section 1.8 will present formulas for the two-dimensional case.

1.1 Acoustic sources and dipoles

Let us start this chapter by having a look at a spherical wave. We consider a function (a signal) $\lambda : \mathbb{R} \to \mathbb{R}$ such that $\lambda(t) = 0$ for all $t < 0$. A function of the time variable that vanishes for $t < 0$ will be always referred to as a **causal function**. We now choose $\mathbf{x}_0 \in \mathbb{R}^3$ and consider the function

$$u(\mathbf{x}, t) := \frac{\lambda(t - c^{-1}|\mathbf{x} - \mathbf{x}_0|)}{4\pi|\mathbf{x} - \mathbf{x}_0|}. \tag{1.1}$$

A more or less boring computation shows that

$$c^{-2}\frac{\partial^2 u}{\partial t^2} = \Delta u \qquad \forall \mathbf{x} \neq \mathbf{x}_0 \quad \forall t > 0,$$

F.-J. Sayas, *Retarded Potentials and Time Domain Boundary Integral Equations*, Springer Series in Computational Mathematics 50,
DOI 10.1007/978-3-319-26645-9_1

as long as $\lambda \in \mathcal{C}^2(\mathbb{R})$, where the Laplace operator Δ is taken in the space variables. (The result is actually true for less smooth λ, but we are not going to worry about regularity at this point.) It is interesting to notice the following facts.

- The function u moves on spherical surfaces. Actually,

$$u(\mathbf{x},t) = \frac{\lambda(t-c^{-1}r)}{4\pi r} \qquad |\mathbf{x}-\mathbf{x}_0| = r. \tag{1.2}$$

  This shows that points on a sphere centered at $\mathbf{x}_0$ perceive the same solution at the same time.
- The previous formula shows also that for a point at distance r of the point source, we need to wait $c^{-1}r$ time units to start perceiving any signal. Apart from this delay, the entire signal is received at speed c (and with a damping factor $4\pi r$). The signal goes through, exactly as emitted.

There are other kinds of solutions of the wave equation that can be understood as traveling on spherical surfaces. If u is a sufficiently smooth solution of the wave equation, so are the three components of ∇u and therefore, so is $\nabla u \cdot \mathbf{n}_0$, where $\mathbf{n}_0$ is a fixed vector. With this idea, and starting in (1.1), we can create a new family of solutions to the wave equation:

$$\begin{aligned} u(\mathbf{x},t) &:= \nabla_{\mathbf{x}_0}\left(\frac{\varphi(t-c^{-1}|\mathbf{x}-\mathbf{x}_0|)}{4\pi|\mathbf{x}-\mathbf{x}_0|}\right)\cdot\mathbf{n}_0 \qquad (1.3)\\ &= -\nabla_{\mathbf{x}}\left(\frac{\varphi(t-c^{-1}|\mathbf{x}-\mathbf{x}_0|)}{4\pi|\mathbf{x}-\mathbf{x}_0|}\right)\cdot\mathbf{n}_0\\ &= \varphi(t-c^{-1}|\mathbf{x}-\mathbf{x}_0|)\frac{(\mathbf{x}-\mathbf{x}_0)\cdot\mathbf{n}_0}{4\pi|\mathbf{x}-\mathbf{x}_0|^3} + c^{-1}\dot\varphi(t-c^{-1}|\mathbf{x}-\mathbf{x}_0|)\frac{(\mathbf{x}-\mathbf{x}_0)\cdot\mathbf{n}_0}{4\pi|\mathbf{x}-\mathbf{x}_0|^2}.\end{aligned}$$

We will assume that $\mathbf{n}_0$ is a unit vector. This formula bears some similitude with (1.1). For instance, at time t, al points on the surface $|x - x_0| = r$ receive information from the signal φ emitted at time $t - c^{-1}r$. However, the points on the surface do not only observe the value $\varphi(t-c^{-1}r)$ but also its trend $\dot\varphi(t-c^{-1}r)$. The main difference between the wave (1.3) and the spherical wave (1.1) is directionality. While points seeing the source in the direction $\mathbf{n}_0$ get to perceive the signal, all points such that $(\mathbf{x}-\mathbf{x}_0)\cdot\mathbf{n}_0 = 0$ are in a deaf spot and miss the entire signal. Actually, if the angle between $\mathbf{x}-\mathbf{x}_0$ and $\mathbf{n}_0$ is θ, then

$$u(\mathbf{x},t) = \frac{1}{4\pi r}\left(\frac{\varphi(t-c^{-1}r)}{r} + \frac{\dot\varphi(t-c^{-1}r)}{c}\right)\cos\theta \tag{1.4}$$

The points $\mathbf{x}_0 \pm r\mathbf{n}_0$ (respective North and South pole of the sphere with axis $\mathbf{n}_0$) get the signal with the same amount of attenuation, but mirrored. The reader is encouraged to check the physical dimensions of all the elements in formulas (1.2) and (1.4) to recognize that the respective transmitted signals (λ and φ) have different dimensions. (We can understand the dimensional mismatch by noticing the differentiation in the space variables in (1.3)–(1.4), which seems to require some kind of compensation.)

Another way of motivating the directional spherical wave (1.3) uses the physical idea of dipole. Take two source points

$$\mathbf{x}_0 \pm \tfrac{h}{2}\mathbf{n}_0,$$

separated a distance h in the direction $\mathbf{n}_0$. The upper point $\mathbf{x}_0 + \frac{h}{2}\mathbf{n}_0$ emits a signal $h^{-1}\varphi$ and simultaneously the point $\mathbf{x}_0 - \frac{h}{2}\mathbf{n}_0$ emits the signal $-h^{-1}\varphi$. The receiver gets to hear the signal

$$\frac{1}{h}\left(\frac{\varphi(t - c^{-1}|\mathbf{x} - \mathbf{x}_0 - \frac{h}{2}\mathbf{n}_0|)}{4\pi|\mathbf{x} - \mathbf{x}_0 - \frac{h}{2}\mathbf{n}_0|} - \frac{\varphi(t - c^{-1}|\mathbf{x} - \mathbf{x}_0 + \frac{h}{2}\mathbf{n}_0|)}{4\pi|\mathbf{x} - \mathbf{x}_0 + \frac{h}{2}\mathbf{n}_0|}\right)$$

which in the limit $h \to 0$ turns into (1.3).

1.2 Acoustic layer potentials

The single layer potential can be understood as the (continuous) superposition of spherical waves (1.1) being emitted from points on a surface Γ:

$$(\mathcal{S} * \lambda)(\mathbf{x}, t) := \int_\Gamma \frac{\lambda(\mathbf{y}, t - c^{-1}|\mathbf{x} - \mathbf{y}|)}{4\pi|\mathbf{x} - \mathbf{y}|} \mathrm{d}\Gamma(\mathbf{y}). \tag{1.5}$$

The causal signal $\lambda(t)$ has been substituted by a density distribution of causal signals $\lambda(\mathbf{y}, t)$, i.e., $\lambda : \Gamma \times \mathbb{R} \to \mathbb{R}$ such that $\lambda(\,\cdot\,, t) \equiv 0$ for $t < 0$. The convolution sign in the notation of this is purely formal for the time being. This also applies to the symbol for the double layer potential that we will define shortly.

The reader who meets this kind of potential expression for the first time is encouraged to have a close look at the relatively bad aspect that it has: there is integration in the space variable $\mathbf{y}$ that somehow got its way into the time variable (through the delay). A particular set of densities is the addition of tensor products of functions of space and time

$$\lambda(\mathbf{y}, t) = \sum_{j=1}^{N} \Phi_j(\mathbf{y})\lambda_j(t),$$

producing simpler propagated signals

$$\sum_{j=1}^{N} \int_\Gamma \frac{\Phi_j(\mathbf{y})\lambda_j(t - c^{-1}|\mathbf{x} - \mathbf{y}|)}{4\pi|\mathbf{x} - \mathbf{y}|} \mathrm{d}\Gamma(\mathbf{y}).$$

Simplifying even more, we can assume that the surface Γ has been subdivided into N panels $\{\Gamma_1, \ldots, \Gamma_N\}$ and Φ_j is just the characteristic function of the panel Γ_j. This is how the single layer potential looks like now:

$$\sum_{j=1}^{N} \int_{\Gamma_j} \frac{\lambda_j(t - c^{-1}|\mathbf{x} - \mathbf{y}|)}{4\pi |\mathbf{x} - \mathbf{y}|} \mathrm{d}\Gamma(\mathbf{y}).$$

In any of the above expressions, it is easy to check that if a point is at a distance r from Γ, it will take $T = c^{-1} r$ time units for the signal to reach the point. Apart from very simple configurations, different points $\mathbf{x}$ will perceive different outputs, since the balance of distances $|\mathbf{x} - \mathbf{y}|$ with the spacial distribution of the density is going to differ depending on the point of view.

Another class of signals we can plug into the potential expression are time-harmonic signals. A noncausal time harmonic signal emitted from Γ would be

$$\mathrm{Re}\,(\lambda(\mathbf{y}) e^{-\imath\omega t}) \qquad \lambda : \Gamma \to \mathbb{C},$$

which is heard as a time harmonic signal

$$\mathrm{Re}\Big(e^{-\imath\omega t} \underbrace{\int_\Gamma \frac{e^{\imath\omega c^{-1}|\mathbf{x}-\mathbf{y}|}}{4\pi |\mathbf{x} - \mathbf{y}|} \lambda(\mathbf{y}) \mathrm{d}\Gamma(\mathbf{y})} \Big).$$

The underbraced expression can be recognized as a single layer potential associated with the Helmholtz equation $\Delta + k^2$, ($k = \omega/c$ is the wave number) which is the equation satisfied by the spatial part of a time harmonic solution to the wave equation.

A double layer potential can be defined with the same idea of superposition. The directionality at the point $\mathbf{y} \in \Gamma$ is given by the unit normal vector $\boldsymbol{\nu}(\mathbf{y})$:

$$\begin{aligned}
(\mathcal{D} * \varphi)(\mathbf{x}, t) :=& \int_\Gamma \nabla_{\mathbf{y}} \left(\frac{\varphi(\mathbf{z}, t - c^{-1}|\mathbf{x} - \mathbf{y}|)}{4\pi |\mathbf{x} - \mathbf{y}|} \right) \Big|_{\mathbf{z}=\mathbf{y}} \cdot \boldsymbol{\nu}(\mathbf{y}) \mathrm{d}\Gamma(\mathbf{y}) \\
=& \int_\Gamma \frac{(\mathbf{x} - \mathbf{y}) \cdot \boldsymbol{\nu}(\mathbf{y})}{4\pi |\mathbf{x} - \mathbf{y}|^3} \big(\varphi(\mathbf{y}, t - c^{-1}|\mathbf{x} - \mathbf{y}|) \\
& \qquad + c^{-1}|\mathbf{x} - \mathbf{y}| \dot{\varphi}(\mathbf{y}, t - c^{-1}|\mathbf{x} - \mathbf{y}|) \big) \mathrm{d}\Gamma(\mathbf{y}).
\end{aligned}$$

Obviously, for this expression to make sense we need an orientable surface with a well-defined normal vector field (almost everywhere, so polyhedra are not a problem).

1.3 Jump relations

Let us try to see some properties of the possible limits of the layer potentials when we get close to the surface.

Continuity of the single layer potentials. A possible way to study the single layer potential is by studying functions of the form

$$w(\mathbf{x}, \hat{\mathbf{x}}, t) := \int_\Gamma \frac{\lambda(\mathbf{y}, t - c^{-1}|\hat{\mathbf{x}} - \mathbf{y}|)}{4\pi|\mathbf{x} - \mathbf{y}|} \mathrm{d}\Gamma(\mathbf{y}), \tag{1.6}$$

since $(\mathcal{S} * \lambda)(\mathbf{x}, t) = w(\mathbf{x}, \mathbf{x}, t)$. Let $\mathbf{z} \in \Gamma$. We first take the limit $\hat{\mathbf{x}} \to \mathbf{z}$ in (1.6) and we obtain (formally at least)

$$\int_\Gamma \frac{\lambda(\mathbf{y}, t - c^{-1}|\mathbf{z} - \mathbf{y}|)}{4\pi|\mathbf{x} - \mathbf{y}|} \mathrm{d}\Gamma(\mathbf{y}). \tag{1.7}$$

In a second step, we recognize in (1.7) the form of a Coulomb potential (the single layer potential for the Laplacian), which is continuous across Γ. This means that

$$\lim_{\mathbf{x} \to \mathbf{z} \in \Gamma} (\mathcal{S} * \lambda)(\mathbf{x}, t) = \int_\Gamma \frac{\lambda(\mathbf{y}, t - c^{-1}|\mathbf{z} - \mathbf{y}|)}{4\pi|\mathbf{z} - \mathbf{y}|} \mathrm{d}\Gamma(\mathbf{y}) =: (\mathcal{V} * \lambda)(\mathbf{z}, t).$$

Discontinuity of the normal derivative of the single layer potential. We next look at directional derivatives of $\mathcal{S} * \lambda$. Let $\boldsymbol{\nu} = \boldsymbol{\nu}(\mathbf{z})$ with $\mathbf{z} \in \Gamma$. Then:

$$\begin{aligned}
\big(\nabla_{\mathbf{x}}(\mathcal{S} * \lambda) \cdot \boldsymbol{\nu}\big)(\mathbf{x}, t) &= -c^{-1} \int_\Gamma \frac{\dot{\lambda}(\mathbf{y}, t - c^{-1}|\mathbf{x} - \mathbf{y}|)}{4\pi|\mathbf{x} - \mathbf{y}|} \frac{(\mathbf{x} - \mathbf{y}) \cdot \boldsymbol{\nu}}{|\mathbf{x} - \mathbf{y}|} \mathrm{d}\Gamma(\mathbf{y}) \\
&\quad - \int_\Gamma \frac{\lambda(\mathbf{y}, t - c^{-1}|\mathbf{x} - \mathbf{y}|)}{4\pi|\mathbf{x} - \mathbf{y}|} \frac{(\mathbf{x} - \mathbf{y}) \cdot \boldsymbol{\nu}}{|\mathbf{x} - \mathbf{y}|^2} \mathrm{d}\Gamma(\mathbf{y}) \\
&=: a(\mathbf{x}, t) + b(\mathbf{x}, t).
\end{aligned}$$

With arguments similar to those we used in the continuity analysis of $\mathcal{S} * \lambda$, we can prove that a is continuous across Γ. We are now going to give a simplified argument demonstrating that

$$b(\mathbf{z} - \varepsilon\boldsymbol{\nu}(\mathbf{z}), t) - b(\mathbf{z} + \varepsilon\boldsymbol{\nu}(\mathbf{z}), t) \stackrel{\varepsilon \to 0}{\longrightarrow} \lambda(\mathbf{z}, t),$$

which is equivalent to showing that the jump of the normal derivative of $\mathcal{S} * \lambda$ across Γ is λ. Note that when $\mathbf{x} \to \mathbf{z} \in \Gamma$, only a neighborhood of $\mathbf{z}$ in Γ is relevant from the point of view of creating a discontinuity in the integral:

$$b(\mathbf{x}, t) = -\int_\Gamma \frac{\lambda(\mathbf{y}, t - c^{-1}|\mathbf{x} - \mathbf{y}|)}{4\pi|\mathbf{x} - \mathbf{y}|} \frac{(\mathbf{x} - \mathbf{y}) \cdot \boldsymbol{\nu}(\mathbf{y})}{|\mathbf{x} - \mathbf{y}|^2} \mathrm{d}\Gamma(\mathbf{y}).$$

To further simplify the exposition, let us assume that Γ is a flat surface around $\mathbf{z}$. After translation, rotation, and localization, we can assume that

$$\mathbf{z} = \mathbf{0} \qquad \boldsymbol{\nu} = (0,0,1) \qquad \Gamma = \{(\mathbf{y},0) \, : \, \mathbf{y} \in \mathbb{R}^2, \ |\mathbf{y}| < R\} = B(0,R) \times \{0\}.$$

If $\mathbf{x} = \mathbf{z} \pm \varepsilon \boldsymbol{\nu}(\mathbf{z}) = \pm\varepsilon(0,0,1)$, then

$$\begin{aligned} b(\mathbf{0} - \varepsilon\boldsymbol{\nu}, t) - b(\mathbf{0} + \varepsilon\boldsymbol{\nu}, t) &= \varepsilon \int_{B(0,R)} \frac{\lambda(\mathbf{y}, t - c^{-1}|(\mathbf{y},\varepsilon)|)}{2\pi |(\mathbf{y},\varepsilon)|^3} \mathrm{d}\mathbf{y} \\ &= \lambda(\mathbf{0}, t) \int_{B(0,R)} \frac{\varepsilon}{2\pi |(\mathbf{y},\varepsilon)|^3} \mathrm{d}\mathbf{y} \\ &\quad + \varepsilon \int_{B(0,R)} \frac{\lambda(\mathbf{y}, t - c^{-1}|(\mathbf{y},\varepsilon)|) - \lambda(\mathbf{0}, t)}{4\pi |(\mathbf{y},\varepsilon)|^3} \mathrm{d}\mathbf{y}. \end{aligned} \tag{1.8}$$

Note that

$$\int_{B(0,R)} \frac{\varepsilon}{2\pi |(\mathbf{y},\varepsilon)|^3} \mathrm{d}\mathbf{y} = \int_0^R \frac{\varepsilon\, r}{\sqrt{(r^2+\varepsilon^2)^3}} \mathrm{d}r = 1 - \frac{\varepsilon}{\sqrt{R^2+\varepsilon^2}} \stackrel{\varepsilon \to 0}{\longrightarrow} 1. \tag{1.9}$$

On the other hand, for smooth λ

$$|\lambda(\mathbf{y}, t - c^{-1}|(\mathbf{y},\varepsilon)|) - \lambda(\mathbf{0}, t)| \le C_1 |\mathbf{y}| + C_2 \varepsilon, \tag{1.10}$$

$$\int_{B(0,R)} \frac{\varepsilon^2}{2\pi |(\mathbf{y},\varepsilon)|^3} \mathrm{d}\mathbf{y} \stackrel{\varepsilon \to 0}{\longrightarrow} 0, \tag{1.11}$$

and

$$\begin{aligned} \int_{B(0,R)} \frac{\varepsilon |\mathbf{y}|}{2\pi |(\mathbf{y},\varepsilon)|^3} \mathrm{d}\mathbf{y} &= \int_0^R \frac{\varepsilon r^2}{\sqrt{(r^2+\varepsilon^2)^3}} \mathrm{d}r \\ &= \varepsilon \log(\sqrt{R^2+\varepsilon^2} + R) - \frac{\varepsilon R}{\sqrt{R^2+\varepsilon^2}} - \varepsilon \log \varepsilon \stackrel{\varepsilon \to 0}{\longrightarrow} 0. \end{aligned} \tag{1.12}$$

Using (1.9), (1.10), (1.11), and (1.12) in (1.8), the result follows. The case of curved boundaries is very similar.

Discontinuity of the double layer potential. The expression for the double layer potential

$$\begin{aligned} (\mathcal{D} * \varphi)(\mathbf{x}, t) = c^{-1} &\int_\Gamma \frac{\dot\varphi(\mathbf{y}, t - c^{-1}|\mathbf{x}-\mathbf{y}|)}{4\pi |\mathbf{x}-\mathbf{y}|} \frac{(\mathbf{x}-\mathbf{y}) \cdot \boldsymbol{\nu}(\mathbf{y})}{|\mathbf{x}-\mathbf{y}|} \mathrm{d}\Gamma(\mathbf{y}) \\ + &\int_\Gamma \frac{\varphi(\mathbf{y}, t - c^{-1}|\mathbf{x}-\mathbf{y}|)}{4\pi |\mathbf{x}-\mathbf{y}|} \frac{(\mathbf{x}-\mathbf{y}) \cdot \boldsymbol{\nu}(\mathbf{y})}{|\mathbf{x}-\mathbf{y}|^2} \mathrm{d}\Gamma(\mathbf{y}) \end{aligned}$$

definitely resembles that of the directional derivative of the single layer potential. We can recognize two terms: the first one is continuous and in the second one, we can use exactly the same arguments to prove that for every $\mathbf{z} \in \Gamma$ such that Γ is flat around $\mathbf{z}$

$$(\mathcal{D} * \varphi)(\mathbf{z} + h\boldsymbol{\nu}(\mathbf{z}), t) - (\mathcal{D} * \varphi)(\mathbf{z} - h\boldsymbol{\nu}(\mathbf{z}), t) \overset{\varepsilon \to 0}{\longrightarrow} \varphi(\mathbf{z}, t).$$

Note that the sign of the jump is the opposite to the one of the normal derivative of $\mathcal{S} * \lambda$.

Continuity of the normal derivative of the double layer potential. Assuming more regularity for the density φ, it is possible to show that the normal derivative of $\mathcal{D} * \varphi$ is continuous across smooth points of Γ. The proof is more involved (tangential integration by parts is involved and finite part integrals make their appearance) and requires a certain amount of patience. Because we will take a different point of view, using Laplace transform techniques and basing our results on well-established properties of layer potentials for elliptic problems, we will just accept this result for the time being.

1.4 A Calderón type calculus

The structure of the boundary integral calculus for the wave equation is very similar to that of elliptic operators, so those accustomed to the many formulas (Green representation theorem, boundary integral identities, Calderón projector, etc) of the boundary integral calculus will recognize here exactly the same basic structure. The main difference is at the analytic level: spaces are much less clear and the theory requires quite some effort to be developed. The boundary integral calculus can be derived in several ways. My favorite is the following. It develops from three concepts:

- a uniqueness theorem for transmission problems,
- a concept of single layer operator,
- a concept of double layer operator.

(The three concepts can be grouped in one: an existence and uniqueness theorem for transmission problems.) Once these elements have been established, the representation theorem (Green's Theorem for steady state problems, Kirchhoff's formula for waves) is a direct consequence of these elements. The boundary integral operators are the averages of the Cauchy data of layer operators and they yield a collection of integral identities satisfied by interior and exterior solutions. Two sound mathematically oriented references on boundary integral equations are the monographs of Hsiao and Wendland [55] and the more numerically oriented presentation of Sauter and Schwab [80]. The many intricacies of the theory of Calderón projectors, boundary integral operators and potentials, are thoroughly explained in McLean's celebrated book [68].

We are going to informally expose this theory. We will need Chapters 2 and 3 to develop a rigorous theory for the main building blocks. The geometric layout is composed of a bounded domain Ω^-, with Lipschitz boundary Γ and exterior $\Omega^+ := \mathbb{R}^d \setminus \overline{\Omega^-}$ (that is supposed to be connected). The restriction (trace) of a function u to the boundary Γ from the interior and exterior of Γ will be denoted $\gamma^- u$ and $\gamma^+ u$, respectively. The normal derivative (with the normal vector pointing outwards) from inside and outside are $\partial_\nu^- u$ and $\partial_\nu^+ u$. Jumps of these two quantities across the interface Γ are denoted

$$[\![\gamma u]\!] := \gamma^- u - \gamma^+ u, \qquad [\![\partial_\nu u]\!] := \partial_\nu^- u - \partial_\nu^+ u.$$

Averages are denoted with double curly brackets

$$\{\!\{\gamma u\}\!\} := \tfrac{1}{2}(\gamma^- u + \gamma^+ u), \qquad \{\!\{\partial_\nu u\}\!\} := \tfrac{1}{2}(\partial_\nu^- u + \partial_\nu^+ u).$$

In the background of this theory there is a class of functions $u(\mathbf{x}, t)$ for which we can take second time derivatives, spatial Laplacian, traces and normal derivatives on the boundary and initial values at $t = 0$. For the moment let us refer to these functions as admissible functions.

The uniqueness result. The first key ingredient is a uniqueness result for a kind of transmission problem of the wave equation. It can be informally stated as follows: if an admissible function u satisfies

$$\begin{aligned}
c^{-2} u_{tt} &= \Delta u && \text{in } \mathbb{R}^d \setminus \Gamma \times (0, \infty),\\
[\![\gamma u]\!] &= 0 && \text{on } \Gamma \times (0, \infty),\\
[\![\partial_\nu u]\!] &= 0 && \text{on } \Gamma \times (0, \infty),\\
u(\,\cdot\,, 0) &= 0 && \text{in } \mathbb{R}^d \setminus \Gamma,\\
u_t(\,\cdot\,, 0) &= 0 && \text{in } \mathbb{R}^d \setminus \Gamma,
\end{aligned}$$

then u is necessarily zero.

Those used to frequency domain problems will be wondering where the radiation condition is. This can be dealt with in several ways, such as demanding finite energy for each time, or asking for bounded spatial support for each time, etc. At this level, we assume that the class of admissible functions includes this radiation condition. When we develop the correct theoretical frame, radiation will be part of causality and will not have to be expressed as a separate condition.

A single layer potential. For a function $\lambda : \Gamma \times (0, \infty) \to \mathbb{R}$ in a certain class of functions, there exists an admissible function $u := \mathcal{S} * \lambda$ such that

$$\begin{aligned}
c^{-2} u_{tt} &= \Delta u && \text{in } \mathbb{R}^d \setminus \Gamma \times (0, \infty),\\
[\![\gamma u]\!] &= 0 && \text{on } \Gamma \times (0, \infty),
\end{aligned}$$

$$\begin{aligned} [\![\partial_\nu u]\!] &= \lambda && \text{on } \Gamma \times (0,\infty), \\ u(\,\cdot\,,0) &= 0 && \text{in } \mathbb{R}^d \setminus \Gamma, \\ u_t(\,\cdot\,,0) &= 0 && \text{in } \mathbb{R}^d \setminus \Gamma. \end{aligned}$$

Obviously, $\mathcal{S} * \lambda$ is the unique solution of this problem.

A double layer potential. For a function $\varphi : \Gamma \times (0,\infty) \to \mathbb{R}$ in a certain class, there exists an admissible function $u := \mathcal{D} * \varphi$ such that

$$\begin{aligned} c^{-2} u_{tt} &= \Delta u && \text{in } \mathbb{R}^d \setminus \Gamma \times (0,\infty), \\ [\![\gamma u]\!] &= -\varphi && \text{on } \Gamma \times (0,\infty), \\ [\![\partial_\nu u]\!] &= 0 && \text{on } \Gamma \times (0,\infty), \\ u(\,\cdot\,,0) &= 0 && \text{in } \mathbb{R}^d \setminus \Gamma, \\ u_t(\,\cdot\,,0) &= 0 && \text{in } \mathbb{R}^d \setminus \Gamma. \end{aligned}$$

There is an inherent compatibility condition between the three classes of functions where u, λ, and ϕ take values. It can be expressed as follows: given u in the class of pressure (wave) fields, the quantities $\lambda := [\![\partial_\nu u]\!]$ and $\varphi := [\![\gamma u]\!]$ can be used as respective inputs of the single and double layer potentials.

First consequence: Kirchhoff's formula. If u is a solution of the wave equation around Γ

$$\begin{aligned} c^{-2} u_{tt} &= \Delta u && \text{in } \mathbb{R}^d \setminus \Gamma \times (0,\infty), \\ u(\,\cdot\,,0) &= 0 && \text{in } \mathbb{R}^d \setminus \Gamma, \\ u_t(\,\cdot\,,0) &= 0 && \text{in } \mathbb{R}^d \setminus \Gamma, \end{aligned}$$

then

$$u = \mathcal{S} * [\![\partial_\nu u]\!] - \mathcal{D} * [\![\gamma u]\!]. \tag{1.13}$$

This is a direct consequence of the definitions of layer potentials and the uniqueness theorem for transmission problems.

New definitions: boundary integral operators. The properties of potentials

$$[\![\gamma(\mathcal{S} * \lambda)]\!] = 0 \qquad \text{and} \qquad [\![\partial_\nu(\mathcal{D} * \varphi)]\!] = 0 \tag{1.14}$$

allow us to define the following four operators:

$$\begin{aligned} \mathcal{V} * \lambda &:= \{\!\{\gamma(\mathcal{S} * \lambda)\}\!\} = \gamma^-(\mathcal{S} * \lambda) = \gamma^+(\mathcal{S} * \lambda), \\ \mathcal{K}^t * \lambda &:= \{\!\{\partial_\nu(\mathcal{S} * \lambda)\}\!\}, \end{aligned}$$

$$\begin{aligned} \mathcal{K} * \varphi &:= \{\!\{\gamma(\mathcal{D} * \varphi)\}\!\}, \\ \mathcal{W} * \varphi &:= -\{\!\{\partial_\nu(\mathcal{D} * \varphi)\}\!\} = -\partial_\nu^-(\mathcal{D} * \varphi) = -\partial_\nu^+(\mathcal{D} * \varphi). \end{aligned}$$

The superscript t in $\mathcal{K}^t$ is not the time variable, but a sort of transposition symbol, that is difficult to explain at this moment. We will come back to this issue in Chapter 2. Since

$$[\![\partial_\nu(\mathcal{S} * \lambda)]\!] = \lambda \qquad \text{and} \qquad [\![\gamma(\mathcal{D} * \varphi)]\!] = -\varphi,$$

the definitions imply that

$$\partial_\nu^\pm(\mathcal{S} * \lambda) = \mp\tfrac{1}{2}\lambda + \mathcal{K}^t * \lambda \qquad \text{and} \qquad \gamma^\pm(\mathcal{D} * \varphi) = \pm\tfrac{1}{2}\varphi + \mathcal{K} * \varphi. \tag{1.15}$$

The collection of all these formulas is often referred to as **the jump relations** of potentials.

Boundary integral identities. Starting at the representation theorem (Kirchhoff's formula)

$$u = \mathcal{S} * [\![\partial_\nu u]\!] - \mathcal{D} * [\![\gamma u]\!]$$

and using the jump relations, we can write, for instance,

$$\begin{bmatrix} \{\!\{\gamma u\}\!\} \\ \{\!\{\partial_\nu u\}\!\} \end{bmatrix} = \begin{bmatrix} -\mathcal{K} & \mathcal{V} \\ \mathcal{W} & \mathcal{K}^t \end{bmatrix} * \begin{bmatrix} [\![\gamma u]\!] \\ [\![\partial_\nu u]\!] \end{bmatrix}. \tag{1.16}$$

Exterior solutions: direct method. The previous presentation was carried out for solutions of transmission problems. The reader might wonder what to do when we only have an exterior solution at our disposal, i.e., a solution of

$$\begin{aligned} c^{-2}u_{tt} &= \Delta u && \text{in } \Omega^+ \times (0,\infty), \\ u(\,\cdot\,,0) &= 0 && \text{in } \Omega^+, \\ u_t(\,\cdot\,,0) &= 0 && \text{in } \Omega^+. \end{aligned}$$

The simplest thing to do is to consider that $u \equiv 0$ in $\Omega^- \times (0,\infty)$ naturally completes the exterior solution. Then

$$[\![\gamma u]\!] = -\gamma^+ u, \quad \{\!\{\gamma u\}\!\} = \tfrac{1}{2}\gamma^+ u, \quad [\![\partial_\nu u]\!] = -\partial_\nu^+ u, \quad \{\!\{\partial_\nu u\}\!\} = \tfrac{1}{2}\partial_\nu^+ u.$$

Therefore, Kirchhoff's formula (1.13) for this u is rewritten as

$$u = \mathcal{D} * \gamma^+ u - \mathcal{S} * \partial_\nu^+ u, \tag{1.17}$$

while the integral identities (1.16) become

$$\frac{1}{2}\begin{bmatrix}\gamma^+ u\\ \partial_\nu^+ u\end{bmatrix} = \begin{bmatrix}\mathcal{K} & -\mathcal{V}\\ -\mathcal{W} & -\mathcal{K}^t\end{bmatrix} * \begin{bmatrix}\gamma^+ u\\ \partial_\nu^+ u\end{bmatrix}, \tag{1.18}$$

or, equivalently,

$$\begin{bmatrix}\frac{\delta_0}{2} - \mathcal{K} & \mathcal{V}\\ \mathcal{W} & \frac{\delta_0}{2} + \mathcal{K}^t\end{bmatrix} * \begin{bmatrix}\gamma^+ u\\ \partial_\nu^+ u\end{bmatrix} = \begin{bmatrix}0\\ 0\end{bmatrix}. \tag{1.19}$$

In the latter expression we have used the formal notation $\delta_0 * \xi = \xi$ (convolution with the Dirac delta is the identity) in order to keep convolutional notation for the entire matrix of operators. Some people prefer reordering the columns in the above expression to get

$$\begin{bmatrix}\mathcal{V} & \frac{\delta_0}{2} - \mathcal{K}\\ \frac{\delta_0}{2} + \mathcal{K}^t & \mathcal{W}\end{bmatrix} * \begin{bmatrix}\partial_\nu^+ u\\ \gamma^+ u\end{bmatrix} = \begin{bmatrix}0\\ 0\end{bmatrix},$$

emphasizing some kind of symmetry of the system that is not clear at the present moment.

It is interesting to remark that Kirchhoff's formula for the exterior solution also represents the zero solution inside (that is how we got to it) and therefore

$$\mathcal{D} * \gamma^+ u - \mathcal{S} * \partial_\nu^+ u \equiv 0 \qquad \text{in } \Omega^-.$$

1.5 Scattering problems

The geometric layout of this section will be exactly the same as the one of the previous section: a bounded domain Ω^- with Lipschitz boundary Γ (this is the scatterer) and connected exterior Ω^+ (Figure 1.1).

Incident waves. In a scattering problem there is something called an incident wave that is supposed to be known. Possible incident waves are spherical waves

$$u^{\rm inc}(\mathbf{x}, t) = \frac{\lambda(t - c^{-1}|\mathbf{x} - \mathbf{x}_0|)}{4\pi|\mathbf{x} - \mathbf{x}_0|},$$

emitted from a point source $\mathbf{x}_0 \not\in \overline{\Omega^-}$ and with causal signal λ. Their directional counterparts

$$u^{\rm inc}(\mathbf{x}, t) = \nabla_{\mathbf{x}_0}\left(\frac{\varphi(t - c^{-1}|\mathbf{x} - \mathbf{x}_0|)}{4\pi|\mathbf{x} - \mathbf{x}_0|}\right) \cdot \mathbf{n}_0$$

Fig. 1.1 A cartoon for the geometric setting of this entire book. A Lipschitz obstacle is hit by an incident wave. The scattered wave lives in the exterior domain Ω^-, even if potentials are actually defined on both sides of Γ. Multiple scatterers are covered by this theory.

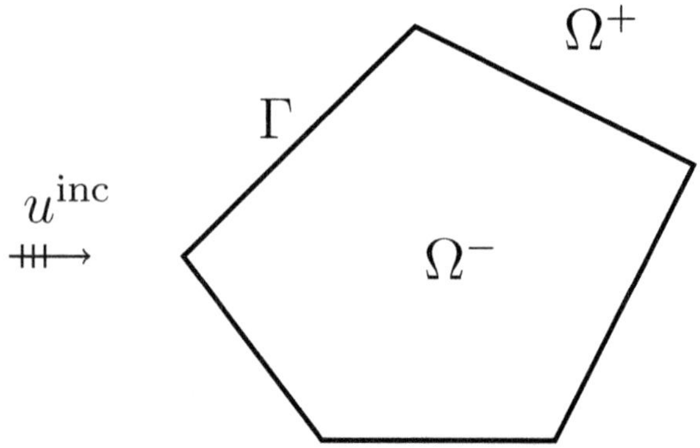

are also valid incident waves. Superposition of several of the above are also acceptable incident waves. These waves are compactly supported at all times. Plane waves,

$$u^{\rm inc}(\mathbf{x},t) := \lambda(c^{-1}\mathbf{x}\cdot\mathbf{d} - t) \qquad \text{where} \qquad |\mathbf{d}| = 1,$$

make for good non-compactly supported incident waves. For the placement of the obstacle to be feasible we need that at time zero the supports of $u^{\rm inc}(\,\cdot\,,0)$ and $u^{\rm inc}_t(\,\cdot\,,0)$ do not intersect $\overline{\Omega^-}$. Apart from the compactness of their support, the nature of spherical and plane waves is very different, since, when seen in free space $\mathbb{R}^d$, the first ones are solutions of a nonhomogeneous wave equation (there is a singular source placed at $\mathbf{x}_0$) but have vanishing initial condition, while plane waves are unforced solutions of the wave equation corresponding to a particular set of non-compactly supported initial conditions.

In all cases, it is customary to consider that an incident wave is a solution of a wave propagation problem in free space:

$$\begin{aligned} c^{-2}u^{\rm inc}_{tt} &= \Delta u^{\rm inc} + f && \text{in } \mathbb{R}^d \times (0,\infty),\\ u^{\rm inc}(\,\cdot\,,0) &= u_0 && \text{in } \mathbb{R}^d,\\ u^{\rm inc}_t(\,\cdot\,,0) &= v_0 && \text{in } \mathbb{R}^d. \end{aligned}$$

Physical placement of the obstacle at time $t = 0$ requires that the supports of u_0 and v_0 do not intersect $\overline{\Omega^-}$. Source terms at any given time $f(\,\cdot\,,t)$, should be set apart from the scatterer as well.

Scattering by obstacles. The influence of the simplest scatterers on the wave field is modeled by a boundary condition. For instance, a sound-soft obstacle induces a Dirichlet boundary condition:

$$\begin{aligned} c^{-2}u^{\rm tot}_{tt} &= \Delta u^{\rm tot} + f && \text{in } \Omega^+ \times (0,\infty),\\ \gamma u^{\rm tot} &= 0 && \text{on } \Gamma \times (0,\infty), \end{aligned}$$

$$u^{\text{tot}}(\,\cdot\,,0) = u_0 \qquad \text{in } \Omega^+,$$
$$u_t^{\text{tot}}(\,\cdot\,,0) = v_0 \qquad \text{in } \Omega^+.$$

The source term and initial conditions are the same as for the incident wave. A sound-hard obstacle is modeled with a Neumann boundary condition instead

$$\partial_\nu u^{\text{tot}} = 0 \qquad \text{on } \Gamma \times (0,\infty).$$

Simple absorbing boundary conditions are modeled with the dynamic boundary conditions

$$\partial_\nu u^{\text{tot}} - \alpha c^{-1} \gamma u_t = 0 \qquad \text{on } \Gamma \times (0,\infty),$$

where $\alpha : \Gamma \to [0,\infty)$ plays the role of a surface impedance function.

The scattered wave. For practical (computational) purposes, the decomposition

$$u^{\text{tot}} = u^{\text{inc}} + u^{\text{scat}}$$

is often used. Note that this is a nonphysical decomposition. The scattered wave field is just the difference between what is happening (u^{tot}) and what would happen if the obstacle was not present (u^{inc}). While the scattered wave will respect some basic causality principles and will travel at the correct speed, it is a nonphysical object which will cross the obstacle unphysically, because it has to compensate for the fact that the incident wave does not perceive the obstacle while, obviously, the total field does. Our future unknown will be the scattered wave field $u := u^{\text{scat}} = u^{\text{tot}} - u^{\text{inc}}$ that satisfies a homogeneous equation, with homogeneous initial conditions but nonvanishing boundary conditions:

$$\begin{aligned} c^{-2} u_{tt} &= \Delta u && \text{in } \Omega^+ \times (0,\infty), \\ \mathrm{BC}(u) &= -\mathrm{BC}(u^{\text{inc}}) && \text{on } \Gamma \times (0,\infty), \\ u(\,\cdot\,,0) &= 0 && \text{in } \Omega^+, \\ u_t(\,\cdot\,,0) &= 0 && \text{in } \Omega^+. \end{aligned}$$

Other possible scattering problems include penetrable scatterers with different material properties. Some methods for these problems combine integral equations for the exterior scattered field with volume formulations for the interior of the scatterer. Mathematically speaking, they do not differ in essence from the kind of problems we meet when building transparent boundary conditions for wave propagation problems.

1.6 Integral formulations in scattering

Consider again the problem of scattering by a sound-soft obstacle:

$$\begin{aligned} c^{-2}u_{tt} &= \Delta u && \text{in } \Omega^+ \times (0,\infty), \\ \gamma u &= \beta && \text{on } \Gamma \times (0,\infty), \\ u(\,\cdot\,,0) &= 0 && \text{in } \Omega^+, \\ u_t(\,\cdot\,,0) &= 0 && \text{in } \Omega^+. \end{aligned}$$

The incident wave is only perceived as a boundary condition

$$\beta := -\gamma u^{\text{inc}} : \Gamma \times (0,\infty) \to \mathbb{R}.$$

Indirect formulation. Since single layer potentials are solutions of the wave equation with homogeneous initial conditions, we are allowed to propose

$$u := \mathcal{S} * \lambda \tag{1.20}$$

as a possible solution of the scattering problem, for a causal density to be determined. Recalling that we have denoted $\mathcal{V} * \lambda := \gamma(\mathcal{S} * \lambda)$, it follows that in order for $u = \mathcal{S} * \lambda$ to be a solution, we need

$$\mathcal{V} * \lambda = \beta. \tag{1.21}$$

In due time, we will see that (1.21) is a uniquely solvable equation. With it we will determine the density λ and the representation formula (1.20) will deliver the solution to the scattering problem. Because single layer potentials are continuous across the interface Γ, we are naturally attaching a continuous extension of the solution to the interior of the scatterer, so with (1.20)–(1.21) we are actually solving

$$\begin{aligned} c^{-2}u_{tt} &= \Delta u && \text{in } \mathbb{R}^d \setminus \Gamma \times (0,\infty), \\ \gamma^{\pm} u &= \beta && \text{on } \Gamma \times (0,\infty), \\ u(\,\cdot\,,0) &= 0 && \text{in } \mathbb{R}^d \setminus \Gamma, \\ u_t(\,\cdot\,,0) &= 0 && \text{in } \mathbb{R}^d \setminus \Gamma. \end{aligned}$$

These are two separate wave equations (one in Ω^+ and another one in Ω^-) that have no connection apart from sharing the same boundary condition (this is not a transmission problem). When $\beta = -\gamma u^{\text{inc}}$ it follows that the interior solution is just $-u^{\text{inc}}$.

Direct formulation. Another approach consists of taking Kirchhoff's formula as the integral representation formula. Therefore, we have (recall (1.17))

$$u = \mathcal{D} * \gamma^+ u - \mathcal{S} * \partial_\nu^+ u.$$

Setting $\lambda := \partial_\nu^+ u$ to be our boundary unknown and substituting the boundary condition, it follows that

$$u = \mathcal{D} * \beta - \mathcal{S} * \lambda. \tag{1.22}$$

The first of the integral identities (1.18) or (1.19)

$$\mathcal{V} * \lambda = -\tfrac{1}{2}\beta + \mathcal{K} * \beta \tag{1.23}$$

can be used as an integral equation in order to try to find λ and obtain the scattered wave-field with it. While the pair of integral equation and integral representation for the indirect formulation (1.20)–(1.21) is definitely simpler, the pair (1.22)–(1.23) has the advantage of computing a physical quantity on the boundary. The practical choice in frequency domain problems is almost invariably the direct method, but the situation is much less clear in the time domain, for complicated reasons we will not deal with. There is also the question of whether we should strive for formulations where the boundary unknown appears under the action of an operator of the second kind (identity plus integral operator). I will not deal with this question either, since the choices in the time domain are still quite open. (A recent contribution [45] seems to have opened the way for a theoretical study of second kind formulations.)

1.7 Galerkin semidiscretization

We are going to do some work on the equation

$$\mathcal{V} * \lambda = \beta \tag{1.24}$$

that appeared in Section 1.6. In strong integral form, this is

$$\int_\Gamma \frac{\lambda(\mathbf{y}, t - c^{-1}|\mathbf{x}-\mathbf{y}|)}{4\pi|\mathbf{x}-\mathbf{y}|} \mathrm{d}\Gamma(\mathbf{y}) = \beta(\mathbf{x}, t) \qquad \mathbf{x} \in \Gamma \qquad t > 0. \tag{1.25}$$

The unknown is a causal function $\lambda : \Gamma \times \mathbb{R} \to \mathbb{R}$. For (semi)discretization let us choose a finite dimensional space $X^h \subset L^\infty(\Gamma)$ with basis $\{\Phi_1, \dots, \Phi_N\}$ and let us look for

$$\lambda^h := \sum_{j=1}^N \lambda_j(t)\Phi_j(\mathbf{x}), \tag{1.26}$$

where the unknowns are N causal functions $\lambda_j : \mathbb{R} \to \mathbb{R}$ of the time variable. Because we are drastically reducing the set of possible densities we cannot expect that the integral equation (1.24) stands any chance of having a solution of the form (1.26). Instead, we will look for some averages (moments) of left- and right-hand side to coincide, namely,

$$\int_\Gamma \Phi_i(\mathbf{x})\big(\mathcal{V} * \lambda^h\big)(\mathbf{x},t)\mathrm{d}\Gamma(\mathbf{x}) = \int_\Gamma \Phi_i(\mathbf{x})\,\beta(\mathbf{x},t)\mathrm{d}\Gamma(\mathbf{x}) \qquad i = 1,\ldots,N, \qquad t > 0.$$

This is equivalent to the following system

$$\sum_{j=1}^N \int_\Gamma \int_\Gamma \frac{\Phi_i(\mathbf{x})\Phi_j(\mathbf{y})}{4\pi|\mathbf{x}-\mathbf{y}|}\lambda_j(t - c^{-1}|\mathbf{x}-\mathbf{y}|)\,\mathrm{d}\Gamma(\mathbf{x})\mathrm{d}\Gamma(\mathbf{y}) = \int_\Gamma \Phi_i(\mathbf{x})\,\beta(\mathbf{x},t)\mathrm{d}\Gamma(\mathbf{x}) \tag{1.27}$$

$$i = 1,\ldots,N, \qquad t > 0.$$

This is a system of delay equations with delays integrated over distances between points. In Chapter 4 we will see that this system can be considered as a system of convolution equations that can be discretized with the convolution quadrature method.

A second option consists of using a time-Galerkin discretization of the system above. This is equivalent to decomposing each λ_j as a linear combination of shape functions and testing the equations above in the following way: oddly enough, we multiply them *by the derivative of the shape functions* and possibly some fixed time-dependent weight, and we then integrate in the time variable. The inherent coercivity principle is explained in Section 3.7. To explain this more clearly, let us assume that we have a time-grid with constant time-step

$$0 = t_0 < t_1 < t_2 < \ldots < t_n < \ldots \qquad t_n = \kappa\, n,$$

and that we use piecewise constant functions in time. This corresponds to choosing the basis of characteristic functions

$$\chi_{(t_{n-1},t_n)}(t) \qquad n \geq 1.$$

The fully discrete unknown now has the form

$$\sum_{j=1}^N \sum_{m=1}^\infty \lambda_j^m\, \chi_{(t_{m-1},t_m)}(t)\Phi_j(\mathbf{x}), \qquad \lambda_j^m \in \mathbb{R}.$$

Each of the equations (1.27) is multiplied by $\dot\chi_{(t_{n-1},t_n)}$ and integrated from 0 to ∞. Since $\dot\chi_{(t_{n-1},t_n)} = -\delta_{t_n} + \delta_{t_{n-1}}$, this is equivalent to setting up a system

$$\sum_{j=1}^N \sum_{m=1}^n \lambda_j^m(-A_{ij}^{n-m} + A_{ij}^{n-1-m}) = -\beta_i^n + \beta_i^{n-1} \tag{1.28}$$

with

$$
\begin{aligned}
A_{ij}^{n-m} &= \int_\Gamma \int_\Gamma \frac{\Phi_i(\mathbf{x})\Phi_j(\mathbf{y})}{4\pi|\mathbf{x}-\mathbf{y}|} \chi_{(t_{m-1},t_m)}(t_n - c^{-1}|\mathbf{x}-\mathbf{y}|)\mathrm{d}\Gamma(\mathbf{x})\mathrm{d}\Gamma(\mathbf{y}) \\
&= \int_\Gamma \int_\Gamma \frac{\Phi_i(\mathbf{x})\Phi_j(\mathbf{y})}{4\pi|\mathbf{x}-\mathbf{y}|} \chi_{(t_n-t_m,t_n-t_{m-1})}(c^{-1}|\mathbf{x}-\mathbf{y}|)\mathrm{d}\Gamma(\mathbf{x})\mathrm{d}\Gamma(\mathbf{y}) \\
&= \int_\Gamma \int_\Gamma \frac{\Phi_i(\mathbf{x})\Phi_j(\mathbf{y})}{4\pi|\mathbf{x}-\mathbf{y}|} \chi_{(t_{n-m},t_{n-m+1})}(c^{-1}|\mathbf{x}-\mathbf{y}|)\mathrm{d}\Gamma(\mathbf{x})\mathrm{d}\Gamma(\mathbf{y}) \\
&= \iint_{\Gamma_c^{n-m}} \frac{\Phi_i(\mathbf{x})\Phi_j(\mathbf{y})}{4\pi|\mathbf{x}-\mathbf{y}|} \mathrm{d}\Gamma(\mathbf{x})\mathrm{d}\Gamma(\mathbf{y}),
\end{aligned}
$$

where

$$
\Gamma_c^\ell := \{(\mathbf{x},\mathbf{y}) \in \Gamma \times \Gamma \,:\, ct_\ell \le |\mathbf{x}-\mathbf{y}| \le c\,t_{\ell+1}\}, \qquad \ell \ge 0,
$$

and

$$
\beta_i^n = \int_\Gamma \Phi_i(\mathbf{x})\beta(\mathbf{x},t_n)\mathrm{d}\Gamma(\mathbf{x}).
$$

The system (1.28) has a block triangular structure with some Toeplitz character to it. (This character, as well as the convolutional structure in discrete time, is heavily dependent on us having chosen a constant time-step.) We can start by solving for $n = 1$

$$
\sum_{j=1}^N A_{ij}^0 \lambda_j^1 = \beta_i^1, \qquad i = 1,\ldots,N.
$$

Next we solve for $n = 2$

$$
\sum_{j=1}^N A_{ij}^0 \lambda_j^2 + \sum_{j=1}^N (A_{ij}^1 - A_{ij}^0)\lambda_j^1 = \beta_i^2 - \beta_i^1,
$$

for $n = 3$,

$$
\sum_{j=1}^N A_{ij}^0 \lambda_j^3 + \sum_{j=1}^N (A_{ij}^1 - A_{ij}^0)\lambda_j^2 + \sum_{j=1}^N (A_{ij}^2 - A_{ij}^1)\lambda_j^0 = \beta_i^3 - \beta_i^2,
$$

and so on. Each time step requires the solution of a linear system with the same matrix. (If piecewise polynomial discontinuous functions are used for space discretization, the matrix A^0 is actually very sparse.) The sets Γ^ℓ are empty for large enough ℓ (depending on the size of Γ, the speed of waves c, and the time step κ) and therefore the system (1.28) has a finite tail.

The main difficulty of this full Galerkin approach is related to the precise computation of the matrices A^n. Thinking of a piecewise constant approximation in the space variable, we need to compute integrals

$$\iint_{\Gamma_{i,j}^{\ell}} \frac{1}{|\mathbf{x}-\mathbf{y}|} \mathrm{d}\Gamma(\mathbf{x})\mathrm{d}\Gamma(\mathbf{y}), \qquad \Gamma_{i,j}^{\ell} := \{(\mathbf{x},\mathbf{y}) \,:\, \mathbf{x} \in \Gamma_i,\ \mathbf{y} \in \Gamma_j,\ ct_\ell \le |\mathbf{x}-\mathbf{y}| \le ct_{\ell+1}\},$$

where a weakly singular integrand is integrated on domains with quite exotic shape. This is not easily done and it has been frequently reported that the method is extremely sensitive to small errors (both numerical and human) in computation of these integrals.

1.8 Two-dimensional waves

Single layer potential. The layer potentials in two dimensions are considerably more complicated. The single layer potential admits the expression:

$$\begin{aligned}(\mathcal{S} * \lambda)(\mathbf{x},t) &:= \frac{1}{2\pi} \int_\Gamma \int_0^{t-c^{-1}|\mathbf{x}-\mathbf{y}|} \frac{\lambda(\mathbf{y},\tau)}{\sqrt{(t-\tau)^2 - c^{-2}|\mathbf{x}-\mathbf{y}|^2}} \mathrm{d}\Gamma(\mathbf{y})\,\mathrm{d}\tau \\ &= \int_\Gamma \int_0^t G_2(t-\tau, |\mathbf{x}-\mathbf{y}|)\, \lambda(\mathbf{y},\tau)\, \mathrm{d}\Gamma(\mathbf{y})\mathrm{d}\tau\end{aligned}$$

where

$$G_2(t,r) := \frac{H(t-c^{-1}r)}{2\pi\sqrt{t^2 - c^{-2}r^2}},$$

and H is the Heaviside function. Unlike in the three-dimensional case, there is a space-and-time integral in the definition, since the two-dimensional wave kernel is a function. Compare this with the three-dimensional kernel that can be formally expressed as

$$G_3(t,r) := \frac{\delta(t-c^{-1}r)}{4\pi r}.$$

The Dirac delta distribution creates the pure delay (no integration over the past) of this kernel.

Why? There is a simple argument that shows how the two-dimensional potential comes up. Assume that the causal signal $\lambda(t)$ is simultaneously emitted by all points in the z axis. A point $(x, y, 0) \equiv (r\cos\theta, r\sin\theta, 0)$ then receives the signal

$$\int_{-\infty}^{\infty} \frac{\lambda(t - c^{-1}\sqrt{r^2 + z^2})}{4\pi\sqrt{r^2 + z^2}} \mathrm{d}z = 2\int_{0}^{\infty} \frac{\lambda(t - c^{-1}\sqrt{r^2 + z^2})}{4\pi\sqrt{r^2 + z^2}} \mathrm{d}z.$$

The change of variables (changing the third dimension z to a fictitious time τ)

$$(0, \infty) \ni z \longleftrightarrow \tau := t - c^{-1}\sqrt{r^2 + z^2} \in (-\infty, t - c^{-1}r),$$
$$\frac{\mathrm{d}z}{\sqrt{z^2 + r^2}} = \frac{\mathrm{d}\tau}{zc^{-1}} = \frac{\mathrm{d}\tau}{\sqrt{(t-\tau)^2 - c^{-2}r^2}}$$

transforms the last integral to

$$\int_{-\infty}^{t-c^{-1}r} \frac{\lambda(\tau)}{2\pi\sqrt{(t-\tau)^2 - c^{-2}r^2}} \mathrm{d}\tau = \int_{0}^{t-c^{-1}r} \frac{\lambda(\tau)}{2\pi\sqrt{(t-\tau)^2 - c^{-2}r^2}} \mathrm{d}\tau,$$

that is, to the cylindrical (two-dimensional circular) wave from which the layer potential is constructed.

The double layer potential. If we formally take the gradient of G_2

$$\nabla G_2(t, r) = \frac{c^{-1}}{2\pi} \frac{\delta(t - c^{-1}r)}{\sqrt{t^2 - c^{-2}r^2}} (-\nabla r) - \frac{c^{-2}}{2\pi} \frac{H(t - c^{-1}r)}{(t^2 - c^{-2}r^2)^{3/2}} (-\tfrac{1}{2}\nabla r^2)$$

we reach an expression for the double layer potential in two dimensions:

$$\begin{aligned}(\mathcal{D} * \varphi)(\mathbf{x}, t) :=\ & \frac{c^{-1}}{2\pi} \int_{\Gamma} \frac{\varphi(\mathbf{y}, t - c^{-1}|\mathbf{x}-\mathbf{y}|)}{|\mathbf{x}-\mathbf{y}|} \frac{(\mathbf{x}-\mathbf{y}) \cdot \boldsymbol{\nu}(\mathbf{y})}{\sqrt{(t-\tau)^2 - c^{-2}|\mathbf{x}-\mathbf{y}|^2}} \mathrm{d}\Gamma(\mathbf{y}) \\ & - \frac{c^{-2}}{2\pi} \int_{\Gamma} \int_{0}^{t-c^{-1}|\mathbf{x}-\mathbf{y}|} \frac{\varphi(\mathbf{y}, \tau)}{(t-\tau)^2 - c^{-2}|\mathbf{x}-\mathbf{y}|^2} \frac{(\mathbf{x}-\mathbf{y}) \cdot \boldsymbol{\nu}(\mathbf{y})}{\sqrt{(t-\tau)^2 - c^{-2}|\mathbf{x}-\mathbf{y}|^2}} \mathrm{d}\Gamma(\mathbf{y})\mathrm{d}\tau \\ =\ & \int_{\Gamma} \int_{0}^{t} \partial_{\boldsymbol{\nu}(\mathbf{y})} G_2(t - \tau, |\mathbf{x}-\mathbf{y}|)\, \varphi(\mathbf{y}, \tau)\mathrm{d}\Gamma(\mathbf{y})\mathrm{d}\tau.\end{aligned}$$

This one has a circular wave front with no memory plus another one that integrates over the past. Again, it is illustrative to compare with the three-dimensional case:

$$\nabla G_3(t, r) = c^{-1} \frac{\delta'(t - c^{-1}r)}{4\pi r} (-\nabla r) + \frac{\delta(t - c^{-1}r)}{4\pi r^2} (-\nabla r).$$

Chapter 2
From time domain to Laplace domain

A possible theoretical framework for the time domain layer potentials for the wave equation is that of vector-valued distributions. From this moment on, instead of thinking of functions of the space and time variables $u(\mathbf{x}, t)$ we will think of functions of the time variable with values on a space of functions of the space variables, which amounts to considering the functions $u(t) = u(\,\cdot\,, t)$ (we will not change the name). In principle, our distributions will be only allowed to take real values, but the test functions (and the Laplace transforms of the distributions) will take complex values.

Some fast references before we take off. The distributional Laplace transform is introduced in most medium/advanced texts on the theory of distributions. The textbook of Laurent Schwartz [83] is an excellent introduction. The subject of vector-valued distributions is much more involved (and there is the risk of wanting to know and understand it all). François Trèves's monograph on distributions [85] is the standard (but not easy to read) reference. A compendium of this theory, with a different point of view, can be found in volume five of the English language edition of Robert Dautray and Jacques-Louis Lions's ambitious encyclopedia on theoretical and numerical continuum models [35].

Notation. Given a domain $\mathcal{O}$, we will denote

$$(u, v)_{\mathcal{O}} := \int_{\mathcal{O}} u\, v \qquad (\nabla u, \nabla v)_{\mathcal{O}} := \int_{\mathcal{O}} \nabla u \cdot \nabla v,$$

$$\|u\|_{\mathcal{O}}^2 := (u, \overline{u})_{\mathcal{O}}, \qquad \|\nabla u\|_{\mathcal{O}}^2 := (\nabla u, \nabla \overline{u})_{\mathcal{O}}, \qquad \|u\|_{1,\mathcal{O}}^2 := \|u\|_{\mathcal{O}}^2 + \|\nabla u\|_{\mathcal{O}}^2.$$

We will make unannounced use of basic properties of the Sobolev space $H^1(\mathcal{O})$, the trace operator and the concept of weak normal derivative (we will introduce the latter though). If Γ is a Lipschitz curve/surface, we will use the angled bracket

F.-J. Sayas, *Retarded Potentials and Time Domain Boundary Integral Equations*, Springer Series in Computational Mathematics 50,
DOI 10.1007/978-3-319-26645-9_2

$$\langle \xi, \eta \rangle_\Gamma := \int_\Gamma \xi \, \eta \, \mathrm{d}\Gamma$$

to denote the $L^2(\Gamma)$-inner product and its extension as the $H^{-1/2}(\Gamma) \times H^{1/2}(\Gamma)$ duality product. The $H^{\pm 1/2}(\Gamma)$-norms are denoted $\| \cdot \|_{\pm 1/2,\Gamma}$. *As a general rule, all brackets will be considered as bilinear (and not sesquilinear) forms, even if spaces include complex-valued functions.*

2.1 Causal tempered distributions

We first consider the Schwartz class

$$\mathcal{S}(\mathbb{R}) := \{\varphi \in \mathcal{C}^\infty(\mathbb{R}) \, : \, p\,\varphi^{(k)} \in L^\infty(\mathbb{R}) \ \forall k \geq 0, \quad \forall p \in \mathcal{P}(\mathbb{R})\},$$

where $\mathcal{P}(\mathbb{R})$ is the space of polynomials with complex coefficients. It is well known that we can define a metric in $\mathcal{S}(\mathbb{R})$ that makes this space complete and such that convergence $\varphi_n \to \varphi$ with respect to that metric is equivalent to convergence in $L^\infty(\mathbb{R})$ of $p\,\varphi_n^{(k)} \to p\,\varphi^{(k)}$ for all $k \geq 0$ and $p \in \mathcal{P}(\mathbb{R})$. Quite obviously differentiation and multiplication by polynomials are continuous operators in the Schwartz class.

A tempered distribution with values in the Banach space X is a continuous linear map $f : \mathcal{S}(\mathbb{R}) \to X$. The action of f on a general element $\varphi \in \mathcal{S}(\mathbb{R})$ is denoted using angled brackets $\langle f, \varphi \rangle$. A **causal tempered distribution with values in** X (these are the ones we will be paying attention to) is a tempered X-valued distribution such that

$$\langle f, \varphi \rangle = 0 \qquad \forall \varphi \in \mathcal{S}(\mathbb{R}) \text{ such that } \operatorname{supp} \varphi \subset (-\infty, 0).$$

There is no commonly used notation for the set of causal tempered X-valued distributions. To avoid continuous repetition of this long expression, we will write

$$f \in \mathrm{CT}(X).$$

Some very simple examples. If $a \in X$, we can define the following two distributions

$$\langle \delta_0 \otimes a, \varphi \rangle := \varphi(0)\, a, \qquad \langle \dot{\delta}_0 \otimes a, \varphi \rangle := -\dot{\varphi}(0)\, a.$$

Some functions are distributions. Many functions can be considered as causal tempered distributions. For instance, if $f : [0, \infty) \to X$ is a continuous function, such that

$$\|f(t)\|_X \leq C\,(1 + t^m) \qquad t \geq 0,$$

for some positive integer m, we can define

$$\langle f, \varphi \rangle := \int_0^\infty \varphi(t) f(t) \mathrm{d}t. \tag{2.1}$$

(Note that we are giving the same name to f as a function and as a distribution, which is justified by proving that two different functions cannot define the same distribution.) The integral has to be understood in the sense of Bochner, although with the regularity assumed for f, this definition can be done as an improper Riemann integral with values on a Banach space. The definition (2.1) can be used for $f \in L^p((0, \infty), X)$ for any $1 \leq p \leq \infty$. When functions are used as distributions they are automatically extended by zero for values $t < 0$.

Differentiation. If f is a causal tempered X-valued distribution, we define

$$\langle \dot{f}, \varphi \rangle := -\langle f, \dot{\varphi} \rangle,$$

thus obtaining another element $\dot{f} \in \mathrm{CT}(X)$. (This is very simple to prove.) We will sometimes write $\frac{\mathrm{d}}{\mathrm{d}t} f$ instead of $\dot{f}$. For instance,

$$\tfrac{\mathrm{d}}{\mathrm{d}t}(\delta_0 \otimes a) = \dot{\delta}_0 \otimes a.$$

We have to be careful with what we understand by differentiation, since this is an operator that considers f as an entity defined for all $t \in \mathbb{R}$ even if it is zero for negative t. If, for instance, we have a function $f \in \mathcal{C}^1([0, \infty), X)$ such that f is bounded and so is its derivative f', then

$$\dot{f} = f' + \delta_0 \otimes f(0),$$

that is, differentiation in the sense of distributions takes into account the jump at the origin.

Composition with steady-state operators. Consider two Banach spaces X and Y and a bounded linear operator $A : X \to Y$. For $f \in \mathrm{CT}(X)$ we define

$$\langle Af, \varphi \rangle := A \langle f, \varphi \rangle$$

and we obtain that $Af \in \mathrm{CT}(Y)$. (Once more, this is very simple to prove.) In particular, if $X \subset Y$ with continuous embedding, every $f \in \mathrm{CT}(X)$ can be understood as $f \in \mathrm{CT}(Y)$.

2.2 Distributional scattered waves

Let Γ be the boundary of a bounded Lipschitz domain $\Omega^- \subset \mathbb{R}^d$. The exterior domain $\Omega^+ := \mathbb{R}^d \setminus \overline{\Omega^-}$ will always be assumed to be connected, just to avoid some minor annoyances. We will however accept the possibility that Ω^- has several connected components, since nothing is changed by this hypothesis. *From now on, when we deal with the wave equation, we will take the wave speed $c = 1$ to simplify matters.* Once we have introduced the Laplace transform version of all the potentials and operators, it will be clear that changing speed of the waves is equal to rescaling time by c, which is equivalent to scaling by $1/c$ in the Laplace domain.

The domain of the Laplacian. The space

$$H^1_\Delta(\mathbb{R}^d \setminus \Gamma) := \{u \in H^1(\mathbb{R}^d \setminus \Gamma) \; : \; \Delta u \in L^2(\mathbb{R}^d \setminus \Gamma)\}$$

will be the natural environment for the definition of the layer potentials. The Laplace operator Δ that appears in the definition of this space is the Laplacian in the sense of distributions in $\mathbb{R}^d \setminus \Gamma$. To emphasize this fact, we write $\Delta u \in L^2(\mathbb{R}^d \setminus \Gamma)$, although $L^2(\mathbb{R}^d \setminus \Gamma) \cong L^2(\mathbb{R}^d)$.

The space $H^1_\Delta(\mathbb{R}^d \setminus \Gamma)$ is endowed with the norm

$$\|u\|^2_\Delta := \|u\|^2_{\mathbb{R}^d} + \|\nabla u\|^2_{\mathbb{R}^d \setminus \Gamma} + \|\Delta u\|^2_{\mathbb{R}^d \setminus \Gamma},$$

which makes it a Hilbert space. In this set we can define the interior and exterior traces $\gamma^\pm : H^1_\Delta(\mathbb{R}^d \setminus \Gamma) \to H^{1/2}(\Gamma)$ and the interior and exterior normal derivatives $\partial^\pm_\nu : H^1_\Delta(\mathbb{R}^d \setminus \Gamma) \to H^{-1/2}(\Gamma)$. These last ones are defined using Green's formula:

$$\begin{aligned}
\langle \partial_\nu u^-, \gamma^- v\rangle_\Gamma &:= (\Delta u, v)_{\Omega^-} + (\nabla u, \nabla v)_{\Omega^-} \qquad && \forall v \in H^1(\Omega^-),\\
\langle \partial_\nu u^+, \gamma^+ v\rangle_\Gamma &:= -(\Delta u, v)_{\Omega^+} - (\nabla u, \nabla v)_{\Omega^+} \qquad && \forall v \in H^1(\Omega^+).
\end{aligned}$$

The jumps of the trace and normal derivative are also well defined (and define bounded operators) in $H^1_\Delta(\mathbb{R}^d \setminus \Gamma)$:

$$[\![\gamma u]\!] = \gamma^- u - \gamma^+ u, \qquad [\![\partial_\nu u]\!] = \partial^-_\nu u - \partial^+_\nu u.$$

Scattered waves around Γ. The set of solutions of the wave equation that we will care about are

$$u \in \mathrm{CT}(H^1_\Delta(\mathbb{R}^d \setminus \Gamma)) \qquad \text{such that} \quad \ddot{u} = \Delta u. \tag{2.2}$$

The equality above should be understood exactly as follows. We know that $\Delta : H^1_\Delta(\mathbb{R}^d \setminus \Gamma) \to L^2(\mathbb{R}^d \setminus \Gamma)$ is bounded. Therefore, if $u \in \mathrm{CT}(H^1_\Delta(\mathbb{R}^d \setminus \Gamma))$, then $\Delta u \in \mathrm{CT}(L^2(\mathbb{R}^d \setminus \Gamma))$. As mentioned in Section 2.1, the causal tempered $H^1_\Delta(\mathbb{R}^d \setminus \Gamma)$-valued distribution $\ddot{u}$ can be understood as $\ddot{u} \in \mathrm{CT}(L^2(\mathbb{R}^d \setminus \Gamma))$. Therefore, the equality in (2.2) is that of two causal tempered distributions with values in $L^2(\mathbb{R}^d \setminus \Gamma)$.

A remark concerning initial conditions. The equation (2.2) is implicitly imposing homogeneous initial conditions. If we want to represent a (smooth) function $u : [0,\infty) \to H^1_\Delta(\mathbb{R}^d \setminus \Gamma)$ such that $u(0) = u_0$, $u'(0) = v_0$ and $\Delta u(t) = u''(t)$ for all $t \geq 0$ (these time derivatives are strong derivatives), we need to consider a causal distribution u such that

$$\ddot{u} = \Delta u + \delta_0 \otimes v_0 + \dot{\delta}_0 \otimes u_0.$$

This definition includes an extension by zero to $t < 0$ and not a possible extension by solving the wave equation backwards. Once more, in (2.2) we are assuming homogeneous boundary conditions in a sense that is not easy to specify, since there is no time regularity assumed so far.

What we will end up showing. A goal of what follows is to show that solutions of (2.2) are determined (with an explicit formula) by the causal distributions $[\![\gamma u]\!]$ and $[\![\partial_\nu u]\!]$. This is what is known as Kirchhoff's formula in the three-dimensional case. We will get to this in Section 3.5.

The exterior Dirichlet problem. An exterior causal solution of the Dirichlet problem can be defined in several different ways. Dirichlet data is $\beta \in \mathrm{CT}(H^{1/2}(\Gamma))$. We can consider a causal tempered $H^1_\Delta(\Omega^+)$-valued distribution such that

$$\ddot{u} = \Delta u \qquad \text{and} \qquad \gamma u = \beta,$$

with equalities as elements of $\mathrm{CT}(L^2(\Omega^+))$ and $\mathrm{CT}(H^{1/2}(\Gamma))$ respectively. We can also consider u as in (2.2) satisfying the boundary conditions

$$\gamma^+ u = \beta, \qquad \gamma^- u = 0.$$

(This is a way of taking $u(t) = 0$ in Ω^- for all t, as we will see once we show the corresponding existence and uniqueness results.) A third option consists of looking for u as in (2.2) satisfying

$$\gamma^+ u = \gamma^- u = \beta.$$

This will lead to the single layer representation of the exterior solution. You might be inclined to think that this is just fooling around with the interior domain, when all we care about is what happens in Ω^+. The truth is that for layer potentials, there is no inside or outside; what layer potentials see is the domain minus the boundary $\mathbb{R}^d \setminus \Gamma$.

2.3 The Laplace transform

In this section we present the Laplace transform for causal tempered distributions. We note that the Laplace transform can be defined for more general distributions than the ones we are going to deal with. The positive half complex plane is denoted

$$\mathbb{C}_+ := \{s \in \mathbb{C} \, : \, \mathrm{Re}\, s > 0\}.$$

Laplace transform of a causal tempered distribution. Given $f \in \mathrm{CT}(X)$ we define the Laplace transform of f as the function

$$\mathbb{C}_+ \ni s \longmapsto \mathrm{F}(s) := \langle f, \exp(-s\,\cdot\,) \rangle. \tag{2.3}$$

Notice that even if the function $t \mapsto \exp(-s\,t)$ is not an element of the Schwartz class, the definition (2.3) still makes sense because of the causality of f. A fully precise definition of the duality in (2.3) can be done by using a smooth version of the Heaviside function:

$$h \in \mathcal{C}^\infty(\mathbb{R}) \qquad 0 \le h \le 1 \qquad h \equiv 1 \text{ in } [-1/2, \infty), \qquad h \equiv 0 \text{ in } (-\infty, -1].$$

We can then define $\mathrm{F}(s) := \langle f, h \exp(-s\,\cdot\,) \rangle$ and prove that the particular choice of h does not modify the definition of the Laplace transform. The Laplace transform of f is also denoted $\mathcal{L}\{f\}$. It is an analytic function of $s \in \mathbb{C}_+$.

Simple properties. The following two Laplace transforms are straightforward to compute

$$\mathcal{L}\{\delta_0 \otimes a\} \equiv a, \qquad \mathcal{L}\{\dot\delta_0 \otimes a\} = s\,a.$$

It is easy to prove that

$$\mathcal{L}\{\dot f\}(s) = s\mathrm{F}(s). \tag{2.4}$$

Again, for smooth bounded functions $f \, : \, [0,\infty) \to X$, the differentiation formula (2.4) is taking care of the value at zero. If f' is the distribution associated with the function $f' : [0,\infty) \to X$, then (2.4) is just a compact way of writing the much more popular formula (for functions)

$$\mathcal{L}\{f'\}(s) = s\mathrm{F}(s) - f(0).$$

Finally, if $f \in \mathrm{CT}(X)$ and $A \in \mathcal{B}(X, Y)$, then

$$\mathcal{L}\{Af\}(s) = A\,\mathrm{F}(s), \tag{2.5}$$

as can be easily verified.

An important result is the statement of injectivity of the Laplace transform (which is the main reason why it can be used as a transform): *if* $f \in \mathrm{CT}(X)$ *satisfies* $\mathrm{F}(s) = 0$ *for all* $s \in \mathbb{C}_+$, *then* $f = 0$.

2.4 Potentials in the Laplace domain

A first simple use of the Laplace transform gives uniqueness of solution to some wave propagation problems around Γ.

Proposition 2.4.1. *Let*

$$u \in \mathrm{CT}(H^1_\Delta(\mathbb{R}^d \setminus \Gamma)) \quad \textit{such that} \quad \ddot{u} = \Delta u$$

with equality as causal tempered $L^2(\mathbb{R}^d \setminus \Gamma)$*-valued distributions. If*

$$[\![\gamma u]\!] = 0, \qquad [\![\partial_\nu u]\!] = 0, \tag{2.6}$$

—as $H^{1/2}(\Gamma)$*- and* $H^{-1/2}(\Gamma)$*-valued distributions—, then* $u = 0$.

Proof. Note that for all $s \in \mathbb{C}_+$, $\mathrm{U}(s) \in H^1_\Delta(\mathbb{R}^d \setminus \Gamma)$ satisfies

$$s^2\mathrm{U}(s) = \Delta\mathrm{U}(s) \qquad \text{in } \mathbb{R}^d \setminus \Gamma,$$

and

$$[\![\gamma \mathrm{U}(s)]\!] = 0, \qquad [\![\partial_\nu \mathrm{U}(s)]\!] = 0.$$

(To see this, we just need to apply steady-state bounded operators to the Laplace transform and use (2.5).)

Applying Green's Theorem (the definition of the weak normal derivatives), we easily prove that

$$\int_{\mathbb{R}^d\setminus\Gamma} \left(|\nabla\mathrm{U}(s)|^2 + s^2|\mathrm{U}(s)|^2\right) = \langle \partial_\nu^- \mathrm{U}(s), \gamma^-\overline{\mathrm{U}(s)}\rangle_\Gamma - \langle \partial_\nu^+ \mathrm{U}(s), \gamma^+\overline{\mathrm{U}(s)}\rangle_\Gamma = 0.$$

Therefore

$$0 = \mathrm{Re}\left(\bar{s}\int_{\mathbb{R}^d\setminus\Gamma} \left(|\nabla\mathrm{U}(s)|^2 + s^2|\mathrm{U}(s)|^2\right)\right) = (\mathrm{Re}\, s)\left(\|\nabla\mathrm{U}(s)\|^2_{\mathbb{R}^d\setminus\Gamma} + |s|^2\|\mathrm{U}(s)\|^2_{\mathbb{R}^d\setminus\Gamma}\right),$$

which implies that $\mathrm{U}(s) = 0$ for all $s \in \mathbb{C}_+$. The injectivity theorem for the Laplace transform implies that $u = 0$. □

Towards a definition of the single layer potential. Formally, the single layer potential $u = \mathcal{S} * \lambda$ is defined as the solution of a transmission problem. If λ is a causal tempered $H^{-1/2}(\Gamma)$-valued distribution, we can think of finding $u \in \mathrm{CT}(H^1_\Delta(\mathbb{R}^d \setminus \Gamma))$ such that

$$\ddot{u} = \Delta u, \qquad [\![\gamma u]\!] = 0, \qquad [\![\partial_\nu u]\!] = \lambda. \tag{2.7}$$

In the next chapter, when we get to return to the time domain, we will see that problem (2.7) does actually give a correct definition of the retarded single layer potential.

The single layer potential in the Laplace domain. If u is a solution of (2.7), then for all $s \in \mathbb{C}_+$, $\mathrm{U}(s) \in H^1_\Delta(\mathbb{R}^d \setminus \Gamma)$ satisfies

$$\Delta \mathrm{U}(s) - s^2 \mathrm{U}(s) = 0 \quad \text{in } \mathbb{R}^d \setminus \Gamma, \qquad [\![\gamma \mathrm{U}(s)]\!] = 0 \qquad [\![\partial_\nu \mathrm{U}(s)]\!] = \Lambda(s), \tag{2.8}$$

where Λ is the Laplace transform of λ. However, the solution of (2.8) can be expressed using a single layer potential associated with the differential operator $\Delta - s^2$. For smooth enough densities $\lambda : \Gamma \to \mathbb{C}$ ($\lambda \in L^2(\Gamma)$ is enough), we define

$$\mathrm{S}(s)\lambda := \int_\Gamma E(\,\cdot\,, \mathbf{y}; s)\lambda(\mathbf{y})\, \mathrm{d}\Gamma(\mathbf{y}), \tag{2.9}$$

where

$$E(\mathbf{x}, \mathbf{y}; s) := \begin{cases} \frac{\imath}{4} H_0^{(1)}(\imath s|\mathbf{x} - \mathbf{y}|) & (d = 2), \\ \dfrac{e^{-s|\mathbf{x}-\mathbf{y}|}}{4\pi|\mathbf{x} - \mathbf{y}|} & (d = 3), \end{cases} \tag{2.10}$$

is the fundamental solution of $\Delta - s^2$. The integral formula (2.9) can be extended to any $\lambda \in H^{-1/2}(\Gamma)$. For any $s \in \mathbb{C}_+$, $\mathrm{S}(s)$ defines a bounded operator $\mathrm{S}(s) : H^{-1/2}(\Gamma) \to H^1_\Delta(\mathbb{R}^d \setminus \Gamma)$ such that $u = \mathrm{S}(s)\lambda$ is the unique solution of

$$\Delta u - s^2 u = 0 \quad \text{in } \mathbb{R}^d \setminus \Gamma, \quad [\![\gamma u]\!] = 0, \quad [\![\partial_\nu u]\!] = \lambda.$$

Therefore

$$\mathrm{U}(s) = \mathrm{S}(s)\Lambda(s) \qquad s \in \mathbb{C}_+. \tag{2.11}$$

Once again, we will have to wait until the next section to take an inverse Laplace transform in (2.11) and use it to give a definition of the layer potential in the time domain. For correct definitions of layer potentials and boundary integral operators for elliptic problems, the reader is referred to the very elegant presentation of Martin Costabel in [34]. The entire theory is presented in great detail in the often quoted monograph of William McLean on elliptic systems and their boundary integral representations [68].

The single layer operator. If our starting point is the Dirichlet problem:

$$u \in \mathrm{CT}(H^1_\Delta(\mathbb{R}^d \setminus \Gamma)), \qquad \ddot{u} = \Delta u, \qquad \gamma^+ u = \gamma^- u = \beta,$$

where $\beta \in \mathrm{CT}(H^{1/2}(\Gamma))$, we can use the Laplace transform to note that, if such u exists, then

$$\mathrm{U}(s) = \mathrm{S}(s)\Lambda(s),$$

where $\Lambda : \mathbb{C}_+ \to H^{-1/2}(\Gamma)$ is given at each $s \in \mathbb{C}_+$ as the solution of the boundary integral equation

$$\mathrm{V}(s)\Lambda(s) = \mathrm{B}(s),$$

where

$$H^{-1/2}(\Gamma) \ni \lambda \longmapsto \mathrm{V}(s)\lambda := \gamma^\pm \mathrm{S}(s)\lambda.$$

In Section 2.6 we will give bounds (in terms of s) for the norms $\mathrm{S}(s)$, $\mathrm{V}(s)$, and $\mathrm{V}^{-1}(s)$. These will be used to prove estimates in the time domain once we have an inversion formula.

Changing the background velocity. In the previous definitions we have taken $c = 1$. It is actually very simple to modify the speed of waves to another value of c. Taking the Laplace transform of $c^{-2}\ddot{u} = \Delta u$, it is clear that we get to the equation $\Delta \mathrm{U} - (s/c)^2 \mathrm{U} = 0$. Therefore, the layer potential and operator at this velocity are simply $\mathrm{S}(s/c)$ and $\mathrm{V}(s/c)$.

2.5 The energy norm

The key to understanding the single layer potential and operator in the time domain is the correct analysis of the Laplace transformed potential $\mathrm{S}(s)$ and its associated boundary integral operator $\mathrm{V}(s) = \gamma^\pm \mathrm{S}(s)$ for every $s \in \mathbb{C}_+$. For short, we will often write

$$\sigma := \mathrm{Re}\, s.$$

We start with some notation. For $\sigma > 0$, we write

$$\underline{\sigma} := \min\{1, \sigma\},$$

and we consider the norms

$$\|u\|^2_{c,\mathcal{O}} := \|\nabla u\|^2_{\mathcal{O}} + c^2 \|u\|^2_{\mathcal{O}}, \qquad \text{for } c > 0.$$

We also consider the bilinear form

$$a_{s,\mathcal{O}}(u,v) := (\nabla u, \nabla v)_{\mathcal{O}} + s^2 (u,v)_{\mathcal{O}}$$

and note the following simple properties:

$$|a_{s,\mathcal{O}}(u,v)| \le |||u|||_{|s|,\mathcal{O}} |||v|||_{|s|,\mathcal{O}} \qquad \forall u, v \in H^1(\mathcal{O}), \tag{2.12}$$

$$\mathrm{Re}\big(e^{-\imath \operatorname{Arg} s} a_{s,\mathcal{O}}(u,\overline{u})\big) = \frac{\sigma}{|s|} |||u|||^2_{|s|,\mathcal{O}} \qquad \forall u \in H^1(\mathcal{O}). \tag{2.13}$$

Also

$$\underline{\sigma} \|u\|_{1,\mathcal{O}} \le |||u|||_{|s|,\mathcal{O}} \le \frac{|s|}{\underline{\sigma}} \|u\|_{1,\mathcal{O}} \qquad \forall u \in H^1(\mathcal{O}), \tag{2.14}$$

where we have used that

$$\max\{1, |s|\} \min\{1, \mathrm{Re}\, s\} \le |s| \qquad \forall s \in \mathbb{C}_+. \tag{2.15}$$

The following careful lifting property, due to Alain Bamberger and Tuong Ha–Duong, appears (in a slightly different language) in [9]. Its proof relies on classical Sobolev space techniques: use a partition of unity, map to a half space, and use the Fourier representation of Sobolev spaces in free space. We will give a fully detailed proof in Section 2.7.

Proposition 2.5.1. *Let $\mathcal{O}$ be a Lipschitz domain. There exists $C_{\mathcal{O}} > 0$ such that for all $\xi \in H^{1/2}(\partial\mathcal{O})$ and $c > 0$, the solution $u \in H^1(\mathcal{O})$ of the Dirichlet problem*

$$\begin{aligned} -\Delta u + c^2 u &= 0 \qquad \text{in } \mathcal{O}, \\ \gamma u &= \xi \qquad \text{on } \partial\mathcal{O} \end{aligned}$$

can be bounded by

$$|||u|||_{c,\mathcal{O}} \le C_{\mathcal{O}} \max\{1, c\}^{1/2} \|\xi\|_{1/2,\partial\mathcal{O}}.$$

As a consequence, we obtain a bound for the normal derivative of the solution of $\Delta u - s^2 u = 0$ for $s \in \mathbb{C}_+$: [9, 60].

Proposition 2.5.2. *Let $\mathcal{O}$ be a Lipschitz domain and let $C_{\mathcal{O}} > 0$ be the constant of Proposition 2.5.1. If $s \in \mathbb{C}_+$ and $u \in H^1_\Delta(\mathcal{O})$ satisfies $\Delta u - s^2 u = 0$ in $\mathcal{O}$, then*

$$\|\partial_\nu u\|_{-1/2,\partial\mathcal{O}} \le C_{\mathcal{O}} \left(\frac{|s|}{\underline{\sigma}}\right)^{1/2} |||u|||_{|s|,\mathcal{O}}.$$

Proof. Let $\xi \in H^{1/2}(\partial\mathcal{O})$ and consider $v \in H^1(\mathcal{O})$ satisfying

$$\Delta v - |s|^2 v = 0, \qquad \gamma v = \xi.$$

Then,

$$\begin{aligned}
|\langle \partial_\nu u, \xi\rangle_{\partial\mathcal{O}}| &= \left|(\nabla u, \nabla v)_{\mathcal{O}} + s^2(u,v)_{\mathcal{O}}\right| && (\text{since } \Delta u - s^2 u = 0)\\
&= |a_{s,\mathcal{O}}(u,v)| \\
&\le |||u|||_{|s|,\mathcal{O}} |||v|||_{|s|,\mathcal{O}} && (\text{by (2.12)})\\
&\le C_{\mathcal{O}} \max\{1,|s|\}^{1/2} |||u|||_{|s|,\mathcal{O}} \|\xi\|_{1/2,\partial\mathcal{O}} && (\text{by Proposition 2.5.1})\\
&\le C_{\mathcal{O}} \left(\frac{|s|}{\underline{\sigma}}\right)^{1/2} |||u|||_{|s|,\mathcal{O}} \|\xi\|_{1/2,\partial\mathcal{O}}. && (\text{by (2.15)})
\end{aligned}$$

Therefore

$$\|\partial_\nu u\|_{-1/2,\partial\mathcal{O}} = \sup_{0\neq\xi\in H^{1/2}(\partial\mathcal{O})} \frac{|\langle \partial_\nu u, \xi\rangle_{\partial\mathcal{O}}|}{\|\xi\|_{1/2,\partial\mathcal{O}}} \le C_{\mathcal{O}} \left(\frac{|s|}{\underline{\sigma}}\right)^{1/2} |||u|||_{|s|,\mathcal{O}},$$

and the proof is finished. □

2.6 Bounds in the resolvent set

We start with a coercivity estimate [9, 60] for $\mathrm{V}(s)$. A time domain version of it will be given in Section 3.7.

Proposition 2.6.1. *There exists $C_\Gamma > 0$ such that for all $s \in \mathbb{C}_+$*

$$\mathrm{Re}\left(e^{\imath \operatorname{Arg} s}\langle \overline{\lambda}, \mathrm{V}(s)\lambda\rangle_\Gamma\right) \ge C_\Gamma \frac{\sigma\underline{\sigma}}{|s|^2}\|\lambda\|^2_{-1/2,\Gamma} \qquad \forall \lambda \in H^{-1/2}(\Gamma).$$

Therefore $\mathrm{V}(s) : H^{-1/2}(\Gamma) \to H^{1/2}(\Gamma)$ is invertible and

$$\|\mathrm{V}(s)^{-1}\|_{H^{1/2}(\Gamma)\to H^{-1/2}(\Gamma)} \le C_\Gamma^{-1}\frac{|s|^2}{\sigma\underline{\sigma}}.$$

Proof. The proof of this result relies on the results of Section 2.5 together with the understanding of $\mathrm{V}(s)$ in terms the solution of a transmission problem, a technique that goes back (at least) to the pioneering work of Jean-Claude Nédélec and Jacques Planchard [70] for the single layer operator of the three-dimensional Laplace equation. Let $\lambda \in H^{-1/2}(\Gamma)$ and $u = \mathrm{S}(s)\lambda$. Then $u \in H^1_\Delta(\mathbb{R}^d \setminus \Gamma)$ is the unique solution of the transmission problem

$$\Delta u - s^2 u = 0 \quad \text{in } \mathbb{R}^d \setminus \Gamma \qquad [\![\gamma u]\!] = 0, \qquad [\![\partial_\nu u]\!] = \lambda \tag{2.16}$$

and

$$\mathrm{V}(s)\lambda = \gamma^\pm u. \tag{2.17}$$

Therefore

$$\begin{aligned}
\langle \overline{\lambda}, \mathrm{V}(s)\lambda\rangle_\Gamma &= \langle \partial_\nu^- \overline{u}, \gamma^- u\rangle_\Gamma - \langle \partial_\nu^+ \overline{u}, \gamma^+ u\rangle_\Gamma && \text{(by (2.16) and (2.17))}\\
&= (\nabla \overline{u}, \nabla u)_{\mathbb{R}^d\setminus\Gamma} + \overline{s}^2 (\overline{u}, u)_{\mathbb{R}^d\setminus\Gamma} && \text{(since } \Delta\overline{u} - \overline{s}^2\overline{u} = 0)\\
&= a_{\overline{s},\mathbb{R}^d\setminus\Gamma}(\overline{u}, u).
\end{aligned}$$

Then

$$\begin{aligned}
\mathrm{Re}\big(e^{\imath \mathrm{Arg} s}\langle \overline{\lambda}, \mathrm{V}(s)\lambda\rangle_\Gamma\big) &= \mathrm{Re}\big(e^{-\imath \mathrm{Arg}\overline{s}} a_{\overline{s},\mathbb{R}^d\setminus\Gamma}(u, \overline{u})\big)\\
&= \frac{\sigma}{|s|} |\!|\!| u |\!|\!|^2_{|s|,\mathbb{R}^d\setminus\Gamma} && \text{(by (2.13))}\\
&\geq \frac{\sigma}{|s|}\frac{\underline{\sigma}}{|s|}\big(C^2_{\Omega^-}\|\partial_\nu^- u\|^2_{-1/2,\Gamma} + C^2_{\Omega^+}\|\partial_\nu^+ u\|^2_{-1/2,\Gamma}\big) && \text{(by Prop 2.5.2)}\\
&\geq C_\Gamma \frac{\sigma\underline{\sigma}}{|s|^2}\|\lambda\|^2_{-1/2,\Gamma}. && (\lambda = [\![\partial_\nu u]\!])
\end{aligned}$$

This coercivity estimate proves invertibility of $\mathrm{V}(s)$ by a Lax-Milgram argument and a bound for the inverse of $\mathrm{V}(s)$. □

Proposition 2.6.2. *There exists $C_\Gamma > 0$ such that for all $s \in \mathbb{C}_+$*

$$\|\mathrm{S}(s)\lambda\|_{1,\mathbb{R}^d\setminus\Gamma} \leq C_\Gamma \frac{|s|}{\sigma\underline{\sigma}^2}\|\lambda\|_{-1/2,\Gamma} \qquad \forall \lambda \in H^{-1/2}(\Gamma), \tag{2.18}$$

and

$$\|\mathrm{V}(s)\|_{H^{-1/2}(\Gamma)\to H^{1/2}(\Gamma)} \leq C_\Gamma^2 \frac{|s|}{\sigma\underline{\sigma}^2}. \tag{2.19}$$

Finally, there exists $C'_\Gamma > 0$ such that for all $s \in \mathbb{C}_+$,

$$\|\partial_\nu^\pm \mathrm{S}(s)\lambda\|_{-1/2,\Gamma} \leq C'_\Gamma \frac{|s|^{3/2}}{\sigma\underline{\sigma}^{3/2}}\|\lambda\|_{-1/2,\Gamma} \qquad \forall \lambda \in H^{-1/2}(\Gamma). \tag{2.20}$$

Proof. Let $\lambda \in H^{-1/2}(\Gamma)$ and $u = \mathrm{S}(s)\lambda$. Then

$$\begin{aligned}
\frac{\sigma}{|s|}\underline{\sigma}^2 \|u\|^2_{1,\mathbb{R}^d\setminus\Gamma} &\leq \frac{\sigma}{|s|} |\!|\!| u |\!|\!|^2_{|s|,\mathbb{R}^d\setminus\Gamma} && \text{(by (2.14))}\\
&= \operatorname{Re}\left(e^{\imath \operatorname{Arg} s}\langle\overline{\lambda}, \mathrm{V}(s)\lambda\rangle_\Gamma\right) && \text{(proof of Prop 2.6.1)}\\
&\leq \|\lambda\|_{-1/2,\Gamma}\|\gamma u\|_{1/2,\Gamma} && (\gamma u = \mathrm{V}(s)\lambda)\\
&\leq C_\Gamma \|\lambda\|_{-1/2,\Gamma}\|u\|_{1,\mathbb{R}^d\setminus\Gamma}. && (C_\Gamma = \|\gamma\|)
\end{aligned}$$

This proves (2.18) and therefore (2.19). Note finally that

$$\begin{aligned}
\|\partial_\nu^\pm u\|^2_{-1/2,\Gamma} &\leq D_\Gamma \frac{|s|}{\underline{\sigma}} |\!|\!| u |\!|\!|^2_{|s|,\mathbb{R}^d\setminus\Gamma} && \text{(by Proposition 2.5.2)}\\
&\leq C_\Gamma D_\Gamma \frac{|s|}{\underline{\sigma}}\frac{|s|}{\sigma}\|\lambda\|_{-1/2,\Gamma}\|\mathrm{V}(s)\lambda\|_{1/2,\Gamma} && \text{(see above)}\\
&\leq C_\Gamma^2 D_\Gamma \frac{|s|^3}{\sigma^2\underline{\sigma}^3}\|\lambda\|^2_{-1/2,\Gamma}, && \text{(by (2.19))}
\end{aligned}$$

which completes the proof. □

Proposition 2.6.3. *For all $s \in \mathbb{C}_+$,*

$$\|\mathrm{S}(s)\lambda\|_\Delta \leq 2C_\Gamma \frac{|s|^2}{\sigma\underline{\sigma}^3}\|\lambda\|_{-1/2,\Gamma} \qquad \forall\lambda \in H^{-1/2}(\Gamma),$$

where C_Γ is the constant of Proposition 2.6.2.

Proof. Let $u = \mathrm{S}(s)\lambda$. Then, by the arguments of the proof of Proposition 2.6.1 and by Proposition 2.6.2,

$$|\!|\!| u |\!|\!|^2_{|s|,\mathbb{R}^d\setminus\Gamma} \leq \frac{|s|}{\sigma}|\langle\overline{\lambda}, \mathrm{V}(s)\lambda\rangle_\Gamma| \leq C_\Gamma^2 \frac{|s|^2}{\sigma^2\underline{\sigma}^2}\|\lambda\|^2_{-1/2,\Gamma}.$$

Therefore, since $\Delta u = s^2 u$, we can bound

$$\|\Delta u\|_{\mathbb{R}^d\setminus\Gamma} = \|s^2 u\|_{\mathbb{R}^d\setminus\Gamma} \leq |s| |\!|\!| u |\!|\!|_{|s|,\mathbb{R}^d\setminus\Gamma} \leq C_\Gamma \frac{|s|^2}{\sigma\underline{\sigma}}\|\lambda\|_{-1/2,\Gamma}. \tag{2.21}$$

The result is then a consequence of (2.18) and (2.21) with the overestimate $\frac{1}{\underline{\sigma}} + |s| \leq 2\frac{|s|}{\underline{\sigma}^2}$. □

2.7 Appendix: Proof of the lifting lemma

In this section we give a proof of Proposition 2.5.1. This is done in three steps: lifting in the half space, lifting in the half space with compact support, and mapping to a general domain. We will denote

$$\mathbb{R}^d_+ := \mathbb{R}^{d-1} \times (0,\infty), \qquad \partial_\circ \mathbb{R}^d = \mathbb{R}^{d-1} \times \{0\},$$

and $\gamma_\circ : H^1(\mathbb{R}^d_+) \to H^{1/2}(\mathbb{R}^{d-1})$ for the trace operator. We will write $\mathbb{R}^d \ni \mathbf{x} = (\tilde{\mathbf{x}}, x_d) \in \mathbb{R}^{d-1} \times \mathbb{R}$. The first lemma builds a parameter-dependent lifting in the half space.

Lemma 2.7.1. *Let $\xi \in H^{1/2}(\mathbb{R}^{d-1}) \equiv H^{1/2}(\partial_\circ \mathbb{R}^d)$ and let $w \in H^1(\mathbb{R}^d_+)$ be the solution of*

$$-\Delta w + c^2 w = 0 \qquad \text{in } \mathbb{R}^d_+, \tag{2.22a}$$

$$\gamma_\circ w = \xi \qquad \text{on } \partial_\circ \mathbb{R}^d. \tag{2.22b}$$

Then

$$\|\nabla w\|^2_{\mathbb{R}^d_+} + c^2 \|w\|^2_{\mathbb{R}^d_+} \le (2\pi)^{d-1} \max\{1, c\} \|\xi\|^2_{1/2,\mathbb{R}^{d-1}}.$$

Proof. Using a density argument (for smooth functions) it is easy to show that the $(d-1)$-variable Fourier transform

$$\breve{w}(\boldsymbol{\omega}, x_d) := \int_{\mathbb{R}^{d-1}} e^{-\imath \boldsymbol{\omega}\cdot\tilde{\mathbf{x}}} w(\tilde{\mathbf{x}}, x_d) \mathrm{d}\tilde{\mathbf{x}}$$

transforms $L^2(\mathbb{R}^d_+)$ into itself. Again with a density argument, we can prove that the solution of (2.22) is in the anisotropic Sobolev space $\breve{H} := \{v \in L^2(\mathbb{R}^d_+) \,:\, \partial_{x_d} v \in L^2(\mathbb{R}^d_+)\}$ and satisfies

$$-\partial_{x_d x_d} \breve{w} + (|\boldsymbol{\omega}|^2 + c^2)\breve{w} = 0 \qquad \text{in } \mathbb{R}^d_+,$$

$$\breve{\gamma}_\circ \breve{w} = \hat{\xi} \qquad \text{on } \partial_\circ \mathbb{R}^d,$$

where $\breve{\gamma}_\circ : \breve{H} \to L^2(\mathbb{R}^{d-1})$ is the associated one-dimensional trace operator and $\hat{\xi}$ is the Fourier transform of ξ. Therefore

$$\breve{w}(\boldsymbol{\omega}, x_d) = \hat{\xi}(\boldsymbol{\omega}) e^{-x_d(|\boldsymbol{\omega}|^2+c^2)^{1/2}}.$$

In other words, we can prove that

$$w(\tilde{\mathbf{x}}, x_d) = \frac{1}{(2\pi)^{d-1}} \int_{\mathbb{R}^{d-1}} e^{\imath \boldsymbol{\omega}\cdot\tilde{\mathbf{x}}} \hat{\xi}(\boldsymbol{\omega}) e^{-x_d(|\boldsymbol{\omega}|^2+c^2)^{1/2}} \mathrm{d}\boldsymbol{\omega}$$

is the solution operator corresponding to problem (2.22). Using the Plancherel identity corresponding to the transform $w \mapsto \breve{w}$, we can easily justify the following:

$$\begin{aligned}
\|\nabla w\|^2_{\mathbb{R}^d_+} &+ c^2\|w\|^2_{\mathbb{R}^d_+} \\
&= (2\pi)^{d-1}\int_{\mathbb{R}^d_+}\Big(|\partial_{x_d}\breve{w}(\omega,x_d)|^2 + (c^2+|\omega|^2)|\breve{w}(\omega,x_d)|^2\Big)\mathrm{d}\omega\,\mathrm{d}x_d \\
&= (2\pi)^{d-1}\int_{\mathbb{R}^d_+}\Big(2(c^2+|\omega|^2)|\hat{\xi}(\omega)|^2 e^{-2x_d(c^2+|\omega|^2)^{1/2}}\Big)\mathrm{d}\omega\,\mathrm{d}x_d \\
&= (2\pi)^{d-1}\int_{\mathbb{R}^{d-1}}(c^2+|\omega|^2)^{1/2}|\hat{\xi}(\omega)|^2\mathrm{d}\omega \\
&\le (2\pi)^{d-1}\max\{1,c\}\int_{\mathbb{R}^{d-1}}(1+|\omega|^2)^{1/2}|\hat{\xi}(\omega)|^2\mathrm{d}\omega \\
&= (2\pi)^{d-1}\max\{1,c\}\|\xi\|^2_{1/2,\mathbb{R}^{d-1}}.
\end{aligned}$$

Note that the last equality is due to the Fourier transform representation of the space $H^{1/2}(\mathbb{R}^{d-1})$. □

In the second lemma leading towards the proof, we build a local parameter-dependent lifting from compactly supported functions in the hyperplane $x_n = 0$ to functions defined in the half-plane $x_n > 0$ that are compactly supported as well. This lemma is a technical step which is needed for the construction of a lifting in the original domain $\mathcal{O}$ by using a partition of unity and pullbacks (mappings from a reference configuration).

Lemma 2.7.2. *Let $r < 1$ and*

$$H^{1/2}_r(\mathbb{R}^{d-1}) := \{\xi \in H^{1/2}(\mathbb{R}^{d-1}) \ :\ \operatorname{supp}\xi \subset \overline{B(\mathbf{0};r)}\}.$$

Consider the following sets

$$\mho := B(\mathbf{0};1)\times(-1,1)\subset\mathbb{R}^d,\quad \mho_+ := \mho\cap\mathbb{R}^d_+,\quad \partial_\circ\mho := \mho\cap\partial_\circ\mathbb{R}^d = B(\mathbf{0};1)\times\{0\},$$

and the Sobolev space $H^1_\star(\mho_+) := \{u \in H^1(\mho_+)\ :\ \gamma u = 0 \text{ in } \partial\mho_+\setminus\partial_\circ\mho\}$. For every $c > 0$, there exists a bounded operator $\gamma_\circ^+ : H^{1/2}_r(\mathbb{R}^{d-1}) \to H^1_\star(\mho_+)$ satisfying

$$\gamma_\circ\gamma_\circ^+\xi = \xi \qquad \forall \xi \in H^{1/2}_r(\mathbb{R}^{d-1})$$

and

$$\|\nabla v\|^2_{\mho_+} + c^2\|v\|^2_{\mho_+} \le C\max\{1,c\}\|\xi\|^2_{1/2,\mathbb{R}^{d-1}},\qquad v = \gamma_\circ^+\xi$$

where $C > 0$ is independent of c.

Proof. Let $\varphi \in \mathcal{D}(\mathbb{R}^d)$ satisfy: $0 \le \varphi \le 1$, $\varphi \equiv 1$ in $B(\mathbf{0}; r) \times (-1/2, 1/2)$ and $\operatorname{supp}\varphi \subset \mho$. We then define $\gamma_\circ^+\xi := \varphi w$, where w solves (2.22) with $\max\{1, c\}$ instead of c. The rest of the proof is left as an exercise. □

Lemma 2.7.3. *Let $\mathcal{O}$ be a Lipschitz domain. For all $c > 0$ there exists an operator $\gamma^+ : H^{1/2}(\partial\mathcal{O}) \to H^1(\mathcal{O})$ such that*

$$\gamma\gamma^+\xi = \xi \quad \forall \xi \in H^{1/2}(\partial\mathcal{O})$$

and

$$\|\nabla v\|_{\mathcal{O}}^2 + c^2\|v\|_{\mathcal{O}}^2 \le C\max\{1, c\}\|\xi\|_{1/2,\partial\mathcal{O}}^2 \qquad v = \gamma^+\xi. \tag{2.23}$$

Proof. Let $\mho, \mho_+$ and $\partial_\circ\mho$ be as in Lemma 2.7.2. We can then find open sets Ω_j and Lipschitz maps with Lipschitz inverses $\chi_j : \mho \to \Omega_j$ such that $\{\Omega_1, \ldots, \Omega_M\}$ is a cover of $\partial\Omega$ with χ_j mapping $\mho_+$ into $\Omega_j \cap \mathcal{O}$ and $\partial_\circ\mho$ into $\Omega_j \cap \partial\mathcal{O}$. To this covering of $\partial\mathcal{O}$ we associate smooth compactly supported functions $\{\varphi_1, \ldots, \varphi_m\}$ such that $\operatorname{supp}\varphi_j \subset \Omega_j$ and $\sum_j \varphi_j \equiv 1$ in a neighborhood of $\partial\mathcal{O}$. The functions $\tilde{\varphi}_j := \varphi_j \circ \chi_j|_{\partial_\circ\mho}$ are compactly supported in $B(\mathbf{0}; 1) \equiv \partial_\circ\mho_+$ and we can therefore find $r > 0$ such that

$$\operatorname{supp}\tilde{\varphi}_j \subset \overline{B(\mathbf{0}; r)} \quad j = 1, \ldots, M.$$

Therefore, the functions

$$\tilde{\xi}_j := \tilde{\varphi}_j(\xi \circ \chi_j|_{\partial_\circ\mho}) = (\varphi_j\,\xi) \circ \chi_j|_{\partial_\circ\mho}$$

can be extended by zero to $\mathbb{R}^{d-1} \setminus B(\mathbf{0}; 1)$ and are thus elements of the space $H_r^{1/2}(\mathbb{R}^{d-1})$ introduced in Lemma 2.7.2. Finally we define

$$\gamma^+\xi := \sum_{j=1}^{M} (\gamma_\circ^+\tilde{\xi}_j) \circ \Xi_j, \tag{2.24}$$

where $\gamma_\circ^+$ is the lifting of Lemma 2.7.2, $\Xi_j : \Omega_j \cap \mathcal{O} \to \mho_+$ is the inverse of $\chi_j|_{\mho_+}$ and the functions $(\gamma_\circ^+\tilde{\xi}_j) \circ \Xi_j : \Omega_j \cap \mathcal{O} \to \mathbb{R}$ are extended by zero to the rest of $\mathcal{O}$. The rest of the proof is left as an exercise. □

Proof of Proposition 2.5.1. The solution of the Dirichlet problem

$$\begin{aligned} -\Delta u + c^2 u &= 0 \qquad \text{in } \mathcal{O}, \\ \gamma u &= \xi \qquad \text{on } \partial\mathcal{O} \end{aligned}$$

minimizes the quadratic functional $\|\nabla u\|_{\mathcal{O}}^2 + c^2\|u\|_{\mathcal{O}}^2$ among all elements of $H^1(\mathcal{O})$ such that $\gamma u = \xi$. Therefore, the result is a straightforward consequence of Lemma 2.7.3. □

2.8 Exercises

1. (Section 2.1) **Tensor product distributions.** Let $T \in \mathcal{S}'(\mathbb{R})$ be any tempered scalar distribution and let $a \in X$, where X is a Banach space. Show that $u := T \otimes a$ defined by

$$\langle T \otimes a, \varphi \rangle := \langle T, \varphi \rangle_{\mathcal{S}'(\mathbb{R}) \times \mathcal{S}(\mathbb{R})}\, a$$

is a tempered X-valued distribution.

2. (Section 2.1) **A paradoxical example of distributions.**

   (a) Let χ be the characteristic function of the interval $(0, 1)$. Show that the traveling wave function u

$$u(t) := \chi(\cdot - t) : \mathbb{R} \to \mathbb{R}$$

   is a tempered $H^2(\mathbb{R})$-valued distribution, even if as a function it never takes values on $H^2(\mathbb{R})$. **Hint.** Note that $\langle u, \varphi \rangle = \chi * \varphi$, where $*$ denotes convolution in $\mathbb{R}$.

   (b) Extend the previous example to any $\chi \in L^1(\mathbb{R})$ or $\chi \in L^2(\mathbb{R})$.

   (c) Show that

$$\ddot{u} = \partial_x^2 u$$

   as tempered distributions with values in $L^2(\mathbb{R})$. (Here $\partial_x^2 : H^2(\mathbb{R}) \to L^2(\mathbb{R})$ is the distributional derivative that acts as a steady-state operator.) **Hint.** Note that by definition $\langle \partial_x^2 u, \varphi \rangle = \partial_x^2 \langle u, \varphi \rangle$.

3. (Section 2.1) **Standing functions.** Let $\chi \in H^2(\mathbb{R})$ and define $u(t) := \chi$ for all t. Show that u is a tempered $H^2(\mathbb{R})$-valued distribution. Write u as a tensor product. Show that if $\chi \notin H^2(\mathbb{R})$, then u is not an $H^2(\mathbb{R})$-valued distribution.

4. (Section 2.1) **Time-harmonic waves.** Let $\omega > 0$. Consider the function $\phi_\omega(t) := \exp(-\imath \omega t)$ and let $v \in H^1_\Delta(\Omega)$ satisfy $\Delta v + \omega^2 v = 0$. Show that the $H^1_\Delta(\Omega)$-valued distribution $u := \phi_\omega \otimes v$ satisfies the wave equation

$$\ddot{u} = \Delta u,$$

with equality as tempered (noncasual) distributions with values in $L^2(\Omega)$.

5. (Section 2.1) **The passage of a plane wave.** Consider a function $\chi \in L^\infty(\mathbb{R})$, a bounded domain $\Omega \in \mathbb{R}^d$ and a fixed direction $\mathbf{d} \in \mathbb{R}^d$ (with $|\mathbf{d}| = 1$). For $t \in \mathbb{R}$, we define $u(t) : \Omega \to \mathbb{R}$ as

$$u(t)(\mathbf{x}) := \chi(\mathbf{x} \cdot \mathbf{d} - t) \qquad \mathbf{x} \in \Omega.$$

(a) Show that u is an $H^2(\Omega)$-valued tempered distribution. (Note that in principle u is not an $H^2(\mathbb{R}^d)$-valued distribution, unless $d = 1$ and χ satisfies some integrability conditions.)
(b) Show that $\ddot{u} = \Delta u$, with equality as tempered distributions with values in $L^2(\Omega)$.
(c) Give conditions on χ and Ω that ensure that u is causal.

6. (Section 2.1) Show that if $f \in \mathrm{CT}(X)$ is causal and $\operatorname{supp}\dot{f} \subset [T, \infty)$, then $\operatorname{supp} f \subset [T, \infty)$. (**Hint.** The proof is the same as for scalar-valued distributions. Choose a fixed $\varphi_0 \in \mathcal{D}(-\infty, T)$ such that $\int_{-\infty}^T \varphi_0 = 1$, and use the process

$$\mathcal{D}(-\infty, T) \ni \psi \longmapsto \varphi(t) := \int_{-\infty}^t \Big(\psi(\tau) - \Big(\int_{-\infty}^T \psi\Big)\varphi_0(\tau)\Big)\mathrm{d}\tau \in \mathcal{D}(-\infty, T)$$

to show that $\langle f, \psi \rangle = \int_{-\infty}^T \psi \langle f, \varphi_0 \rangle = 0$.)
7. (Section 2.7) Complete the proof of Lemma 2.7.2:

$$\|\nabla(\varphi w)\|^2_{\mho_+} + c^2 \|\varphi w\|^2_{\mho_+} \le 2\|\nabla w\|^2_{\mho_+} + (c^2 + 2\|\nabla \varphi\|^2_{L^\infty(\mho_+)})\|w\|^2_{\mho_+}.$$

This justifies why we can use $\max\{1, c\}$ instead of c in (2.22).
8. (Section 2.7) Complete the proof of Lemma 2.7.3. Use the fact that

$$\|\nabla(w \circ \chi_j)\|^2_{\Omega_j \cap \mathcal{O}} + c^2 \|w \circ \chi_j\|^2_{\Omega_j \cap \mathcal{O}} \le C\Big(\|\nabla w\|^2_{\mho_+} + c^2 \|w\|^2_{\mho_+}\Big) \qquad \forall w \in H^1(\mho_+).$$

Justify that the function defined in (2.24) is well defined as a function in $H^1(\Omega)$, that its trace is ξ, and that the bound (2.23) is satisfied.

Chapter 3
From Laplace domain to time domain

The task of identifying what functions $\mathrm{F}(s)$ are Laplace transforms of causal functions (distributions) is part of what is often called the Paley-Wiener Theorem, which is a collection of results related to holomorphic extensions of the Fourier transform that can be understood as two-sided Laplace transforms. Our presentation will first restrict the kind of symbols (Laplace transforms) that we want to invert. Part of the material that follows is adapted (and modified) from the PhD dissertation of Antonio Laliena [61].

3.1 Inversion of the Laplace transform

A class of symbols. Let X be a Banach space and $\mu \in \mathbb{R}$. We write $\mathrm{F} \in \mathcal{A}(\mu, X)$ when F is an analytic function

$$\mathrm{F} : \mathbb{C}_+ \to X$$

such that

$$\|\mathrm{F}(s)\| \le C_{\mathrm{F}}(\mathrm{Re}\, s)|s|^{\mu} \qquad \forall s \in \mathbb{C}_+,$$

where $C_{\mathrm{F}} : (0,\infty) \to (0,\infty)$ is a nonincreasing function such that

$$C_{\mathrm{F}}(\sigma) \le \frac{C}{\sigma^m} \qquad \forall \sigma \in (0,1], \qquad \text{for some } m \ge 0. \tag{3.1}$$

Since $1 \le |s|/(\mathrm{Re}\, s)$, it is clear that $\mathcal{A}(\mu, X) \subset \mathcal{A}(\mu+\varepsilon, X)$ for all $\varepsilon > 0$.

This class of symbols is a further refinement of a class introduced by Antonio Laliena and myself in [60] motivated by the potentials and operators on the resolvent

F.-J. Sayas, *Retarded Potentials and Time Domain Boundary Integral Equations*, Springer Series in Computational Mathematics 50,
DOI 10.1007/978-3-319-26645-9_3

set (see Section 2.6) and with a view on the analysis of Convolution Quadrature methods (see Chapter 4). The class is more restrictive than the one dealt with by Christian Lubich in his introduction to Convolution Quadrature methods for hyperbolic problems [66]. In Section 2.6, we have seen that

$$\begin{aligned}
\mathrm{S} &\in \mathcal{A}(1, \mathcal{B}(H^{-1/2}(\Gamma), H^1(\mathbb{R}^d \setminus \Gamma)),\\
\mathrm{S} &\in \mathcal{A}(2, \mathcal{B}(H^{-1/2}(\Gamma), H^1_\Delta(\mathbb{R}^d \setminus \Gamma)),\\
\mathrm{V} &\in \mathcal{A}(1, \mathcal{B}(H^{-1/2}(\Gamma), H^{1/2}(\Gamma)),\\
\mathrm{V}^{-1} &\in \mathcal{A}(2, \mathcal{B}(H^{1/2}(\Gamma), H^{-1/2}(\Gamma)),\\
\mathrm{K}^t &\in \mathcal{A}(\tfrac{3}{2}, \mathcal{B}(H^{-1/2}(\Gamma), H^{-1/2}(\Gamma)),
\end{aligned}$$

where we have introduced a new operator

$$\mathrm{K}^t(s) := \{\partial_\nu \mathrm{S}(s)\} = \tfrac{1}{2}(\partial_\nu^- \mathrm{S}(s) + \partial_\nu^+ \mathrm{S}(s)).$$

The operator $\mathrm{K}(s)$ will be defined in Section 3.4. At that time we will give a precise meaning to the transposition superscript t in $\mathrm{K}^t(s)$.

The strong inversion formula. We next deal with the inversion of the Laplace transform. The next paragraphs are rigorous but do not give details on the proofs. The interested reader is suggested to work Exercises 1 through 3 to fill in the details. Let $\mathrm{F} \in \mathcal{A}(\mu, X)$ with $\mu < -1$. For any $\sigma > 0$ we define

$$f(t) := \frac{1}{2\pi\imath}\int_{\sigma-\imath\infty}^{\sigma+\imath\infty} e^{st}\mathrm{F}(s)\mathrm{d}s = \frac{1}{2\pi}\int_{-\infty}^{\infty} e^{(\sigma+\imath\omega)t}\mathrm{F}(\sigma+\imath\omega)\,\mathrm{d}\omega. \tag{3.2}$$

A simple computation shows that $f(t)$ is well defined for all $t \in \mathbb{R}$ and that

$$\|f(t)\| \le \frac{1}{2\pi}C_\mathrm{F}(\sigma)\sigma^{1+\mu}e^{\sigma t}B\Big(\tfrac{1}{2}, -\tfrac{\mu+1}{2}\Big), \tag{3.3}$$

where B is the Euler beta function. A classical (but nontrivial) contour integration argument can then be used to prove that $f(t)$ is actually independent of $\sigma > 0$. Taking the limit as $\sigma \to \infty$ in (3.3), we can prove that $f(t) = 0$ for $t \le 0$. For $t \ge 1$, we can use (3.3) with $\sigma = t^{-1}$ and (3.1) to show that

$$\|f(t)\| \le \frac{1}{2\pi}C_\mathrm{F}(t^{-1})t^{|1+\mu|}B\Big(\tfrac{1}{2}, -\tfrac{\mu+1}{2}\Big) \le C\,t^{m+|1+\mu|}B\Big(\tfrac{1}{2}, -\tfrac{\mu+1}{2}\Big),$$

which ensures polynomial growth of $f(t)$ for large t. The Dominated Convergence Theorem can also be used to prove that f is a continuous function of t. Therefore f is a causal tempered distribution: $f \in \mathrm{CT}(X)$. Finally, it can be proved that $\mathcal{L}\{f\} = \mathrm{F}$. In summary, we have sketched the proof of the following result:

Proposition 3.1.1. *If* $\mathrm{F} \in \mathcal{A}(\mu, X)$ *with* $\mu < -1$, *then* F *is the Laplace transform of a continuous causal function* $f : \mathbb{R} \to X$ *with polynomial growth.*

The general case is a consequence of the formula $\mathcal{L}\{\phi^{(k)}\} = s^k \mathcal{L}\{\phi\}$, given the fact that $\mathrm{F} \in \mathcal{A}(\mu, X)$ implies that $s^{-k}\mathrm{F} \in \mathcal{A}(\mu - k, X)$. Proposition 3.1.1 can then be invoked for the symbol $s^{-k}\mathrm{F}(s)$ for sufficiently large k to obtain the following result:

Proposition 3.1.2. *Let* $\mathrm{F} \in \mathcal{A}(\mu, X)$ *with* $\mu \in \mathbb{R}$ *and let* $k := \min\{0, \lfloor \mu + 1 \rfloor\}$. *Then there exists a continuous causal function* $\phi : \mathbb{R} \to X$ *with polynomial growth such that* $\mathrm{F} = \mathcal{L}\{\phi^{(k)}\}$.

The set of all causal tempered distributions whose Laplace transforms are elements of $\mathcal{A}(\mu, X)$ for some μ can be studied similarly, starting with a simple case and proceeding by differentiation. The key idea is the fact that for a continuous causal function such that

$$\|f(t)\| \leq C(1 + t^m) \qquad \forall t \geq 0,$$

we can use the integral form of the Laplace transform and bound

$$\|\mathrm{F}(s)\| \leq C \int_0^\infty e^{-t\mathrm{Re}\, s}(1 + t^m)\mathrm{d}t = C\left(\frac{1}{\mathrm{Re}\, s} + \frac{m!}{(\mathrm{Re}\, s)^{m+1}}\right).$$

Proposition 3.1.3. *If* $f : \mathbb{R} \to X$ *is a continuous causal function with polynomial growth, then* $\mathrm{F} \in \mathcal{A}(0, X)$. *Therefore the set of symbols*

$$\bigcup_{\mu \in \mathbb{R}} \mathcal{A}(\mu, X)$$

is the set of the Laplace transforms of continuous causal functions $\mathbb{R} \to X$ *with polynomial growth and their distributional derivatives.*

For short, the set of all possible distributional derivatives of continuous causal X-valued functions with polynomial growth will be given the notation $\mathrm{TD}(X)$ (TD as in time domain). This is a nonstandard and temporary notation, waiting to be improved.

3.2 From symbols to convolution operators

Let X and Y be two Banach spaces. Let $a \in \mathrm{TD}(\mathcal{B}(X, Y)) \subset \mathrm{CT}(\mathcal{B}(X, Y))$ be such that $\mathrm{A} = \mathcal{L}\{a\} \in \mathcal{A}(\mu, \mathcal{B}(X, Y))$. Some very general results on the theory of vector-valued distributions can be invoked to prove that given any causal (though not necessarily tempered) X-valued distribution g, the convolution $a * g$ is well defined as a causal Y-valued distribution. Note that even if g also happens to be tempered, it does not follow that $a * g$ is tempered. However, the class $\mathrm{TD}(X)$ is mapped to the class $\mathrm{TD}(Y)$ by convolution with a and we will use this fact to give an alternative (but equivalent) definition of convolution.

If $a \in \mathrm{TD}(\mathcal{B}(X, Y))$ and $g \in \mathrm{TD}(X)$, we define

$$a * g := \mathcal{L}^{-1}\{\mathrm{A}\,\mathrm{G}\}. \tag{3.4}$$

It is easy to see that this definition is meaningful by noticing that if $\mathrm{G} \in \mathcal{A}(\nu, X)$, then

$$\|\mathrm{A}(s)\mathrm{G}(s)\|_Y \leq \|\mathrm{A}(s)\|_{\mathcal{L}(X,Y)}\|\mathrm{G}(s)\|_X \leq C_{\mathrm{A}}(\mathrm{Re}\,s)C_{\mathrm{G}}(\mathrm{Re}\,s)|s|^{\mu+\nu}.$$

One important detail concerns the preservation of some kind of delayed causality: namely, we want to prove that if $g = 0$ in $(-\infty, T)$, then so is $a * g$. For any $T \in \mathbb{R}$, we can define the translated distribution $g_T := g(\,\cdot\, - T)$ by

$$\langle g_T, \varphi\rangle := \langle g, \varphi(\,\cdot\, + T)\rangle.$$

If $g \in \mathrm{CT}(X)$ and $T > 0$, then $g_T \in \mathrm{CT}(X)$ (causality is not lost by a displacement to the right), but in general g_{-T} will not be causal. The assertion that g_{-T} is causal is an equivalent way of saying that $g = 0$ in $(-\infty, T)$.

Proposition 3.2.1. *Let* $T > 0$ *and* $g \in \mathrm{TD}(X)$ *be such that* $\mathrm{supp}\, g \subset [T, \infty)$ *and* $a \in \mathrm{TD}(\mathcal{B}(X, Y))$. *Then* $(a * g)_{-T} = a * g_{-T} \in \mathrm{TD}(Y)$.

Proof. The proof of this result follows from a collection of observations. First of all, note that if $g \in \mathrm{TD}(X)$ and $\mathrm{supp}\, g \in [T, \infty)$, it follows that $g_{-T} \in \mathrm{TD}(X)$. (See exercises.) Next,

$$\|\mathrm{G}(s)\| \leq C_{\mathrm{G}}(\mathrm{Re}\,s)|s|^{\nu} \qquad \forall s \in \mathbb{C}_+. \tag{3.5}$$

Since $g_{-T} \in \mathrm{CT}(X)$, then, by definition of the Laplace transform

$$\mathcal{L}\{g_{-T}\} = \langle g_{-T}, \exp(-\,\cdot\,\, s)\rangle = \langle g, \exp(-s(\,\cdot\, - T))\rangle = e^{sT}\mathrm{G}(s).$$

Given that $g_{-T} \in \mathrm{TD}(X)$, it follows that

$$\|e^{sT}\mathrm{G}(s)\| = e^{\mathrm{Re}\,s\,T}\|\mathrm{G}(s)\| \leq C_{\mathrm{G},T}(\mathrm{Re}\,s)|s|^{\nu'}. \tag{3.6}$$

In fact, comparing (3.5) and (3.6) it is clear that we can take $\nu = \nu'$ and the hypothesis on g_{-T} is just one on the behavior of $e^{\sigma T}C_{\mathrm{G}}(\sigma)$. (It should be noted that displacement does not modify the tempered character of a distribution. Moreover, the characterization of $\mathrm{TD}(X)$ as a space of derivatives of causal functions with polynomial growth makes transparent that what we are looking at is purely the preservation of causality by left displacement. More on this in Section 3.6.)

Therefore $\mathcal{L}\{a * g_{-T}\} = \mathrm{A}(s)\,(e^{sT}\mathrm{G}(s)) = e^{sT}\mathrm{A}(s)\mathrm{G}(s)$ and thus

$$\mathcal{L}\{(a * g_{-T})_T\} = e^{-sT}\mathcal{L}\{a * g_{-T}\} = \mathrm{A}(s)\mathrm{G}(s) \in \mathcal{A}(\mu + \nu, Y).$$

We have finally proved that $a * g_{-T} = (a * g)_{-T}$ and thus $(a * g)_{-T}$ is causal, which is what we wanted to show. □

The following result—taken from [43]—gives a taste of the kind of estimates that can be obtained for convolution operators by working in the Laplace domain. It is based on a result by Christian Lubich [66] restricted to the class of symbols we are working with, and uses a couple of clever observations made by Lehel Banjai.

Proposition 3.2.2. *Let* $\mathrm{A} = \mathcal{L}\{a\} \in \mathcal{A}(\mu, \mathcal{B}(X, Y))$ *with* $\mu \geq 0$ *and let*

$$k := \lfloor \mu + 2 \rfloor \qquad \varepsilon := k - (\mu + 1) \in (0, 1].$$

If $g \in \mathcal{C}^{k-1}(\mathbb{R}, X)$ *is causal and* $g^{(k)}$ *is integrable, then* $a * g \in \mathcal{C}(\mathbb{R}, Y)$ *is causal and*

$$\|(a * g)(t)\| \leq 2^{\mu} C_{\varepsilon}(t) C_{\mathrm{A}}(t^{-1}) \int_0^t \|(\mathcal{P}_k g)(\tau)\| \mathrm{d}\tau, \tag{3.7}$$

where

$$C_{\varepsilon}(t) := \frac{1+\varepsilon}{\pi\varepsilon} \frac{t^{\varepsilon}}{(1+t)^{\varepsilon}},$$

and

$$(\mathcal{P}_k g)(t) := e^{-t}(e^{\cdot} g)^{(k)}(t) = \sum_{\ell=0}^{k} \binom{k}{\ell} g^{(\ell)}(t).$$

Proof. If $g^{(k)}$ is causal and integrable, then it is tempered. Furthermore, it is the derivative of a continuous bounded causal function, and therefore $g^{(k)} \in \mathrm{TD}(X)$. Moreover, $\mathcal{L}\{\mathcal{P}_k g\} = (1 + s)^k \mathrm{G}(s) \in \mathcal{A}(0, X)$ (see Proposition 3.1.3) and therefore $\mathrm{G} \in \mathcal{A}(-k, X)$ and

$$\|(1 + s)^k \mathrm{G}(s)\| \leq \int_0^{\infty} \|(\mathcal{P}_k g)(\tau)\| \mathrm{d}\tau \qquad \forall s \in \mathbb{C}_+. \tag{3.8}$$

A simple bound now shows that $\mathrm{A}\,\mathrm{G} \in \mathcal{A}(\mu - k, Y)$ and because $\mu - k = -(1 + \varepsilon) < -1$, it follows from Proposition 3.1.1 that $a * g$ is continuous and causal. We can bound $(a * g)(t)$ using the strong form of the inversion formula, proceeding as follows:

$$
\begin{aligned}
\|(a*g)(t)\| &\le \frac{e^{\sigma t}}{2\pi}\int_{-\infty}^{\infty}\|\mathrm{A}(\sigma+\imath\omega)\mathrm{G}(\sigma+\imath\omega)\|\mathrm{d}\omega && \text{(by (3.2))}\\
&\le \frac{e^{\sigma t}}{2\pi}C_{\mathrm{A}}(\sigma)\int_{-\infty}^{\infty}\|(\sigma+\imath\omega)^{\mu}\mathrm{G}(\sigma+\imath\omega)\|\mathrm{d}\omega && (\mathrm{A}\in\mathcal{A}(\mu))\\
&\le \frac{e^{\sigma t}}{2\pi}C_{\mathrm{A}}(\sigma)\max_{\mathrm{Re}\,s=\sigma}\|(1+s)^{k}\mathrm{G}(s)\|\int_{-\infty}^{\infty}\frac{|\sigma+\imath\omega|^{\mu}}{|1+\sigma+\imath\omega|^{k}}\mathrm{d}\omega\\
&\le \frac{e^{\sigma t}}{\pi}C_{\mathrm{A}}(\sigma)\int_{0}^{\infty}\frac{|\sigma+\imath\omega|^{\mu}}{|1+\sigma+\imath\omega|^{k}}\mathrm{d}\omega\int_{0}^{\infty}\|(\mathcal{P}_{k}g)(\tau)\|\mathrm{d}\tau. && \text{(by (3.8))}
\end{aligned}
$$

We next bound

$$
\begin{aligned}
C(\sigma):=\int_{0}^{\infty}\frac{|\sigma+\imath\omega|^{\mu}}{|1+\sigma+\imath\omega|^{k}}\mathrm{d}\omega &\le \int_{0}^{\sigma}\frac{(2\sigma)^{\mu}}{(1+\sigma)^{k}}\mathrm{d}\omega+\int_{\sigma}^{\infty}\frac{(2\omega)^{\mu}}{\omega^{k}}\mathrm{d}\omega\\
&=(2\sigma)^{\mu}\sigma(1+\sigma)^{-k}+\frac{2^{\mu}}{k-\mu-1}\sigma^{1-k+\mu}\\
&=2^{\mu}\left(\sigma^{\mu}(1+\sigma)^{1-k}+\varepsilon^{-1}\sigma^{-\varepsilon}\right)\le 2^{\mu}(1+\varepsilon^{-1})\sigma^{-\varepsilon}.
\end{aligned}
$$

Taking $\sigma=t^{-1}$, we can estimate

$$
\|(a*g)(t)\|\le 2^{\mu}C_{\varepsilon}(t)C_{\mathrm{A}}(t^{-1})\int_{0}^{\infty}\|(\mathcal{P}_{k}g)(\tau)\|\mathrm{d}\tau,
$$

which is a bound similar to (3.7) with $\|\mathcal{P}_{k}g\|$ integrated over $(0,\infty)$ instead of $(0,t)$. Let us fix $t>0$ now and consider the function

$$
p(\tau):=\begin{cases} g(\tau), & \tau\le t,\\[4pt] e^{-\tau}\displaystyle\sum_{\ell=0}^{k-1}\frac{(\tau-t)^{\ell}}{\ell!}(e^{\cdot}g)^{(\ell)}(t), & \tau\ge t.\end{cases}
$$

Since p satisfies the same hypotheses as g, $a*p$ has the same properties as $a*g$. Also $p-g$ vanishes in $(-\infty,t)$ and hence, by Proposition 3.2.1, the continuous function $a*(g-p)$ vanishes in $(-\infty,t)$ and therefore

$$
\begin{aligned}
\|(a*g)(t)\|=\|(a*p)(t)\| &\le 2^{\mu}C_{\varepsilon}(t)C_{\mathrm{A}}(t^{-1})\int_{0}^{\infty}\|(\mathcal{P}_{k}p)(\tau)\|\mathrm{d}\tau\\
&=2^{\mu}C_{\varepsilon}(t)C_{\mathrm{A}}(t^{-1})\int_{0}^{t}\|(\mathcal{P}_{k}g)(\tau)\|\mathrm{d}\tau,
\end{aligned}
$$

due to the fact that $\mathcal{P}_{k}p=0$ in (t,∞). □

A word of warning. Passage through the Laplace domain is an elegant and powerful way of studying many convolution operators, but the direct and inverse transforms lose quite a lot of information. For instance, if $I : X \to X$ is the identity operator, then $\delta_0 \otimes I$ defines a distribution whose Laplace transform is the constant operator I, and it is therefore an element of the class $\mathcal{A}(0, \mathcal{B}(X, X))$. It is clear that Proposition 3.2.2 (that needs $k = 2$) is not giving the best possible bound of that style (a sort of Sobolev embedding theorem in one variable). We have also seen an example (Section 2.6) of symbols $\mathrm{V}(s)$ and $\mathrm{V}^{-1}(s)$, mutually inverse, both of positive order, although this fact reflects a deeper relation between the space and time variables.

3.3 Scattering by a sound-soft obstacle

The time domain single layer potential. By Proposition 2.6.3 and the preceding theory, there exists $\mathcal{S} \in \mathrm{TD}(\mathcal{B}(H^{-1/2}(\Gamma), H^1_\Delta(\mathbb{R}^d \setminus \Gamma)))$ such that $\mathcal{L}\{\mathcal{S}\} = \mathrm{S}$. Therefore, given $\lambda \in \mathrm{TD}(H^{-1/2}(\Gamma))$, the convolution

$$u = \mathcal{S} * \lambda$$

defines a causal tempered distribution with values in $H^1_\Delta(\mathbb{R}^d \setminus \Gamma)$, that solves

$$\ddot{u} = \Delta u, \qquad [\![\gamma u]\!] = 0, \qquad [\![\partial_\nu u]\!] = \lambda. \tag{3.9}$$

(This can be seen by taking the Laplace transform $\mathrm{U}(s) = \mathrm{S}(s)\Lambda(s)$ and by applying the properties of $\mathrm{S}(s)$.) Because equation (3.9) has at most a causal tempered solution (see Proposition 2.4.1), $\mathcal{S} * \lambda$ is characterized by the above transmission (and initial value) problem.

Remark. For this presentation of the theory it is not relevant to know what $\mathcal{S}$ is. What can be proved, using smooth densities λ (and a density argument), is that $\mathcal{S} * \lambda$ can be represented in integral form (see Chapter 1). The causal distribution $\mathcal{S}$ for $d = 3$ was found in [59]: to $\varphi \in \mathcal{S}(\mathbb{R})$, it associates the operator

$$H^{-1/2}(\Gamma) \ni \lambda \longmapsto \int_\Gamma \frac{\varphi(\,\cdot\, - \mathbf{y})}{4\pi|\,\cdot\, - \mathbf{y}|} \lambda(\mathbf{y}) \mathrm{d}\Gamma(\mathbf{y}) \in H^1_\Delta(\mathbb{R}^d \setminus \Gamma).$$

The retarded single layer operator. If we define $\mathcal{V} := \gamma \mathcal{S}$ (γ is here a steady-state operator that is applied to the causal tempered distribution $\mathcal{S}$), it follows that $\mathcal{V} * \lambda \in \mathrm{TD}(H^{1/2}(\Gamma))$. We also know that $\mathcal{L}\{\mathcal{V}\} = \mathrm{V}$. An integral expression of $\mathcal{V} * \lambda$ can be obtained by taking the trace of $\mathcal{S} * \lambda$.

Extracting information from the bounds. The following bounds have been proved in Section 2.6:

$$\|\mathrm{S}(s)\|_{H^{-1/2}(\Gamma) \to H^1(\mathbb{R}^d \setminus \Gamma)} \le C \frac{|s|}{\sigma \underline{\sigma}^2} \qquad \|\mathrm{S}(s)\|_{H^{-1/2}(\Gamma) \to H^1_\Delta(\mathbb{R}^d \setminus \Gamma)} \le C \frac{|s|^2}{\sigma \underline{\sigma}^3}$$

$$\|\mathrm{V}(s)\|_{H^{-1/2}(\Gamma)\to H^{1/2}(\Gamma)} \le C\frac{|s|}{\sigma\underline{\sigma}^2} \quad \|\mathrm{V}^{-1}(s)\|_{H^{1/2}(\Gamma)\to H^{-1/2}(\Gamma)} \le C\frac{|s|^2}{\sigma\underline{\sigma}}.$$

The first bound and Proposition 3.2.2 ($k = 3$, $\varepsilon = 1$) can be used to show that for λ smooth enough,

$$\|(\mathcal{S} * \lambda)(t)\|_{1,\mathbb{R}^d} \le C\,\tfrac{t}{1+t}\, t \max\{1, t^2\} \int_0^t \|(\mathcal{P}_3\lambda)(\tau)\|_{-1/2,\Gamma} \mathrm{d}\tau.$$

Similarly (same Proposition, $\mu = 2$, $k = 4$, and $\varepsilon = 1$)

$$\|(\mathcal{V}^{-1} * \beta)(t)\|_{-1/2,\Gamma} \le C\,\tfrac{t}{1+t}\, t \max\{1, t\} \int_0^t \|(\mathcal{P}_4\beta)(\tau)\|_{1/2,\Gamma} \mathrm{d}\tau.$$

These inequalities correspond to the two steps of solving the Dirichlet problem

$$\ddot{u} = \Delta u, \qquad \gamma^+ u = \gamma^- u = \beta$$

with a boundary integral method: we first solve

$$\mathcal{V} * \lambda = g \qquad \Longleftrightarrow \qquad \lambda = \mathcal{V}^{-1} * \beta$$

and then input λ in the potential expression

$$u = \mathcal{S} * \lambda = \mathcal{S} * (\mathcal{V}^{-1} * \beta).$$

Associativity and more. The definition of convolution through the Laplace transform proves that associativity is satisfied:

$$\mathcal{S} * (\mathcal{V}^{-1} * \beta) = (\mathcal{S} * \mathcal{V}^{-1}) * \beta.$$

(Of course, the definition of $\mathcal{S} * \mathcal{V}^{-1}$ has to be done in a similar way.) To study the behavior of the convolution with the $\mathcal{B}(H^{1/2}(\Gamma), H^1(\mathbb{R}^d))$-valued causal distribution $\mathcal{S} * \mathcal{V}^{-1}$, we have to study the symbol $\mathrm{S}(s)\mathrm{V}^{-1}(s)$. In fact, we can prove that

$$\|\mathrm{S}(s)\mathrm{V}^{-1}(s)\|_{H^{1/2}(\Gamma)\to H^1(\mathbb{R}^d)} \le C\frac{|s|^{3/2}}{\sigma\underline{\sigma}^{3/2}}. \tag{3.10}$$

Proposition 3.2.2 ($\mu = 3/2$, $k = 3$, $\varepsilon = 1/2$) then yields:

$$\|u(t)\|_{1,\mathbb{R}^d} = \|(\mathcal{S} * \mathcal{V}^{-1} * \beta)(t)\|_{1,\mathbb{R}^d} \le C\frac{t^{1/2}}{1+t^{1/2}}\, t \max\{1, t^{3/2}\} \int_0^t \|(\mathcal{P}_3\beta)(\tau)\|_{1/2,\Gamma} \mathrm{d}\tau.$$

Proof of (3.10) Given $\xi \in H^{1/2}(\Gamma)$, the function $u := \mathrm{S}(s)\mathrm{V}^{-1}(s)\lambda$ is the solution of the Dirichlet problem

$$\Delta u - s^2 u = 0 \quad \text{in } \mathbb{R}^d \setminus \Gamma, \qquad \gamma^{\pm} u = \xi.$$

To give a bound of the norm of this function we decompose it as $u = u_\xi + (u - u_\xi)$, where

$$\Delta u_\xi - |s|^2 u_\xi = 0 \qquad \text{in } \mathbb{R}^d \setminus \Gamma \qquad \gamma u_\xi = \xi,$$

(this is the lifting of Proposition 2.5.1) and obviously $u - u_\xi \in H_0^1(\mathbb{R}^d \setminus \Gamma)$. Therefore

$$a_s(u, u - u_\xi) = 0. \tag{3.11}$$

From here on, the proof collects a simple coercivity argument and the lifting property Proposition 2.5.1. We first bound using coercivity and boundedness of the bilinear form a_s:

$$\begin{aligned}
\|\!|u - u_\xi\|\!|^2_{|s|,\mathbb{R}^d\setminus\Gamma} &\le \frac{|s|}{\sigma} |a_s(u - u_\xi, u - u_\xi)| && \text{(by (2.13))}\\
&= \frac{|s|}{\sigma} |a_s(-u_\xi, u - u_\xi)| && \text{(by (3.11))}\\
&\le \frac{|s|}{\sigma} \|\!|u_\xi\|\!|_{|s|,\mathbb{R}^d\setminus\Gamma} \|\!|u - u_\xi\|\!|_{|s|,\mathbb{R}^d\setminus\Gamma}. && \text{(by (2.12))}
\end{aligned}$$

Next,

$$\begin{aligned}
\|\!|u\|\!|_{|s|,\mathbb{R}^d\setminus\Gamma} &\le \|\!|u - u_\xi\|\!|_{|s|,\mathbb{R}^d\setminus\Gamma} + \|\!|u_\xi\|\!|_{|s|,\mathbb{R}^d\setminus\Gamma}\\
&\le \frac{2|s|}{\sigma} \|\!|u_\xi\|\!|_{|s|,\mathbb{R}^d\setminus\Gamma} && \text{(see above)}\\
&\le \frac{2|s|}{\sigma} C_\Gamma \max\{1, |s|\}^{1/2} \|\xi\|_{1/2,\Gamma} && \text{(by Proposition 2.5.1)}\\
&\le 2C_\Gamma \frac{|s|^{3/2}}{\sigma \underline{\sigma}^{1/2}} \|\xi\|_{1/2,\Gamma},
\end{aligned}$$

which concludes the proof since $\underline{\sigma} \|u\|_{1,\mathbb{R}^d\setminus\Gamma} \le \|\!|u\|\!|_{|s|,\mathbb{R}^d\setminus\Gamma}$.

Concerning $\mathcal{V}^{-1}$. Just a warning about the notation $\mathcal{V}^{-1}$. In the Laplace domain,

$$\mathrm{V}^{-1}(s)\mathrm{V}(s) = I \qquad \forall s \in \mathbb{C}_+,$$

where $I : H^{-1/2}(\Gamma) \to H^{-1/2}(\Gamma)$ is the identity operator. When we take the inverse Laplace transform, we obtain

$$\mathcal{V}^{-1} * \mathcal{V} = \delta_0 \otimes I,$$

that is, the tempered distribution $\mathcal{V}^{-1}$ is a convolutional inverse of $\mathcal{V}$, which somehow justifies this notation. It is not an inverse of $\mathcal{V} : \mathcal{S}(\mathbb{R}) \to \mathcal{B}(H^{-1/2}(\Gamma), H^{1/2}(\Gamma))$ though.

3.4 The double layer potential

In the Laplace domain, the double layer potential is defined with the integral form

$$H^{1/2}(\Gamma) \ni \xi \longmapsto \mathrm{D}(s)\xi := \int_\Gamma \nabla_{\mathbf{y}} E(\,\cdot\,, \mathbf{y}; s) \cdot \boldsymbol{\nu}(\mathbf{y})\, \xi(\mathbf{y}) \mathrm{d}\Gamma(\mathbf{y}). \tag{3.12}$$

Here E is the fundamental solution of $\Delta - s^2$ (recall (2.10)). A simple computation shows that

$$\nabla_{\mathbf{y}} E(\mathbf{x}, \mathbf{y}; s) \cdot \boldsymbol{\nu}(\mathbf{y}) = \begin{cases} -\dfrac{s}{4} H_1^{(1)}(\imath s|\mathbf{x}-\mathbf{y}|) \dfrac{(\mathbf{x}-\mathbf{y}) \cdot \boldsymbol{\nu}(\mathbf{y})}{|\mathbf{x}-\mathbf{y}|} & (d=2) \\ \dfrac{e^{-s|\mathbf{x}-\mathbf{y}|}}{4\pi|\mathbf{x}-\mathbf{y}|} \dfrac{(\mathbf{x}-\mathbf{y}) \cdot \boldsymbol{\nu}(\mathbf{y})}{|\mathbf{x}-\mathbf{y}|} \Big(s + \dfrac{1}{|\mathbf{x}-\mathbf{y}|} \Big) & (d=3). \end{cases}$$

For $s \in \mathbb{C}_+$, the double layer potential provides the unique solution $u = \mathrm{D}(s)\xi \in H^1_\Delta(\mathbb{R}^d \setminus \Gamma)$ to the transmission problem:

$$\Delta u - s^2 u = 0 \quad \text{in } \mathbb{R}^d \setminus \Gamma, \qquad [\![\gamma u]\!] = -\xi, \qquad [\![\partial_\nu u]\!] = 0. \tag{3.13}$$

The integral operators associated with $\mathrm{D}(s)$ are

$$\mathrm{K}(s) := \{\!\{\gamma \mathrm{D}(s)\}\!\} \qquad \mathrm{W}(s) := \partial_\nu^\pm \mathrm{D}(s).$$

The operators $\mathrm{K}(s)$ and $\mathrm{K}^t(s)$ are real-transposed to each other, i.e.,

$$\langle \mathrm{K}^t(s)\lambda, \xi \rangle_\Gamma = \langle \lambda, \mathrm{K}(s)\xi \rangle_\Gamma \qquad \forall \lambda \in H^{-1/2}(\Gamma), \quad \xi \in H^{1/2}(\Gamma). \tag{3.14}$$

As always, our brackets are bilinear. The complex Hilbert space adjoint of $\mathrm{K}(s)$ is $\mathrm{K}^t(\overline{s})$. The proof of (3.14) follows simple techniques that are common in the manipulation of layer potentials and integral operators for elliptic problems. We next collect some Laplace domain bounds related to these operators [10, 60]. The techniques are very similar to those of the single layer operator (Section 2.6), and the reader is encouraged to move on and try to find what are the associated time domain bounds.

Proposition 3.4.1. *There exist generic constants (all of them denoted here by C_Γ) such that for all $s \in \mathbb{C}_+$:*

$$\mathrm{Re}\left(e^{-\imath \mathrm{Arg}\, s}\langle \mathrm{W}(s)\xi, \overline{\xi}\rangle_\Gamma\right) \ge C_\Gamma \frac{\sigma\underline{\sigma}^2}{|s|}\|\xi\|^2_{1/2,\Gamma} \qquad \forall \xi \in H^{1/2}(\Gamma), \tag{3.15}$$

$$\|\mathrm{W}(s)\|_{H^{1/2}(\Gamma)\to H^{-1/2}(\Gamma)} \le C_\Gamma \frac{|s|^2}{\sigma\underline{\sigma}}, \tag{3.16}$$

$$\|\mathrm{W}^{-1}(s)\|_{H^{-1/2}(\Gamma)\to H^{1/2}(\Gamma)} \le C_\Gamma \frac{|s|}{\sigma\underline{\sigma}^2}, \tag{3.17}$$

$$\|\mathrm{D}(s)\|_{H^{1/2}(\Gamma)\to H^1(\mathbb{R}^d\setminus\Gamma)} \le C_\Gamma \frac{|s|^{3/2}}{\sigma\underline{\sigma}^{3/2}}, \tag{3.18}$$

$$\|\mathrm{D}(s)\|_{H^{1/2}(\Gamma)\to H^1_\Delta(\mathbb{R}^d\setminus\Gamma)} \le C_\Gamma \frac{|s|^{5/2}}{\sigma\underline{\sigma}^{5/2}}. \tag{3.19}$$

Proof. Let $\xi \in H^{1/2}(\Gamma)$ and $u := \mathrm{D}(s)\xi$. Since $[\![\gamma u]\!] = -\xi$ and $\partial_\nu^\pm u = -\mathrm{W}(s)\xi$, the definition of the weak normal derivatives yields

$$\langle \mathrm{W}(s)\xi, \overline{\xi}\rangle_\Gamma = \langle -\partial_\nu u, -[\![\gamma \overline{u}]\!]\rangle_\Gamma = \langle \partial_\nu^- u, \gamma^-\overline{u}\rangle_\Gamma - \langle \partial_\nu^+ u, \gamma^+\overline{u}\rangle_\Gamma = a_s(u,\overline{u}), \tag{3.20}$$

and therefore

$$\begin{aligned}
\mathrm{Re}\left(e^{-\imath \mathrm{Arg}\, s}\langle \mathrm{W}(s)\xi, \overline{\xi}\rangle_\Gamma\right) &= \frac{\sigma}{|s|}|\!|\!| u |\!|\!|^2_{|s|,\mathbb{R}^d\setminus\Gamma} && \text{(coercivity of } a_s\text{: (2.13))}\\
&\ge \frac{\sigma\underline{\sigma}^2}{|s|}\|u\|^2_{1,\mathbb{R}^d\setminus\Gamma} && \text{(by (2.14))}\\
&\ge \|\gamma\|^{-2}\frac{\sigma\underline{\sigma}^2}{|s|}\|\xi\|^2_{1/2,\Gamma}, && \text{(trace theorem)}
\end{aligned}$$

which proves (3.15) and therefore (3.17). Going back to (3.20), and using the bound for the normal derivative given in Proposition 2.5.2, it follows that

$$|\!|\!| u |\!|\!|^2_{|s|,\mathbb{R}^d\setminus\Gamma} \le \frac{|s|}{\sigma}\|\partial_\nu u\|_{-1/2,\Gamma}\|\xi\|_{1/2,\Gamma} \le C_\Gamma \frac{|s|^{3/2}}{\sigma\underline{\sigma}^{1/2}}|\!|\!| u |\!|\!|_{|s|,\mathbb{R}^d\setminus\Gamma}\|\xi\|_{1/2,\Gamma}.$$

Therefore, since

$$\|u\|_{1,\mathbb{R}^d\setminus\Gamma} \le \frac{1}{\underline{\sigma}}|\!|\!| u |\!|\!|_{|s|,\mathbb{R}^d\setminus\Gamma} \qquad \|\Delta u\|_{\mathbb{R}^d\setminus\Gamma} = |s|\,\|su\|_{\mathbb{R}^d\setminus\Gamma} \le |s|\,|\!|\!| u |\!|\!|_{|s|,\mathbb{R}^d\setminus\Gamma},$$

and

$$\|\mathrm{W}(s)\xi\|_{-1/2,\Gamma} = \|\partial_\nu u\|_{-1/2,\Gamma} \le C_\Gamma \frac{|s|^{1/2}}{\underline{\sigma}^{1/2}} |\!|\!| u |\!|\!|_{|s|,\mathbb{R}^d\setminus\Gamma},$$

the proof of (3.16), (3.18), and (3.19) follows readily. □

As a direct consequence of the general theory we have established for our class of symbols in the Laplace domain, we can define time domain operator-valued distributions

$$\begin{aligned} \mathcal{D} &\in \mathrm{TD}(\mathcal{B}(H^{1/2}(\Gamma), H^1_\Delta(\mathbb{R}^d \setminus \Gamma))),\\ \mathcal{K} &\in \mathrm{TD}(\mathcal{B}(H^{1/2}(\Gamma), H^{1/2}(\Gamma))),\\ \mathcal{W} &\in \mathrm{TD}(\mathcal{B}(H^{1/2}(\Gamma), H^{-1/2}(\Gamma))),\\ \mathcal{W}^{-1} &\in \mathrm{TD}(\mathcal{B}(H^{-1/2}(\Gamma), H^{1/2}(\Gamma))). \end{aligned}$$

Recalling (3.13), if $\xi \in \mathrm{TD}(H^{1/2}(\Gamma)$, then $u := \mathcal{D}*\xi$ is the unique (causal tempered) solution of

$$\ddot{u} = \Delta u, \qquad [\![\gamma u]\!] = -\xi, \qquad [\![\partial_\nu u]\!] = 0. \tag{3.21}$$

3.5 Kirchhoff's formula

The previous results leave us ready to derive the full Calderón (boundary integral) calculus for the wave equation. It all starts with the weak version of Kirchhoff's formula:

Proposition 3.5.1. *Causal solutions of the wave equation*

$$u \in \mathrm{TD}(H^1_\Delta(\mathbb{R}^d \setminus \Gamma)) \qquad \ddot{u} = \Delta u \tag{3.22}$$

(equality as causal tempered distributions with values in $L^2(\mathbb{R}^d \setminus \Gamma)$) can be represented as

$$u = \mathcal{S} * [\![\partial_\nu u]\!] - \mathcal{D} * [\![\gamma u]\!]. \tag{3.23}$$

Proof. As already mentioned in Chapter 1, the keys to this representation are a uniqueness theorem for transmission problems (that is the role played by Proposition 2.4.1) and the interpretation of layer potentials as solutions of transmission problems, i.e., (3.9) for the single layer potential and (3.21) for the double layer potential. Given u satisfying (3.22) we can define

$$\varphi := [\![\gamma u]\!] \in \mathrm{TD}(H^{1/2}(\Gamma)) \quad \text{and} \quad \lambda := [\![\partial_\nu u]\!] \in \mathrm{TD}(H^{-1/2}(\Gamma))$$

Table 3.1 The jump relations of the layer potentials in two tables. The table of jumps is what defines the layer potentials. The table of averages is the definition of the boundary integral operators.

	$[\![\gamma\cdot]\!]$	$[\![\partial_\nu\cdot]\!]$
$\mathcal{S}*\lambda$	0	λ
$\mathcal{D}*\varphi$	$-\varphi$	0

	$\{\!\{\gamma\cdot\}\!\}$	$\{\!\{\partial_\nu\cdot\}\!\}$
$\mathcal{S}*\lambda$	$\mathcal{V}*\lambda$	$\mathcal{K}^t*\lambda$
$\mathcal{D}*\varphi$	$\mathcal{K}*\varphi$	$-\mathcal{W}*\varphi$

and note that the distribution

$$v := u - \mathcal{S}*\lambda + \mathcal{D}*\varphi \in \mathrm{TD}(H^1_\Delta(\mathbb{R}^d\setminus\Gamma))$$

is a causal solution of

$$\ddot{v} = \Delta v \qquad [\![\gamma v]\!] = 0 \qquad [\![\partial_\nu v]\!] = 0,$$

and therefore $v = 0$. □

Several useful formulas follow from the representation (3.23) and the definition of the integral operators (Table 3.1). For instance, for any solution of (3.22), we can write

$$\begin{bmatrix} \{\!\{\gamma u\}\!\} \\ \{\!\{\partial_\nu u\}\!\} \end{bmatrix} = \begin{bmatrix} \mathcal{V} & -\mathcal{K} \\ \mathcal{K}^t & \mathcal{W} \end{bmatrix} * \begin{bmatrix} [\![\partial_\nu u]\!] \\ [\![\gamma u]\!] \end{bmatrix}.$$

3.6 Another look at causality

There are several aspects of causality in the wave equation that are more or less hidden in the Laplace transform. We have already seen that if data are delayed (their support starts at time $T > 0$), so is the solution. (This was Proposition 3.2.1, which was just a simple statement on the convolutional character of all our operators.) We next see how finite speed of propagation is also observable in the Laplace transform. It all follows from the following simple abstract result, which was also hidden in the proof of Proposition 3.2.1:

Proposition 3.6.1. *Let* $\mathrm{F} \in \mathcal{A}(\mu, X)$, $\mathrm{F} = \mathcal{L}\{f\}$ *and*

$$(0,\infty) \ni \sigma \longmapsto C(\sigma) := \sup_{\mathrm{Re}\, s \geq \sigma} \|s^{-\mu}\mathrm{F}(s)\|.$$

If for some $T > 0$*, the function* $C(\sigma)e^{\sigma T}$ *is nonincreasing, then* f_{-T} *is causal.*

Proof. Note that the function $C(\sigma)$ is well defined, nonincreasing, and admits a rational bound as $\sigma \to 0$. (This is just a simple consequence of the fact that $C \leq C_F$.) If we define $G(s) := e^{sT}\mathrm{G}(s)$, the above hypotheses imply that $\mathrm{G} \in \mathcal{A}(\mu, X)$. Then let $g \in \mathcal{L}^{-1}\{\mathrm{G}\} \in \mathrm{TD}(X)$; it is simple to see that

$$\mathcal{L}\{g_T\} = e^{-sT}\mathrm{G}(s) = \mathrm{F}(s),$$

which means that $g_T = f$ and therefore $f_{-T} = g$ is causal. □

Another way of checking the hypothesis of Proposition 3.6.1 is the following: if we can write

$$\|\mathrm{F}(s)\| \leq e^{-T\,\mathrm{Re}\,s}\, D_\mathrm{F}(\mathrm{Re}\,s)\,|s|^\mu \qquad \forall s \in \mathbb{C}_+,$$

for some $T > 0$, where D_F is nonincreasing and $D_\mathrm{F}(\sigma) \leq C\sigma^{-m}$ for $\sigma \in (0,1)$, then, f_{-T} is causal, that is, the temporal support of f is contained in $[T, \infty)$.

Given the intimate relationship between causality and finite speed of propagation, it is not surprising that this delayed causality property shows how distance from the scatterer Γ gives a delay in the arrival of the solution.

Proposition 3.6.2. *Let $B \subset \mathbb{R}^d$ be such that $\overline{B} \cap \Gamma = \emptyset$ and let*

$$T := \mathrm{dist}(\overline{B}, \Gamma) := \inf\{|\mathbf{x} - \mathbf{y}| \,:\, \mathbf{x} \in B,\ \mathbf{y} \in \Gamma\} > 0.$$

If

$$u \in \mathrm{TD}(H^1_\Delta(\mathbb{R}^d \setminus \Gamma)) \qquad \ddot{u} = \Delta u,$$

then $u_{-T}|_B$ is causal, that is, the temporal support of the distribution $u|_B$ is contained in $[T, \infty)$.

Proof. Thanks to Kirchhoff's formula, we only need to check that this delay property is satisfied by single and double layer potentials. We will only give full details for the single layer potential in three dimensions. The arguments in two dimensions are very similar. The proof of the double layer potential is even simpler.

We are going to show that $s \in \mathbb{C}_+$

$$\sup_{\mathbf{x}\in B} \left\| \frac{e^{-s|\mathbf{x}-\,\cdot\,|}}{4\pi|\mathbf{x}-\,\cdot\,|} \right\|_{1/2,\Gamma} \leq e^{-\sigma T}\alpha(T)\frac{|s|^{1/2}}{\underline{\sigma}^{1/2}}, \qquad \sigma = \mathrm{Re}\,s. \tag{3.24}$$

Since for $\mathbf{x} \notin \Gamma$, the layer potential $\mathrm{S}(s)$ admits the form

$$(\mathrm{S}(s)\lambda)(\mathbf{x}) = \left\langle \lambda, \frac{e^{-s|\mathbf{x}-\,\cdot\,|}}{4\pi|\mathbf{x}-\,\cdot\,|} \right\rangle_\Gamma,$$

($\mathrm{S}(s)\lambda$ is actually a $\mathcal{C}^\infty$ function in $\mathbb{R}^d \setminus \Gamma$), the bound (3.24) gives

$$\|\mathrm{S}(s)\lambda\|_{H^{-1/2}(\Gamma)\to L^2(B)} \le C e^{-\sigma T}\alpha(T)\,\frac{|s|^{1/2}}{\underline{\sigma}^{1/2}},$$

and therefore Proposition 3.6.1 shows that $(\mathcal{S} * \lambda)|_B$, i.e., the left composition with the embedding operator $H^1(\mathbb{R}^d \setminus \Gamma) \to L^2(B)$, is supported in $[T,\infty)$.

To show (3.24) let us first denote

$$f := \frac{e^{-s|\mathbf{x}-\,\cdot\,|}}{|\mathbf{x}-\,\cdot\,|} \;:\; \Gamma \to \mathbb{C}.$$

Note that

$$\|f\|_\Gamma \le |\Gamma|^{1/2}\|f\|_{L^\infty(\Gamma)} \le \frac{|\Gamma|^{1/2}}{T}e^{-\sigma T}, \tag{3.25}$$

and that

$$\|f\|^2_{1,\Gamma} = \|f\|^2_\Gamma + \|\nabla_\Gamma f\|^2_\Gamma \le |\Gamma|\left(\|f\|^2_{L^\infty(\Gamma)} + \|\nabla f\|^2_{L^\infty(\Gamma)}\right) \tag{3.26}$$

(the first gradient is tangential and the second one is not). Since

$$\nabla f(\mathbf{y}) = \frac{e^{-s|\mathbf{x}-\mathbf{y}|}}{|\mathbf{x}-\mathbf{y}|}\left(s + \frac{1}{|\mathbf{x}-\mathbf{y}|}\right)\frac{\mathbf{x}-\mathbf{y}}{|\mathbf{x}-\mathbf{y}|},$$

it is clear that

$$\|\nabla f\|_{L^\infty(\Gamma)} \le \frac{e^{-\sigma T}}{T}\left(|s| + \frac{1}{T}\right). \tag{3.27}$$

Plugging (3.25) and (3.27) in (3.26) and bounding $|s| + \eta \le 2|s|\max\{1,\eta\}/\underline{\sigma}$, we can bound

$$\|f\|_{1,\Gamma} \le |\Gamma|^{1/2}\frac{e^{-\sigma T}}{T}\left(1 + \frac{1}{T}\right)\frac{|s|}{\underline{\sigma}},$$

Finally, using an interpolation property of Sobolev spaces,

$$\|f\|_{1/2,\Gamma} \le \|f\|^{1/2}_\Gamma\|f\|^{1/2}_{1,\Gamma}$$

we prove (3.24). □

Incident waves. Similar bounds can be used to prove that incident (spherical and plane) waves, with very weak smoothness conditions on the transmitted signal, can be used to provide data for scattering problems.

3.7 Coercivity in the time domain

The two coercivity properties we have obtained in the Laplace domain, namely (Proposition 2.6.1)

$$\mathrm{Re}\left(e^{\imath \mathrm{Arg}\, s}\langle \overline{\lambda}, \mathrm{V}(s)\lambda\rangle_\Gamma\right) \geq C_\Gamma \frac{\sigma \underline{\sigma}}{|s|^2}\|\lambda\|^2_{-1/2,\Gamma} \qquad \forall \lambda \in H^{-1/2}(\Gamma). \tag{3.28}$$

and (Proposition 3.4.1)

$$\mathrm{Re}\left(e^{-\imath \mathrm{Arg}\, s}\langle \mathrm{W}(s)\xi, \overline{\xi}\rangle_\Gamma\right) \geq C_\Gamma \frac{\sigma \underline{\sigma}^2}{|s|}\|\xi\|^2_{1/2,\Gamma} \qquad \forall \xi \in H^{1/2}(\Gamma),$$

can be rewritten as some kind of coercivity properties for the associated boundary integral operators in the time domain.

Plancherel's Formula. Consider a real Hilbert space X, whose inner product is denoted $\langle\,\cdot\,,\,\cdot\,\rangle_X$. Associated with X, there is a complex Hilbert space $X + \imath X$: its corresponding inner product can be constructed by separating real and imaginary parts. (We will keep the bracket bilinear instead of sesquilinear though.) Given $f \in L^2(\mathbb{R}, X)$, its Fourier transform $\mathcal{F}\{f\} \in L^2(\mathbb{R}, X + \imath X)$ is an extension of the operator (defined, for instance, for compactly supported functions):

$$\mathcal{F}\{f\}(\eta) = \int_{-\infty}^{\infty} e^{-\imath\eta t} f(t)\mathrm{d}t$$

(the integral is in the sense of Bochner in X). Plancherel's Formula is the statement

$$\int_{-\infty}^{\infty} \langle \overline{\mathcal{F}\{f\}(\eta)}, \mathcal{F}\{g\}(\eta)\rangle_X \mathrm{d}\eta = \frac{1}{2\pi}\int_{-\infty}^{\infty} \langle f(t), g(t)\rangle_X \mathrm{d}t,$$

which is actually in the basis of the possibility of defining $\mathcal{F}$ for functions in $L^2(\mathbb{R}, X)$. When X is the pivotal space of a Gelfand triple

$$Y \subset X \cong X' \subset Y'$$

(like $H^{1/2}(\Gamma) \subset L^2(\Gamma) \subset H^{-1/2}(\Gamma)$), the result can be extended for $f \in L^2(\mathbb{R}, Y')$ and $g \in L^2(\mathbb{R}, Y)$.

Laplace and Fourier. If $f \in L^2((0,\infty),X)$ (extended by zero for negative t), we can easily write

$$\mathrm{F}(s) = \int_0^\infty e^{-st} f(t)\mathrm{d}t = \mathcal{F}\{e^{-\sigma t} f(t)\}(\eta) \qquad s = \sigma + \imath\eta.$$

Therefore if $\mathrm{A} \in \mathcal{A}(\mu', Y')$ and $\mathrm{B} \in \mathcal{A}(\mu, Y)$, with $\mu + \mu' < -1$, we can integrate the duality

$$\langle \overline{\mathrm{A}(s)}, \mathrm{B}(s)\rangle_X$$

along a line $\mathrm{Re}\, s = \sigma$ (the fact that $\mu + \mu' < -1$ allows us to do that), and obtain

$$\int_{\sigma-\imath\infty}^{\sigma+\imath\infty} \langle \overline{\mathrm{A}(s)}, \mathrm{B}(s)\rangle_X \mathrm{d}s = \frac{1}{2\pi} \int_0^\infty e^{-2\sigma t} \langle a(t), b(t)\rangle_X \mathrm{d}t.$$

Coercivity for the single layer operator. We can then integrate the following expression, equivalent to (3.28),

$$\begin{aligned} \mathrm{Re}\left(\langle \overline{\Lambda(s)}, \mathrm{V}(s) s \Lambda(s)\rangle_\Gamma \right) &\geq C_\Gamma \sigma \underline{\sigma} \| s^{-1/2}\Lambda(s)\|^2_{-1/2,\Gamma} \\ &= C_\Gamma \sigma \underline{\sigma} (\overline{s^{-1/2}\Lambda(s)}, s^{-1/2}\Lambda(s))_{-1/2,\Gamma}. \end{aligned} \qquad (3.29)$$

(The bracket in the right-hand side is the $H^{-1/2}(\Gamma)$-inner product.) Formally, we obtain

$$\int_0^\infty e^{-2\sigma t} \langle \lambda(t), (\mathcal{V} * \dot{\lambda})(t)\rangle_\Gamma \mathrm{d}t \geq C_\Gamma \sigma \underline{\sigma} \int_0^\infty e^{-2\sigma t} \| \partial^{-1/2}\lambda(t)\|^2_{-1/2,\Gamma} \mathrm{d}t, \qquad (3.30)$$

where $\partial^{-1/2}\lambda = \mathcal{L}^{-1}\{s^{-1/2}\Lambda(s)\}$.

We can keep track of the symbols, starting with

$$\Lambda \in \mathcal{A}(\mu, H^{-1/2}(\Gamma)),$$

so that

$$s\mathrm{V}(s)\Lambda(s) \in \mathcal{A}(\mu+2, H^{1/2}(\Gamma)), \qquad s^{-1/2}\Lambda(s) \in \mathcal{A}(\mu - \tfrac{1}{2}, H^{-1/2}(\Gamma)).$$

The requirements for integration in the left- and right-hand sides of (3.29) are different. For the left-hand side we need $2\mu + 2 < -1$, that is, $\mu < -3/2$. For the right-hand side, it is enough with $2\mu - 1 < -1$ ($\mu < 0$). This mismatch in time regularity was already noticed in the seminal paper of Bamberger and Ha-Duong [9]. In that work, some anisotropic Sobolev spaces are introduced by freezing a particular value of σ. The kind of densities they deal with are those causal distributions (that admit Laplace transform) such that $e^{-\sigma\,\cdot}\,\dot{\lambda} \in L^2((0,\infty), H^{-1/2}(\Gamma))$ and

$e^{-\sigma\cdot}\,\mathcal{V}*\dot{\lambda}\in L^2((0,\infty),H^{1/2}(\Gamma))$. With that Hilbert space structure, (3.30) can be used as the starting point of a stability analysis for Galerkin methods for equations

$$\mathcal{V}*\lambda=\beta,$$

based on the weak formulation

$$\int_0^\infty e^{-2\sigma t}\langle\mu(t),(\mathcal{V}*\dot{\lambda})(t)\rangle_\Gamma \mathrm{d}t=\int_0^\infty e^{-2\sigma t}\langle\mu,\dot{\beta}(t)\rangle_\Gamma \mathrm{d}t.$$

Note that the type of Galerkin methods that we introduced in Section 1.7 did not include the exponential factor. To the best of my knowledge, there is no theoretical support for the elimination of this weight but experiments show that it is not needed.

3.8 Exercises

1. (Section 3.1) In the following exercises we fill in the gaps of the proof of Proposition 3.1.1. Let $\mathrm{F}:\mathbb{C}_+\to X$ satisfy

$$\|\mathrm{F}(s)\|\le C_\mathrm{F}(\mathrm{Re}\,s)|s|^\mu \qquad \forall s\in\mathbb{C}_+$$

where $C_\mathrm{F}:(0,\infty)\to(0,\infty)$ is nonincreasing, and consider the functions (for $\sigma>0$)

$$f_\sigma(t):=\frac{1}{2\pi\imath}\int_{\sigma-\imath\infty}^{\sigma+\imath\infty}e^{st}\mathrm{F}(s)\mathrm{d}s=\frac{1}{2\pi}\int_{-\infty}^{\infty}e^{\sigma+\imath\omega}\mathrm{F}(\sigma+\imath\omega)\mathrm{d}\omega.$$

(a) Show that f_σ is well defined and

$$\|f_\sigma(t)\|\le\frac{1}{2\pi}C_\mathrm{F}(\sigma)\sigma^{1+\mu}e^{\sigma t}\int_{-\infty}^{\infty}(1+\omega^2)^{\mu/2}\mathrm{d}\omega.$$

(b) Use the Dominated Convergence Theorem to show that f_σ is a continuous function of t.

(c) For fixed $0<\sigma_1<\sigma_2$ and $t\in\mathbb{R}$ we consider the quantities

$$\varepsilon_R^\pm:=\int_{\sigma_1\pm\imath R}^{\sigma_2\pm\imath R}e^{st}\mathrm{F}(s)\mathrm{d}s=e^{\pm R\imath t}\int_{\sigma_1}^{\sigma_2}e^{\omega t}\mathrm{F}(\omega\pm\imath R)\mathrm{d}\omega.$$

Show that $\lim_{R\to\infty}\varepsilon_R=0$. Use this and Cauchy's Homology Theorem to show that $f_{\sigma_1}(t)=f_{\sigma_2}(t)$. We can then write $f(t):=f_\sigma(t)$.

(d) Use the previous result and the bound of (a) to show that when $t < 0, f(t) = 0$.

(e) Show that if $C_F(1/\sigma)$ has tempered growth as $\sigma \to \infty$, then $f : \mathbb{R} \to X$ defines a causal tempered distribution. (**Hint.** As suggested in Section 3.1, take $\sigma = 1/t$ in the bound of (a).)

2. (Section 3.1) Assume that $\mathrm{F} : \mathbb{C}_+ \to X$ satisfies

$$\|\mathrm{F}(s)\| \le C_{\mathrm{F}}(\mathrm{Re}\, s)|s|^{\mu} \qquad \forall s \in \mathbb{C}_+$$

where $C_{\mathrm{F}} : (0,\infty) \to (0,\infty)$ is nonincreasing. Show that

$$\frac{1}{2\pi\imath}\int_{\sigma-\imath\infty}^{\sigma+\imath\infty} \frac{1}{z-s}\mathrm{F}(s)\mathrm{d}s = \begin{cases} 0 & 0 < \mathrm{Re}\, z < \sigma \\ \mathrm{F}(z) & 0 < \sigma < \mathrm{Re}\, z, \end{cases}$$

You can proceed as follows:

(a) If $\mathrm{Re}\, z < \sigma$, show that the integral

$$\frac{1}{2\pi\imath}\int_{\sigma-\imath\infty}^{\sigma+\imath\infty} \frac{1}{z-s}\mathrm{F}(s)\mathrm{d}s$$

does not depend on σ.

(b) Using (a) and taking $\sigma \to \infty$, the first part of the result is proved.

(c) When $0 < \sigma < \mathrm{Re}\, z$, take $\sigma' > \mathrm{Re}\, z$, use Cauchy's Representation Formula in a closed rectangular contour with vertices $\sigma \pm \imath R$ and $\sigma' \pm \imath R$ (with R large enough so that z is enclosed by the contour) and a limiting argument to show that

$$\frac{1}{2\pi\imath}\int_{\sigma-\imath\infty}^{\sigma+\imath\infty} \frac{1}{z-s}\mathrm{F}(s)\mathrm{d}s = F(z) + \frac{1}{2\pi\imath}\int_{\sigma'-\imath\infty}^{\sigma'+\imath\infty} \frac{1}{z-s}\mathrm{F}(s)\mathrm{d}s.$$

Then use (b) to prove the result.

3. (Section 3.1) Let F satisfy the same hypotheses and in the previous two problems. Show then that the Laplace transform of

$$f(t) := \frac{1}{2\pi\imath}\int_{\sigma-\imath\infty}^{\sigma+\imath\infty} e^{st}\mathrm{F}(s)\mathrm{d}s.$$

is F. (**Hint.** Use Problem 2.)

4. (Section 3.2) Show that if $g \in \mathrm{TD}(X)$ and $\mathrm{supp}\, g \subset [T,\infty)$, then $g_{-T} \in \mathrm{TD}(X)$. This can be done in the time domain, by writing $g = \phi^{(k)}$, where $\phi : \mathbb{R} \to X$ is a continuous causal function with polynomial growth (Proposition 3.1.3) and then showing that $\mathrm{supp}\, \phi \subset [T,\infty)$.

5. (Section 3.4) Using Propositions 3.4.1 and 3.2.2, give bounds for

$$\|(\mathcal{D} * \xi)(t)\|_{1,\mathbb{R}^d}, \qquad \|(\mathcal{W} * \xi)(t)\|_{-1/2,\Gamma}, \quad \text{and} \quad \|(\mathcal{W}^{-1} * h)(t)\|_{1/2,\Gamma}.$$

6. (Section 3.4). Find a bound of $\|\mathrm{D}(s)\mathrm{W}^{-1}(s)\|_{H^{-1/2}(\Gamma)\to H^1(\mathbb{R}^d\setminus\Gamma)}$ and use it to find bounds for $\|(\mathcal{D} * \mathcal{W}^{-1} * h)(t)\|_{1,\mathbb{R}^d\setminus\Gamma}$. As a hint, prove that if $u = \mathrm{D}(s)\mathrm{W}^{-1}(s)h$ where $h \in H^{-1/2}(\Gamma)$, then

$$a_{s,\mathbb{R}^d}(u, v) = \langle h, \gamma^+ v - \gamma^- v\rangle_\Gamma \quad \forall v \in H^1(\mathbb{R}^d \setminus \Gamma).$$

7. (Sections 3.3, 3.4, and 3.5) The exterior Dirichlet to Neumann operator $\mathrm{DtN}^+(s) : H^{1/2}(\Gamma) \to H^{-1/2}(\Gamma)$ is defined by

$$\mathrm{DtN}^+(s)\xi := \partial_\nu u \qquad \text{where} \qquad \left[\begin{array}{ll} \Delta u - s^2 u = 0 & \text{in } \Omega_+, \\ \gamma^+ u = \xi. & \end{array}\right.$$

Its inverse is the Neumann to Dirichlet operator $\mathrm{NtD}^+(s)$.

(a) Prove the following coercivity estimates

$$-\mathrm{Re}\left(e^{-\imath \mathrm{Arg}\, s}\langle \mathrm{DtN}^+(s)\xi, \overline{\xi}\rangle_\Gamma\right) \ge C_\Gamma \frac{\sigma \underline{\sigma}^2}{|s|^2}\|\xi\|^2_{1/2,\Gamma} \forall \xi \in H^{1/2}(\Gamma), \qquad \forall s \in \mathbb{C}_+,$$

$$-\mathrm{Re}\left(e^{\imath \mathrm{Arg}\, s}\langle \overline{\lambda}, \mathrm{NtD}^+(s)\lambda\rangle_\Gamma\right) \ge C_\Gamma \frac{\sigma \underline{\sigma}}{|s|^2}\|\lambda\|^2_{1/2,\Gamma} \forall \lambda \in H^{-1/2}(\Gamma), \qquad \forall s \in \mathbb{C}_+.$$

(b) Use the previous estimates to give bounds for

$$\|\mathrm{DtN}^+(s)\|_{H^{1/2}(\Gamma)\to H^{-1/2}(\Gamma)} \qquad \text{and} \qquad \|\mathrm{NtD}^+(s)\|_{H^{-1/2}(\Gamma)\to H^{1/2}(\Gamma)}.$$

(c) Write down the corresponding estimates for the time domain Dirichlet to Neumann and Neumann to Dirichlet operators.
(d) Show that

$$\mathrm{DtN}^+(s) = (-\tfrac{1}{2}\mathrm{I} + \mathrm{K}^t(s))\mathrm{V}^{-1}(s) = \mathrm{V}^{-1}(s)(-\tfrac{1}{2}\mathrm{I} + \mathrm{K}(s))$$

(**Hint.** For the first identity use a single layer potential representation. For the second one, use a direct boundary integral representation of the exterior solution.)

8. (Section 3.6) Complete the proof of Proposition 3.6.2 by providing estimates for the double layer potential in three dimensions and for the single and double layer potentials in two dimensions. Note that in the two dimensional case you will need to use estimates for the behavior of the Hankel functions $H_0^{(1)}$ and $H_1^{(1)}$.
9. (Section 3.7) Using the coercivity estimate of the form

$$\mathrm{Re}\,\langle \mathrm{W}(s)\xi, \overline{s\xi}\rangle_\Gamma \ge C_\Gamma \sigma \underline{\sigma}^2 \|\xi\|^2_{1/2,\Gamma} \qquad \forall \xi \in H^{1/2}(\Gamma),$$

derive a formal coercivity estimate in the time domain for the convolution operator $\xi \mapsto \mathcal{W} * \xi$. Write down what the corresponding variational formulation for the problem $\mathcal{W} * \xi = h$ would be.

Chapter 4
Convolution Quadrature

Convolution quadrature (CQ) is a discretization technique for causal convolutions and convolution equations. Coming from a mathematical argument that might seem bizarre at the beginning, this method ends up using data in the time domain but the Laplace transform of the operator. This mixture of Laplace and time domain looks somewhat unnatural but yields a general type of methods that can be easily used in black-box fashion. The method and much of its initial development are due to Christian Lubich. We will try not to be too heavy on notation for sequences by writing (a_n) to denote the sequence $(a_n)_{n\geq 0}$.

4.1 Discrete convolutions with CQ

The aim of this first section is to introduce, in a simple and non-rigorous way, one of the convolution quadrature methods, based upon the backwards Euler formula. Let $f : [0,\infty) \to \mathcal{B}(X,Y)$ be a (causal) operator-valued function for which the Laplace transform

$$\mathcal{L}\{f\}(s) = \mathrm{F}(s) := \int_0^\infty e^{-st} f(t)\mathrm{d}t$$

exists. Although this is not important at this point of our exposition, let us assume that $\mathrm{F}(s)$ exists for all $s \in \mathbb{C}_+$ and that it decays fast enough so that the inversion formula

$$f(t) = \frac{1}{2\pi\imath} \int_{\sigma-\imath\infty}^{\sigma+\imath\infty} e^{st}\mathrm{F}(s)\mathrm{d}s$$

F.-J. Sayas, *Retarded Potentials and Time Domain Boundary Integral Equations*, Springer Series in Computational Mathematics 50,
DOI 10.1007/978-3-319-26645-9_4

holds for all $\sigma > 0$ (see Section 3.1). Let now $g : [0, \infty) \to X$. We first aim to approximate the convolution $f * g : [0, \infty) \to Y$

$$(f * g)(t) := \int_0^t f(\tau)\, g(t-\tau) \mathrm{d}\tau.$$

At least formally (we are not going to verify conditions to apply Fubini's Theorem yet)

$$\begin{aligned}(f * g)(t) &= \int_0^t f(\tau)\, g(t-\tau)\, \mathrm{d}\tau \\ &= \int_0^t \left(\frac{1}{2\pi\imath} \int_{\sigma-\imath\infty}^{\sigma+\imath\infty} e^{s\tau}\, \mathrm{F}(s) \mathrm{d}s \right) g(t-\tau) \mathrm{d}\tau \\ &= \frac{1}{2\pi\imath} \int_{\sigma-\imath\infty}^{\sigma+\imath\infty} \mathrm{F}(s) \left(\int_0^t e^{s\tau} g(t-\tau)\, \mathrm{d}\tau \right) \mathrm{d}s.\end{aligned}$$

Note now that the function

$$y(t; s) := \int_0^t e^{s\tau} g(t-\tau) \mathrm{d}\tau \tag{4.1}$$

is the unique solution (with values in X) of the initial value problem

$$\begin{bmatrix} y' = s\,y + g \text{ in } [0, \infty), \\ y(0) = 0. \end{bmatrix} \tag{4.2}$$

The expression (4.1) is just the variation of constants formula for the initial value problem (4.2).

Let us now consider a constant time-step grid in $[0, \infty)$

$$0 = t_0 < t_1 < \ldots < t_n < \ldots, \qquad t_n := n\,\kappa.$$

We apply the backwards Euler method to (4.2)

$$\frac{y_n - y_{n-1}}{\kappa} = s\,y_n + g_n, \qquad g_n := g(t_n).$$

Therefore

$$y_n = y_n(s) = \frac{1}{1-\kappa\, s} y_{n-1} + \frac{\kappa}{1-\kappa\, s} g_n.$$

Instead of taking $y_0 = 0$, we will take $y_{-1} = 0$ as starting value (which fits with the idea of causality we have been exploring in previous chapters). Then, elementary algebra shows that

$$y_n(s) = \kappa \sum_{m=0}^{n} \frac{1}{(1-\kappa\, s)^{m+1}} g_{n-m},$$

which is a discrete version of

$$\int_0^{t_n} e^{s\tau} g(t_n - \tau) \mathrm{d}\tau.$$

Hence

$$\begin{aligned}
(f * g)(t_n) &= \frac{1}{2\pi\imath} \int_{\sigma-\imath\infty}^{\sigma+\imath\infty} \mathrm{F}(s) y(t_n; s) \mathrm{d}s \\
&\approx \frac{1}{2\pi\imath} \int_{\sigma-\imath\infty}^{\sigma+\imath\infty} \mathrm{F}(s) y_n(s) \mathrm{d}s \\
&= \sum_{m=0}^{n} \left(\frac{\kappa}{2\pi\imath} \int_{\sigma-\imath\infty}^{\sigma+\imath\infty} \frac{1}{(1-\kappa\, s)^{m+1}} \mathrm{F}(s) \mathrm{d}s \right) g_{n-m}.
\end{aligned}$$

We now turn our attention to what lies inside the bracket in the last expression

$$\begin{aligned}
\omega_m(\kappa) :&= \frac{\kappa}{2\pi\imath} \int_{\sigma-\imath\infty}^{\sigma+\imath\infty} \frac{1}{(1-\kappa\, s)^{m+1}} \mathrm{F}(s) \mathrm{d}s \\
&= \frac{(-1)^m}{\kappa^m} \frac{-1}{2\pi\imath} \int_{\sigma-\imath\infty}^{\sigma+\imath\infty} \frac{1}{(s-\frac{1}{\kappa})^{m+1}} \mathrm{F}(s) \mathrm{d}s \\
&= \frac{(-1)^m}{\kappa^m} \frac{1}{m!} \mathrm{F}^{(m)}\left(\frac{1}{\kappa}\right) \\
&= \frac{1}{m!} \frac{\mathrm{d}^m}{\mathrm{d}\zeta^m} \left(\mathrm{F}\left(\frac{1-\zeta}{\kappa}\right)\right)\bigg|_{\zeta=0}.
\end{aligned}$$

We have thus approximated the convolution $f * g$ at the points t_n by a discrete convolution

$$(f * g)(t_n) \approx \sum_{m=0}^{n} \omega_m(\kappa) g_{n-m}, \qquad g_n := g(t_n) \tag{4.3}$$

where

$$\mathrm{F}\left(\frac{1-\zeta}{\kappa}\right) = \sum_{m=0}^{\infty} \omega_m(\kappa) \zeta^m. \tag{4.4}$$

From (4.3) and (4.4), we may gather that the only things we need in order to apply the method to a particular convolution $f * g$ (even if it does not give a good result!) are the values of g at the discrete time values and the possibility of obtaining the Taylor expansion (4.4).

Generalization. Assume that $p(\zeta)$ is analytic in a neighborhood of the origin and that F is such that we can expand

$$\mathrm{F}\left(\frac{p(\zeta)}{k}\right) = \sum_{m=0}^{\infty} \omega_m(k)\zeta^m. \tag{4.5}$$

In this case, (4.3) defines a new method to approximate convolutions. From the point of view of approximation we will demand that $p(\zeta) \approx \log \zeta$ in a sense to be determined and that p is analytic near the origin. However, at this stage this is not needed. Since $\mathrm{F}(s)$ is defined on $\mathbb{C}_+$ (for the kind of symbols we are interested in), and we need the method for $\kappa \to 0$ (κ is the time-step and we want it to be small), a reasonable hypothesis is

$$\mathrm{Re}\, p(\zeta) > 0, \qquad |\zeta| < c_0,$$

for some $c_0 > 0$.

Composition of discrete convolutions. Assume that we have two different functions f_1 and f_2 and we have expanded

$$\mathrm{F}_1\left(\frac{p(\zeta)}{\kappa}\right) = \sum_{m=0}^{\infty} \omega_m^1(\kappa)\zeta^m, \qquad \mathrm{F}_2\left(\frac{p(\zeta)}{\kappa}\right) = \sum_{m=0}^{\infty} \omega_m^2(\kappa)\zeta^m.$$

It is then simple to see that for any sequence (g_n),

$$\sum_{m=0}^{n} \omega_m^1(\kappa)\left(\sum_{\ell=0}^{n-m} \omega_\ell^2(\kappa) g_{n-m-\ell}\right) = \sum_{m=0}^{n} \omega_m^{12}(\kappa) g_{n-m}, \qquad \forall n,$$

where

$$\omega_m^{12}(\kappa) := \sum_{\ell=0}^{m} \omega_\ell^1(\kappa)\omega_{m-\ell}^2(\kappa)$$

is the convolution of the discrete sequences $(\omega_m^1(\kappa))$ and $(\omega_m^2(\kappa))$ and at the same time the sequence of weights associated with the symbol $s \mapsto \mathrm{F}_1(s)\mathrm{F}_2(s)$.

4.2 The ζ transform

Let X be a Banach space and $\mathbf{h} := (h_n)_{n=0}^{\infty}$ be any sequence of elements of X. Consider the formal series

$$\mathrm{H}(\zeta) := \sum_{m=0}^{\infty} h_m \, \zeta^m.$$

Since it is a power series in the variable ζ, when it converges, it does so in an open ball in the complex plane centered at the origin and possibly in some points of its boundary. In that case, the elements of the sequence can be recovered by noticing that

$$h_m = \frac{1}{m!} \mathrm{H}^{(m)}(0),$$

since H is analytic in a neighborhood of the origin. In any case, we do not demand that this series converges. Constant series

$$\mathrm{H}(\zeta) \equiv h_0$$

correspond to almost trivial sequences $(h_0, 0, \ldots, 0, \ldots)$.

If the convolution of two sequences $\mathbf{h}^1$ and $\mathbf{h}^2$ is well defined,

$$(\mathbf{h}^1 * \mathbf{h}^2)_n = \sum_{m=0}^{n} h^1_m h^2_{n-m},$$

then the ζ-series associated with this convolution is simply the Cauchy product of the formal series $\mathrm{H}_1(\zeta)$ and $\mathrm{H}_2(\zeta)$. The concern for the correct definition of the convolution product is not referred to convergence issues but to the fact that the products $h^1_m h^2_{m-n}$ make sense, which is the case when:

(a) $\mathbf{h}^1$ takes values in $\mathcal{B}(X,Y)$ and $\mathbf{h}^2$ in $\mathcal{B}(Z,X)$ or,
(b) $\mathbf{h}^1$ takes values in $\mathcal{B}(X,Y)$ and $\mathbf{h}^2$ takes values in X.

Let $(a_n) \subset \mathcal{B}(X,Y)$ and let a_0 be invertible. Consider the sequence defined by the recurrence

$$a_0^{\mathrm{inv}} = a_0^{-1}$$

$$a_n^{\mathrm{inv}} = -\left(\sum_{m=0}^{n-1} a_m^{\mathrm{inv}} a_{n-m} \right) a_0^{-1}$$

and let

$$\mathrm{A}(\zeta) = \sum_{n=0}^{\infty} a_n \zeta^n, \qquad \mathrm{A}^{-1}(\zeta) = \sum_{n=0}^{\infty} a_n^{\mathrm{inv}} \zeta^n.$$

Then

$$\mathrm{A}^{-1}(\zeta)\mathrm{A}(\zeta) = I_X, \qquad \mathrm{A}(\zeta)\mathrm{A}^{-1}(\zeta) = I_Y$$

where I_X and I_Y are the respective identity operators in X and Y and, at the same time, the corresponding constant series.

In case $\mathrm{A}(\zeta)$ is convergent —that is, when it defines an analytic function $B(0,r) \to \mathcal{B}(X,Y)$— then if $\mathrm{A}(0) = a_0$ is invertible, $\mathrm{A}(\zeta)^{-1}$ exists for ζ sufficiently small and $\zeta \mapsto \mathrm{A}(\zeta)^{-1}$ is again analytic. The corresponding Taylor series is the same as the formal inversion of the series.

4.3 Convolution equations

Assume now that we are interested in solving a convolution equation

$$f * g = h,$$

where f and h are known: f is the operator part of the equation, h is the right-hand side. If

$$\mathrm{F}\left(\frac{p(\zeta)}{\kappa}\right) = \sum_{m=0}^{\infty} \omega_m(\kappa)\zeta^m$$

then we can try to solve the following triangular system of equations

$$\sum_{m=0}^{n} \omega_m(\kappa) g_{n-m} = h(t_n), \qquad n \geq 0.$$

Obviously this is equivalent to solving

$$\omega_0(\kappa) g_n = h(t_n) - \sum_{m=1}^{n} \omega_m(\kappa) g_{n-m}, \tag{4.6}$$

which is possible if $\omega_0(\kappa)$ is invertible. In the case

$$f : [0,\infty) \to \mathcal{B}(X,Y), \qquad h : [0,\infty) \to Y$$

the method provides a sequence (g_n) of elements of X, that we expect to approximate the sequence $(g(t_n))$ if this last one exists. All the steps in (4.6) involve solving an operator equation with data in Y and solution in X, always with the operator $\omega_0(\kappa)$.

Proposition 4.3.1. *Assume that* $\mathrm{F}^{-1}(s)$ *exists for all* $s \in \mathbb{C}_+$ *and that*

$$\mathrm{F}\left(\frac{p(\zeta)}{\kappa}\right) = \sum_{m=0}^{\infty} \omega_m(\kappa)\zeta^m, \qquad \mathrm{F}^{-1}\left(\frac{p(\zeta)}{\kappa}\right) = \sum_{m=0}^{\infty} \omega_m^{\mathrm{inv}}(\kappa)\zeta^m.$$

Then the sequence defined by (4.6) *coincides with the sequence*

$$g_n := \sum_{m=0}^{n} \omega_m^{\mathrm{inv}}(\kappa) h(t_{n-m}).$$

Proof. Since $\mathrm{F}(\frac{1}{\kappa}p(0)) = \omega_0(\kappa)$ is invertible if $\operatorname{Re} p(0) > 0$, then (4.6) defines a sequence (g_n). On the other hand

$$\mathrm{F}\left(\frac{p(\zeta)}{\kappa}\right)\mathrm{F}^{-1}\left(\frac{p(\zeta)}{\kappa}\right) = \mathrm{I}, \qquad |\zeta| \approx 0.$$

This implies that

$$\sum_{m=0}^{n} \omega_m(\kappa)\left[\sum_{\ell=0}^{n-m} \omega_\ell^{\mathrm{inv}}(\kappa) h(t_{n-m-\ell})\right] = h(t_n), \qquad \forall n.$$

By uniqueness of the equations (4.6), this sequence has to be (g_n). □

Let $p(\zeta)$ be a rational function

$$p(\zeta) = \frac{\alpha(\zeta)}{\beta(\zeta)} = \frac{\alpha_0 + \alpha_1\zeta + \ldots + \alpha_N\zeta^N}{\beta_0 + \beta_1\zeta + \ldots + \beta_N\zeta^N}, \tag{4.7}$$

with

$$\alpha_0\,\beta_0 \neq 0.$$

We consider an implicit N-step method to discretize

$$y' = f(t, y)$$

defined by the implicit equations

$$\begin{aligned} &\alpha_0 y_n + \alpha_1 y_{n-1} + \ldots + \alpha_N y_{n-N} && (4.8)\\ &\quad = \kappa\Big(\beta_0 f(t_n, y_n) + \beta_1 f(t_{n-1}, y_{n-1}) + \ldots + \beta_N f(t_{n-N}, y_{n-N})\Big), && n \geq 0. \end{aligned}$$

Then we can prove two results:

Proposition 4.3.2. *If we derive the CQ-method by approximating*

$$y(t_n;s) = \int_0^{t_n} e^{s\tau} g(t_n-\tau)\mathrm{d}\tau$$

with the multistep method (4.8) *for the equation*

$$\begin{bmatrix} y' = s\,y + g \text{ in } [0,\infty), \\ y(0) = 0, \end{bmatrix}$$

and zero starting values

$$y_{-1} = \ldots = y_{-N} = 0,$$

then we obtain the CQ method associated with p.

Proof. Specialized to the equation $y' = s\,y + g$, the multistep method is

$$\begin{aligned} &\alpha_0 y_n + \alpha_1 y_{n-1} + \ldots + \alpha_N y_{n-N} \\ &\qquad = \kappa\, s \left[\beta_0 y_n + \beta_1 y_{n-1} + \ldots + \beta_N y_{n-N}\right] + \kappa\big(\beta_0 g_n + \beta_1 g_{n-1} + \ldots + \beta_N g_{n-N}\big). \end{aligned}$$

The sequence is well defined if and only if $\alpha_0 - \kappa\, s\,\beta_0 \neq 0$, that is,

$$\tfrac{1}{\kappa}p(0) \neq s.$$

Taking as starting values $y{=}-1{=}\ldots{=}y_{-N}{=}0$ and using $g_{-1}{=}\ldots{=}g_{-N} :=0$, this corresponds to

$$\alpha(\zeta)\mathrm{Y}(\zeta) = \kappa\, s\,\beta(\zeta)\mathrm{Y}(\zeta) + \kappa\mathrm{G}(\zeta).$$

The formal series $\alpha(\zeta) - \kappa\, s\,\beta(\zeta)$ is invertible and

$$\mathrm{Y}(\zeta) = \mathrm{Y}(\zeta;s) = \big(\alpha(\zeta) - \kappa s\beta(\zeta)\big)^{-1}\kappa\,\beta(\zeta)\mathrm{G}(\zeta) = \big(\tfrac{1}{\kappa}p(\zeta) - s\big)^{-1}\mathrm{G}(\zeta).$$

Notice that $p(\zeta)$ is holomorphic near the origin and that $\frac{1}{\kappa}p(\zeta) - s$ is holomorphic and nonvanishing near $\zeta = 0$.

For ζ small

$$\frac{1}{2\pi\imath}\int_{\sigma-\imath\infty}^{\sigma+\imath\infty} \big(\tfrac{1}{\kappa}p(\zeta) - s\big)^{-1}\mathrm{F}(s)\mathrm{d}s = \mathrm{F}\left(\tfrac{1}{\kappa}p(\zeta)\right)$$

and, up to convergence questions, that is essentially all we wanted to show. □

Proposition 4.3.3. *If we apply the CQ method based on*

$$p(\zeta) = \frac{\alpha(\zeta)}{\beta(\zeta)} = \frac{\alpha_0 + \alpha_1\zeta + \ldots + \alpha_N\zeta^N}{\beta_0 + \beta_1\zeta + \ldots + \beta_N\zeta^N}$$

to an equation

$$(\delta_0 \otimes A) * g + (\delta_0' \otimes B) * g = h,$$

(where A and B are bounded operators) we obtain the same result as when applying the multistep method to the implicit differential equation

$$Ag + \tfrac{\mathrm{d}}{\mathrm{d}t}(Bg) = h.$$

4.4 A more abstract point of view

The convolution quadrature is actually a convolution at the continuous level. Let us start again from the beginning. We need p defined in a neighborhood of the origin such that $\mathrm{Re}\, p(\zeta) > 0$ for $|\zeta|$ small enough.

Assume that $\mathrm{F} : \mathbb{C}^+ \to \mathcal{B}(X,Y)$ is analytic. Then, the sequence of weights $(\omega_m(\kappa))$ is well defined with the Taylor expansion

$$\mathrm{F}\Big(\frac{p(\zeta)}{\kappa}\Big) = \sum_{m=0}^{\infty} \omega_m(\kappa)\zeta^m. \tag{4.9}$$

We can then define the $\mathcal{B}(X,Y)$-valued causal (possibly non-tempered) distribution:

$$f_\kappa := \sum_{m=0}^{\infty} \delta_{t_m} \otimes \omega_m(\kappa). \tag{4.10}$$

In this case, it is possible to apply the convolution of f_κ with any causal X-valued distribution g. This is how:

$$f_\kappa * g = \sum_{m=0}^{\infty} \omega_m(\kappa) g(\,\cdot\, - t_m).$$

If $g : \mathbb{R} \to X$ is a continuous causal function (as such, $g(0) = 0$), then

$$(f_\kappa * g)(t) = \sum_{m=0}^{\infty} \omega_m(\kappa) g(t - t_m) = \sum_{m=0}^{\lfloor t/\kappa \rfloor} \omega_m(\kappa) g(t - t_m)$$

is a causal Y-valued continuous function. We can then evaluate this convolution at the mesh points

$$(f_\kappa * g)(t_n) = \sum_{m=0}^{n} \omega_m(\kappa) g(t_{n-m}).$$

If we take the Laplace transform (assuming we can do it) in (4.10) we obtain

$$\begin{aligned} \mathrm{F}_\kappa(s) &= \sum_{m=0}^{\infty} e^{-st_m} \omega_m(\kappa) = \sum_{m=0}^{\infty} (e^{-s\kappa})^m \omega_m(\kappa) \\ &= \mathrm{F}\Big(\frac{p(e^{-s\kappa})}{\kappa}\Big). \qquad \text{(by (4.9))} \end{aligned}$$

We will see how, under some hypotheses on $p(\zeta)$, the approximation in the Laplace domain

$$\mathrm{F}(s) \longmapsto \mathrm{F}\big(\tfrac{p(e^{-s\kappa})}{\kappa}\big)$$

defines a transformation of symbols of the classes we studied in Chapter 3. This will justify that f_κ is a causal tempered distribution (and more) and that convolution quadrature can be viewed as discrete convolution (with sequences and ζ-transforms) or as continuous convolution (with distributions and Laplace transforms).

4.5 Symbol approximation by CQ

The building function. We need $p : U \to \mathbb{C}$ holomorphic in U, where $\{\zeta \in \mathbb{C} : |\zeta| \le 1\} \subset U$. Note that this precludes the existence of any kind of poles of p in the closed unit ball. Two main assumptions are made:

(a) $p : B(0,1) \to \mathbb{C}_+$.
(b) There exist $q \ge 1$, $C_0 > 0$ and $\kappa_0 > 0$ such that

$$\left| \frac{p(e^{-\kappa})}{\kappa} - 1 \right| \le C_0 \kappa^q \qquad \forall \kappa \le \kappa_0.$$

In particular $p(1) = 0$ and $p'(1) = -1$.

Unfortunately (because q is going to be the optimal order of the method), the previous hypotheses impose that $q \le 2$. (In the numerical ODE literature, this result is known as Dahlquist's Second Barrier.) Only three properties of this function will be used. They are presented in the following result. We will keep the notation of Chapters 2 and 3:

$$\sigma := \mathrm{Re}\, s \qquad \underline{\sigma} := \min\{1, \sigma\}.$$

The proof of this result uses standard techniques of complex analysis. It is included in Section 4.8.

Proposition 4.5.1. *If p satisfies the above properties and $\kappa_{\max} > 0$, there exist three positive constants C_1, C_2, C_3 such that for all $s \in \mathbb{C}_+$*

$$|p(e^{-s})| \le C_1 |s| \qquad |p(e^{-s}) - s| \le C_2 |s|^{q+1} \tag{4.11}$$

and

$$C_3 \underline{\sigma} \le \mathrm{Re}\left(\frac{p(e^{-s\kappa})}{\kappa}\right) \qquad \forall \kappa \le \kappa_{\max}. \tag{4.12}$$

The constant C_3 depends on $\kappa_{\max}$.

The next result shows that the discrete convolution operator f_κ is actually a convolution operator. Given a symbol F, we can define the approximate symbol

$$\mathrm{F}_\kappa(s) := \mathrm{F}\left(\frac{p(e^{-s\kappa})}{\kappa}\right).$$

The analysis of this new symbol and of its approximation properties to the original symbol $\mathrm{F}(s)$ are sketched next. Note that $\mathrm{F}_\kappa(s + \imath\kappa^{-1}) = \mathrm{F}_\kappa(s)$. This periodicity property reflects the fact that f_κ is discrete. Recall (Section 3.1) that the assumption $\mathrm{F} \in \mathcal{A}(\mu, X)$ means that we can bound

$$\|\mathrm{F}(s)\| \le C_\mathrm{F}(\mathrm{Re}\, s)|s|^\mu \qquad \forall s \in \mathbb{C}_+,$$

where C_F is nonincreasing and can be bounded $C_\mathrm{F}(\sigma) \le C\sigma^{-m}$ as σ tends to zero.

Proposition 4.5.2. *For $\mathrm{F} \in \mathcal{A}(\mu, X)$ with $\mu \ge 0$,*

$$\|\mathrm{F}_\kappa(s)\| \le C_1^\mu C_\mathrm{F}(C_3 \underline{\sigma})|s|^\mu.$$

Therefore F_κ is a symbol of order μ with a bounding function independent of κ and $f_\kappa \in \mathrm{TD}(X)$. Consequently

$$\|\mathrm{F}(s) - \mathrm{F}_\kappa(s)\| \le (1 + C_1^\mu) C_\mathrm{F}(\underline{C_3\, \sigma})|\sigma|^\mu \qquad \forall s \in \mathbb{C}_+. \tag{4.13}$$

Proof. Applying the definition, it follows that

$$\|\mathrm{F}_\kappa(s)\| \le C_{\mathrm{F}}\Bigg(\underbrace{\mathrm{Re}\left(\frac{p(e^{-s\kappa})}{\kappa}\right)}_{\ge C_3\underline{\sigma}}\Bigg)\left|\frac{p(e^{-s\kappa})}{\kappa}\right|^{\mu} \qquad \text{(by (4.12))}$$

$$\le C_{\mathrm{F}}(C_3\underline{\sigma})\left|\frac{p(e^{-s\kappa})}{s\kappa}\right|^{\mu}|s|^{\mu} \qquad (C_{\mathrm{F}}\text{ is nonincreasing})$$

$$\le C_{\mathrm{F}}(C_3\underline{\sigma})C_1^{\mu}|s|^{\mu}. \qquad \text{(by (4.11))}$$

The final bound follows from simple arguments:

$$\begin{aligned}|s|^{-\mu}\|\mathrm{F}(s)-\mathrm{F}_\kappa(s)\| &\le C_{\mathrm{F}}(\sigma)+C_1^{\mu}C_{\mathrm{F}}(C_3\underline{\sigma})\\ &\le C_{\mathrm{F}}(\underline{\sigma})+C_1^{\mu}C_{\mathrm{F}}(C_3\underline{\sigma})\\ &\le (1+C_1^{\mu})C_{\mathrm{F}}(\min\{1,C_3\}\,\underline{\sigma}).\end{aligned}$$

□

The remainder of this section is devoted to proving estimates for the difference $\mathrm{F}-\mathrm{F}_\kappa$. Propositions 4.5.5 and 4.5.6 will be the keys to prove estimates for CQ (this will be done in Section 4.6) with different kinds of regularity for the data. We start the process by proving that the classes $\mathcal{A}(\mu,X)$ are closed under differentiation.

Proposition 4.5.3. *If* $\mathrm{F}\in\mathcal{A}(\mu,X)$*, then* $\mathrm{F}'\in\mathcal{A}(\mu,X)$*. When* $\mu\ge 0$ *we can bound*

$$\|\mathrm{F}'(s)\| \le \left(\tfrac{3}{2}\right)^{\mu}\tfrac{2}{\sigma}C_{\mathrm{F}}(\tfrac{\sigma}{2})|s|^{\mu}.$$

Proof. Let $\Xi(s) := \{z\in\mathbb{C} : |z-s| = \sigma/2\}$ with positive orientation. We can then write

$$\mathrm{F}'(s) = \frac{1}{2\pi\imath}\int_{\Xi(s)}(z-s)^{-2}\mathrm{F}(z)\mathrm{d}z,$$

and bound using the fact that $|\Xi(s)| = \pi\sigma$, $\mathrm{Re}\,z\ge\sigma/2$ and $|z|\le\frac{3}{2}|s|$ for all $z\in\Xi(s)$. □

Proposition 4.5.4. *Let* $\mathrm{F}\in\mathcal{A}(\mu,X)$ *with* $\mu\ge 0$*. Then*

$$\|\mathrm{F}(s)-\mathrm{F}_\kappa(s)\| \le A\,D_{\mathrm{F}}(\sigma)\underline{\sigma}^{-1}\,\kappa^{q}|s|^{\mu+q+1} \qquad \forall s\in\mathbb{C}_+,$$

where

$$D_{\mathrm{F}}(\sigma) := C_{\mathrm{F}}(\tfrac{1}{2}\underline{C_3}\,\underline{\sigma})$$

The constant A depends on $p(\zeta)$ *(through the quantities of Proposition 4.5.1) and on* μ*.*

Proof. Recall that

$$\mathrm{F}(s) - \mathrm{F}_\kappa(s) = \mathrm{F}(s) - \mathrm{F}\left(\tfrac{p(e^{-s\kappa})}{\kappa}\right)$$

and let $\Xi_\kappa(s) := \{\alpha s + (1-\alpha)\frac{p(e^{-s\kappa})}{\kappa} \,:\, 0 \le \alpha \le 1\}$, be the straight line path than joins s and $\frac{p(e^{-s\kappa})}{\kappa}$. For $z \in \Xi_\kappa(s)$, we can bound

$$\operatorname{Re} z \ge \min\{\sigma, \operatorname{Re}\left(\frac{p(s^{-s\kappa})}{\kappa}\right)\} \ge \min\{\sigma, C_3\underline{\sigma}\} \ge \min\{1, C_3\}\underline{\sigma} = \underline{C_3\sigma}, \tag{4.14}$$

and

$$|z| \le \max\{|s|, \left|\tfrac{p(e^{-s\kappa})}{\kappa}\right|\} \le |s| \max\{1, C_1\}. \tag{4.15}$$

Using (4.14) and (4.15) in Proposition 4.5.3, it follows that

$$\begin{aligned} \|\mathrm{F}'(z)\| &\le \left(\tfrac{3}{2}\right)^\mu \frac{2}{\operatorname{Re} z} C_{\mathrm{F}}\left(\tfrac{1}{2}\operatorname{Re} z\right) |z|^\mu \\ &\le \tfrac{2}{\underline{C_3}} \left(\tfrac{3}{2}\max\{1, C_1\}\right)^\mu \tfrac{1}{\underline{\sigma}} C_{\mathrm{F}}\left(\tfrac{1}{2}\underline{C_3\sigma}\right) |s|^\mu \qquad \forall z \in \Xi_\kappa(s). \end{aligned}$$

This bound, the inequality

$$\left| s - \frac{p(e^{-s\kappa})}{\kappa} \right| = \kappa^{-1} \left| s\kappa - p(e^{-s\kappa}) \right| \le C_2 \kappa^q |s|^{q+1},$$

and

$$\|\mathrm{F}(s) - \mathrm{F}\left(\tfrac{p(e^{-s\kappa})}{\kappa}\right)\| \le \sqrt{2} \left| s - \frac{p(e^{-s\kappa})}{\kappa} \right| \max_{z \in \Xi_\kappa(s)} \|\mathrm{F}'(z)\|$$

(see Exercises) prove the result. □

Proposition 4.5.5 (L^1 estimates). *Let $r > 0$ and $\mathrm{F} \in \mathcal{A}(\mu, X)$ with $\mu \ge 0$. Then*

$$\|(\,\cdot\,)^{-1-\mu-r}(\mathrm{F} - \mathrm{F}_\kappa)\|_{L^1(\sigma + \imath\mathbb{R}, X)} \le C\, D_{\mathrm{F}}(\sigma)\underline{\sigma}^{-1} \max\{\underline{\sigma}^{-r}, \sigma^{q-r}\}\, E(\kappa),$$

where

$$E(\kappa) := \kappa^{r\frac{q}{q+1}} + \delta_{r,q+1}\kappa^q |\log \kappa| + \kappa^q$$

and the constant C depends on p, μ, and r.

Proof. We can gather the approximation properties proved in Propositions 4.5.2 and 4.5.4 in the inequality

$$\|s^{-1-\mu-r}(\mathrm{F}(s)-\mathrm{F}_\kappa(s))\| \le B\, D_{\mathrm{F}}(\sigma)\begin{cases} |s|^{-r-1}, \\ \underline{\sigma}^{-1}|s|^{q-r}\kappa^q, \end{cases} \tag{4.16}$$

where the constant B depends on μ and on $p(\zeta)$. If $s=\sigma+\imath\eta$ we can always bound

$$\tfrac{1}{\sqrt{2}}\underline{\sigma}(1+|\eta|) \le |s| \le \max\{1,\sigma\}(1+|\eta|).$$

Therefore

$$\begin{aligned} |s|^{-r-1} &\le 2^{\frac{r+1}{2}}\,\underline{\sigma}^{-r-1}(1+|\eta|)^{-r-1}, && \\ |s|^{q-r} &\le 2^{\frac{r-q}{2}}\underline{\sigma}^{q-r}(1+|\eta|)^{q-r}, && q-r\le 0, \\ |s|^{q-r} &\le \max\{1,\sigma\}^{q-r}(1+|\eta|)^{q-r}, && q-r>0. \end{aligned}$$

The last two inequalities can be collected in the following condensed expressions:

$$\begin{aligned} |s|^{q-r} &\le 2^{\frac{r-q}{2}}\max\{\underline{\sigma}^{-r},\sigma^{q-r}\}\,(1+|\eta|)^{q-r}, \\ |s|^{-r-1} &\le 2^{\frac{r+1}{2}}\,\underline{\sigma}^{-1}\max\{\underline{\sigma}^{-r},\sigma^{q-r}\}\,(1+|\eta|)^{-r-1}. \end{aligned}$$

Bringing these inequalities to (4.16) and collecting constants again, we can write

$$\begin{aligned} \|s^{-1-\mu-r}(\mathrm{F}(s)-\mathrm{F}_\kappa(s))\| \le{}& C\,D_{\mathrm{F}}(\sigma)\underline{\sigma}^{-1}\max\{\underline{\sigma}^{-r},\sigma^{q-r}\} \\ &\times\min\{(1+|\eta|)^{-r-1},\kappa^q(1+|\eta|)^{q-r}\}, \end{aligned}$$

with C depending on B (and thus on p and μ) and on the index r. Since for $r,q>0$,

$$\int_{-\infty}^{\infty}\min\{(1+|\eta|)^{-r-1},\kappa^q(1+|\eta|)^{q-r}\}d\eta \le C\times E(\kappa),$$

the result follows readily. □

Proposition 4.5.6 (Pointwise estimates). *Let* $\mathrm{F}\in\mathcal{A}(\mu,X)$ *with* $\mu\ge 0$.

(a) *If* r,ε *satisfy*

$$0\le\varepsilon\le r-(q+1),$$

then

$$\|\mathrm{F}(s)-\mathrm{F}_\kappa(s)\| \le C|s|^{\mu+r-\varepsilon}D_{\mathrm{F}}(\sigma)\underline{\sigma}^{-1}\sigma^{q+1-r+\varepsilon}\kappa^q \qquad \forall s\in\mathbb{C}_+.$$

(b) *If* r, ε *satisfy*

$$0 \le \varepsilon \le r \le q+1,$$

then

$$\|\mathrm{F}(s) - \mathrm{F}_\kappa(s)\| \le C|s|^{\mu+r-\varepsilon} D_\mathrm{F}(\sigma)\underline{\sigma}^{-1}\,\kappa^{(r-\varepsilon)\frac{q}{q+1}} \qquad \forall s \in \mathbb{C}_+.$$

The constant C *depends on* μ, p, *and* r.

Proof. Using Propositions 4.5.2 and 4.5.4 (see also (4.16)) we can easily bound

$$\begin{aligned}\|\mathrm{F}(s) - \mathrm{F}_\kappa(s)\| &\le B\,|s|^\mu D_\mathrm{F}(\sigma)\min\{1, |s|^{q+1}\kappa^q\underline{\sigma}^{-1}\} \\ &= \underbrace{B|s|^\mu D_\mathrm{F}(\sigma)\underline{\sigma}^{-1}}_{=\Xi}\min\{1, \sigma, |s|^{q+1}\kappa^q\}.\end{aligned}$$

With the hypotheses (a) for the parameters, we continue the bounds as

$$\|\mathrm{F}(s) - \mathrm{F}_\kappa(s)\| \le \Xi|s|^{q+1}\kappa^q \le \Xi|s|^{r-\varepsilon}\sigma^{q+1-(r-\varepsilon)}\kappa^q,$$

due to the fact that $q+1-r+\varepsilon \le 0$.

In the case of (b), we have that $(r-\varepsilon)/(q+1) \le 1$, and therefore

$$\min\{1, \alpha\} \le \alpha^{\frac{r-\varepsilon}{q+1}} \qquad \forall \alpha > 0.$$

We proceed as before to bound

$$\|\mathrm{F}(s) - \mathrm{F}_\kappa(s)\| \le \Xi\min\{1, |s|^{q+1}\kappa^q\} \le \Xi|s|^{r-\varepsilon}\kappa^{(r-\varepsilon)\frac{q}{q+1}},$$

which finishes the proof. □

4.6 Convergence of CQ

Proposition 4.6.1 (Uniform convergence with L^1 regularity). *Let* $\mathrm{F} \in \mathcal{A}(\mu, \mathcal{B}(X,Y))$ *with* $\mu \ge 0$ *and define* $D_\mathrm{F}(\sigma) := C_\mathrm{F}(\frac{1}{2}\underline{C}_3\,\underline{\sigma})$. *Let* g *be a causal* X*-valued function such that*

$$g \in \mathcal{C}^{k-1}(\mathbb{R}, X), \qquad g^{(k)} \in L^1(\mathbb{R}, X), \qquad \textit{with} \quad k > \mu + 1.$$

Then

$$\|(f * g)(t) - (f_\kappa * g)(t)\| \le D\,\mathrm{error}(\kappa)\,h(t)\int_0^t \|g^{(k)}(\tau)\|\mathrm{d}\tau,$$

where D depends only on μ, k, and p,

$$\mathrm{error}(\kappa) := \kappa^{(k-\mu-1)\frac{q}{q+1}} + \delta_{k,\mu+q+2}\kappa^q|\log\kappa| + \kappa^q$$

and

$$h(t) := \begin{cases} t^{k-\mu}D_{\mathrm{F}}(t^{-1}), & t \geq 1, \\ D_{\mathrm{F}}(1)t^{k-\mu-1-q}, & t \leq 1 \quad \text{and} \quad k \geq \mu+1+q, \\ D_{\mathrm{F}}(1), & t \leq 1 \quad \text{and} \quad \mu+1 < k < \mu+1+q. \end{cases}$$

Optimal convergence (order q) is obtained for $k > \mu + q + 2$.

Proof. The proof uses techniques that we have already met in the analysis of convolution operators. (See Section 3.2 and especially Proposition 3.2.2.) The Laplace transform of the error $f * g - f_\kappa * g$ is

$$(\mathrm{F} - \mathrm{F}_\kappa)\,\mathrm{G} \in \mathcal{A}(\mu - k, Y) \qquad \mu - k < -1.$$

(To see this, use Proposition 4.5.2 for the term $\mathrm{F} - \mathrm{F}_\kappa$ and recall the proof that $\mathrm{G} \in \mathcal{A}(-k, X)$ given in Proposition 3.2.2.) Since $\mu - k < -1$ we can apply the strong inversion formula for the Laplace transform (see Section 3.1 and in particular (3.2)), so that

$$\begin{aligned} \|(f * g)(t) - (f_\kappa * g)(t)\| &\leq \frac{e^{\sigma t}}{2\pi}\|(\mathrm{F} - \mathrm{F}_\kappa)\mathrm{G}\|_{L^1(\sigma+\imath\mathbb{R},Y)} \\ &\leq \frac{e^{\sigma t}}{2\pi}\|(\,\cdot\,)^{-k}(\mathrm{F} - \mathrm{F}_\kappa)\|_{L^1(\sigma+\imath\mathbb{R},\mathcal{L}(X,Y))} \sup_{\mathrm{Re}\,s=\sigma}\|s^k\mathrm{G}(s)\|. \end{aligned}$$

Using the integral formula for the Laplace transform of $s^k\mathrm{G}(s)$, it follows that

$$\sup_{\mathrm{Re}\,s=\sigma}\|s^k\mathrm{G}(s)\| \leq \int_0^\infty e^{-\sigma\tau}\|g^{(k)}(\tau)\|\mathrm{d}\tau \leq \int_0^\infty \|g^{(k)}(\tau)\|\mathrm{d}\tau.$$

Using Proposition 4.5.5 with $r = k - 1 - \mu$ and noticing that $\mathrm{error}(\kappa) = E(\kappa)$ for those parameters, we arrive at an error bound, proportional to

$$e^{\sigma t}D_{\mathrm{F}}(\sigma)\underline{\sigma}^{-1}\max\{\underline{\sigma}^{-k+1+\mu}, \sigma^{q-k+1+\mu}\}\mathrm{error}(\kappa)\int_0^\infty \|g^{(k)}(\tau)\|\mathrm{d}\tau.$$

The dependence on t of the error constants is obtained by taking $\sigma = t^{-1}$ or $\sigma = 1$ (this last value when $t \geq 1$ and $k - \mu - 1 - q < 0$).

To change the integration interval for $\|g^{(k)}\|$ from $(0, \infty)$ to $(0, t)$ we can use the causality argument of Proposition 3.2.2. (See Exercises.) □

Remark. Note that the error term in Proposition 4.6.1 can be split into three different cases:

(a) If $\mu + 1 < k < \mu + q + 2$, then

$$\text{error}(\kappa) = \kappa^{\alpha \frac{q}{q+1}} \qquad \alpha := k - \mu - 1 \in (0, q+1).$$

(b) If $k = \mu + q + 2$ (which can only happen if μ is an integer), then $\text{error}(\kappa) = \kappa^q |\log \kappa|$.

(c) Finally, for $k > \mu + q + 2$, the error is of optimal order κ^q.

Proposition 4.6.2 (Uniform estimates with L^2 regularity). *Let* $\mathrm{F} \in \mathcal{A}(\mu, \mathcal{B}(X, Y))$ *with* $\mu \geq 0$ *and define* $D_{\mathrm{F}}(\sigma) := C_{\mathrm{F}}(\frac{1}{2}\underline{C}_3\,\underline{\sigma})$. *Let* g *be a causal* X*-valued function such that*

$$g \in \mathcal{C}^{k-1}(\mathbb{R}, X), \qquad g^{(k)} \in L^2(\mathbb{R}, X).$$

(a) *If* $k > \mu + q + \frac{3}{2}$, *then*

$$\|(f * g)(t) - (f_\kappa * g)(t)\| \leq \kappa^q\, h(t) \left(\int_0^t \|g^{(k)}(\tau)\|^2 \mathrm{d}\tau \right)^{1/2},$$

where

$$h_1(t) = C \frac{2^{\alpha/4}}{\sqrt{\alpha}} D_{\mathrm{F}}(t^{-1}) \max\{1, t\} t^{\alpha} \qquad \alpha = k - \mu - q - \tfrac{3}{2}.$$

(b) *If* $\mu + q + \frac{3}{2} \geq k > \mu + \frac{1}{2}$, *then*

$$\|(f * g)(t) - (f_\kappa * g)(t)\| \leq \kappa^{\beta \frac{q}{q+1}} |\log \kappa|^{1/2}\, h(t) \left(\int_0^t \|g^{(k)}(\tau)\|^2 \mathrm{d}\tau \right)^{1/2},$$

where

$$h_2(t) = C \frac{2^{\beta/4}}{\sqrt{\beta}} D_{\mathrm{F}}(t^{-1}) \max\{1, t^{\beta+1}\} \qquad \beta = k - \mu - \tfrac{1}{2} \in (0, q+1].$$

The constant C *depends on* k, μ, *and* p.

Proof. Let $e_\kappa := f * g - f_\kappa * g$ and assume that g is smooth causal and compactly supported. Using the inversion formula for the Laplace transform (using the inversion contour $\operatorname{Re} s = \sigma$), we can bound

$$\|e_\kappa(t)\| \leq \frac{e^{\sigma t}}{2\pi} \left(\int_{-\infty}^{\infty} \frac{\mathrm{d}\eta}{|\sigma + \imath\eta|^{\frac{1}{2}+\varepsilon}} \right)^{1/2} \sup_{\operatorname{Re} s = \sigma} |s|^{-k+\frac{1}{2}+\varepsilon} \|\mathrm{F}(s) - \mathrm{F}_\kappa(s)\|$$

$$\Big(\int_{-\infty}^{\infty} \|(\sigma+\imath\eta)^k \mathrm{G}(\sigma+\imath\eta)\|^2 \mathrm{d}\eta\Big)^{1/2}$$
$$\leq \frac{e^{\sigma t}}{2\pi}\Big(\frac{2^{\varepsilon+\frac{1}{2}}}{\varepsilon\sigma^{2\varepsilon}}\Big)^{1/2} \sup_{\mathrm{Re}\, s=\sigma} |s|^{-k+\frac{1}{2}+\varepsilon}\|\mathrm{F}(s)-\mathrm{F}_\kappa(s)\|\Big(\frac{1}{2\pi}\int_0^\infty \|g^{(k)}(\tau)\|^2 \mathrm{d}\tau\Big)^{1/2},$$

where in the last inequality we have used Plancherel's identity (see Section 3.7) and, for obvious reasons, we need to take $\varepsilon > 0$.

In case (a), we apply Proposition 4.5.6(a) with

$$r = k - \mu - \tfrac{1}{2} > q+1, \qquad \varepsilon = \tfrac{1}{2}(r-(q+1)) = \tfrac{1}{2}\alpha > 0$$

so that

$$\mu + r - \varepsilon = k - \tfrac{1}{2} - \varepsilon \qquad \text{and} \qquad q + 1 - r + \varepsilon = -\tfrac{1}{2}\alpha.$$

Then

$$\frac{2^{\frac{\varepsilon}{2}+\frac{1}{4}}}{\sigma^\varepsilon\sqrt{\varepsilon}} \sup_{\mathrm{Re}\, s=\sigma} |s|^{-k+\frac{1}{2}+\varepsilon}\|\mathrm{F}(s)-\mathrm{F}_\kappa(s)\| \leq \frac{2^{3/4}}{\sqrt{\alpha}}\frac{2^{\frac{1}{2}(\alpha+1)}}{\sigma^\alpha} D_{\mathrm{F}}(\sigma)\underline{\sigma}^{-1}\kappa^q.$$

The result follows now from taking $\sigma = 1/t$, simplifying, using a density argument and causality to generalize for any g and restrict the integral in the right-hand side to $(0,t)$.

Part (b) is left to the reader as an exercise. □

4.7 An example: back to scattering problems

Let us then go back to the problem of scattering by a sound soft obstacle using a single layer potential. Data is a causal function β with values in $H^{1/2}(\Gamma)$. An integral equation

$$\mathcal{V} * \lambda = \beta \qquad \Longleftrightarrow \qquad \lambda = \mathcal{V}^{-1} * \beta$$

is solved to find a density λ, and then a potential is computed using λ:

$$u = \mathcal{S} * \lambda = \mathcal{S} * \mathcal{V}^{-1} * \beta.$$

Recall that (Proposition 2.6.1)

$$\|\mathrm{V}^{-1}(s)\| \leq C\frac{|s|^2}{\sigma\underline{\sigma}}$$

and (Section 3.3)

$$\|S(s)V^{-1}(s)\|_{H^{1/2}(\Gamma)\to H^1(\mathbb{R}^d\setminus\Gamma)} \leq C\frac{|s|^{3/2}}{\sigma\underline{\sigma}^{3/2}}.$$

Because we are first solving a convolution equation, we do not have access to the full function $\lambda_\kappa := (\mathcal{V}^{-1})_\kappa * \beta$ but only to its values at the fixed time steps $t_n = \kappa\, n$. The error bound (Proposition 4.6.1) is applied to V^{-1} and to $S\,V^{-1}$ separately. For the BDF2-based method ($q = 2$) the errors

$$\|\lambda(t_n) - \lambda_\kappa(t_n)\|_{-1/2,\Gamma}$$

are bounded by a product of three quantities:

$$\text{error}_\lambda(\kappa) \times \text{behavior}_\lambda(t_n) \times \int_0^{t_n} \|\beta^{(k)}(\tau)\| \mathrm{d}\tau.$$

The results for possible values of k are displayed in Table 4.1.

The errors

$$\|u(t_n) - u_\kappa(t_n)\|_{1,\Omega+}$$

are bounded again by a triple product

$$\text{error}_u(\kappa) \times \text{behavior}_u(t_n) \times \int_0^{t} \|g^{(k)}(\tau)\| \mathrm{d}\tau,$$

We present the different bounds in Table 4.2.

Table 4.1 The CQ errors for the density

k	$\text{error}_\lambda(\kappa)$	$\text{behavior}_\lambda(t)$ for $t \geq 1$	$\text{behavior}_\lambda(t)$ for $t \leq 1$
4	$\kappa^{2/3}$	t^4	1
5	$\kappa^{4/3}$	t^5	1
6	$\kappa^2\lvert\log\kappa\rvert$	t^6	t
7	κ^2	t^7	t^2

Table 4.2 The CQ errors for the potential

k	$\text{error}_u(\kappa)$	$\text{behavior}_u(t)$ for $t \geq 1$	$\text{behavior}_\lambda(t)$ for $t \leq 1$
3	$\kappa^{1/3}$	t^4	1
4	κ	t^5	1
5	$\kappa^{5/3}$	t^6	$t^{1/2}$
6	κ^2	t^7	$t^{3/2}$

4.8 Appendix: A-acceptable functions

This section is devoted to proving some technical results that were gathered in Proposition 5.3.1. Let us start by recalling the hypotheses on p. We assume that there is an open set U such that $\{\zeta \in \mathbb{C} : |\zeta| \leq 1\} \subset U$ and that $p : U \to \mathbb{C}$ is holomorphic and satisfies

(a) $p : B(0,1) \to \mathbb{C}_+$.
(b) There exist $q \geq 1$, $C_0 > 0$ and $\kappa_0 > 0$ such that

$$\left| \frac{p(e^{-\kappa})}{\kappa} - 1 \right| \leq C_0 \kappa^q \qquad \forall \kappa \leq \kappa_0.$$

In particular $p(1) = 0$ and $p'(1) = -1$.

Note that the functions $s \mapsto p(e^{-s})$ and $s \mapsto s_\kappa = \kappa^{-1} p(e^{-\kappa s})$ are holomorphic $\mathbb{C}_+ \to \mathbb{C}_+$. Functions satisfying these conditions are called A-acceptable.

Proposition 4.8.1. *There exists $C_1 > 0$ such that*

$$|p(e^{-s})| \leq C_1 \, |s| \qquad \forall s \in \mathbb{C}_+.$$

Proof. It is not complicated to prove that

$$\left| \frac{e^{-s} - 1}{s} \right| \leq 2 \qquad \forall s \in \mathbb{C}_+. \tag{4.17}$$

Since the function $p(\zeta)(1-\zeta)^{-1}$ is holomorphic in U, we can bound

$$\left| \frac{p(e^{-s})}{s} \right| \leq 2 \left| \frac{p(e^{-s})}{e^{-s} - 1} \right| \leq 2 \max_{|\zeta| \leq 1} \left| \frac{p(\zeta)}{\zeta - 1} \right|,$$

which completes the proof. □

Proposition 4.8.2. *There exists $C_2 > 0$ such that*

$$|p(e^{-s}) - s| \leq C_2 |s|^{q+1} \qquad \forall s \in \mathbb{C}_+.$$

Proof. The hypotheses of p imply that the function

$$h(s) := \frac{p(e^{-s}) - s}{s^{q+1}}$$

is holomorphic in a neighborhood of the origin and therefore, there exist $\rho > 0$ and $\widetilde{C} > 0$ such that

$$\left| \frac{p(e^{-s}) - s}{s^{q+1}} \right| \le \widetilde{C}, \quad when \quad |s| \le \rho.$$

For $s \in \mathbb{C}_+$ such that $|s| \ge \rho$, we can bound using Proposition 4.8.1

$$|p(e^{-s}) - s| \le (C_1 + 1)|s| \le (C_1 + 1)\rho^{-q}|s|^{q+1}.$$

This finishes the proof. □

Proposition 4.8.3. *There exists* $C_3 > 0$ *such that*

$$\operatorname{Re} p(e^{-s}) \ge C_3 \underline{\sigma} \qquad \forall s \in \mathbb{C}_+.$$

Proof. By the Cauchy Representation Formula

$$p(z) = \frac{1}{2\pi\imath} \int_{|\zeta|=1} \frac{p(\zeta)}{\zeta - z} \mathrm{d}\zeta = \frac{1}{2\pi} \int_0^{2\pi} \frac{p(e^{\imath\theta})}{e^{\imath\theta} - z} e^{\imath\theta} \mathrm{d}\theta \qquad \forall z \in B(0;1)$$

and therefore

$$\operatorname{Re} p(re^{\imath\phi}) = \frac{1}{2\pi} \int_0^{2\pi} \operatorname{Re} p(e^{\imath\theta}) \frac{1 - r^2}{1 - 2r\cos(\phi - \theta) + r^2} \mathrm{d}\theta \qquad \forall r < 1, \quad \forall \phi.$$

(This is just the well-known Poisson formula.) We now use the fact that $\operatorname{Re} p(e^{\imath\theta}) \ge 0$ (this follows from the hypotheses on p) and

$$\frac{1 - r^2}{1 - 2r\cos(\phi - \theta) + r^2} \ge \frac{1 - r^2}{1 + 2r + r^2} = \frac{1 - r}{1 + r}$$

to bound

$$\operatorname{Re} p(r\, e^{\imath\phi}) \ge \frac{1 - r}{1 + r} \underbrace{\frac{1}{2\pi} \int_0^{2\pi} \operatorname{Re} p(e^{\imath\theta}) \mathrm{d}\theta}_{=:C_p > 0}.$$

Applying this to the point $\zeta = e^{-s} = e^{-\sigma} e^{-\imath\omega}$ for $s \in \mathbb{C}_+$, it follows that

$$\operatorname{Re} p(e^{-s}) \ge C_p \frac{1 - e^{-\sigma}}{1 + e^{-\sigma}}.$$

Finally

$$\frac{1 - e^{-\sigma}}{1 + e^{-\sigma}} = \tanh(\sigma/2) \ge \tanh(1/2) \min\{1, \sigma\} = \tanh(1/2)\underline{\sigma},$$

which completes the proof. □

Proposition 4.8.4. *For all* $\kappa \le \kappa_{\max}$

$$\mathrm{Re}\left(\frac{p(e^{-\kappa s})}{\kappa}\right) \ge \frac{C_3}{\max\{1,\kappa_{\max}\}}\underline{\sigma} \qquad \forall s \in \mathbb{C}_+.$$

Proof. We just apply Proposition 4.8.4 to bound

$$\begin{aligned}\mathrm{Re}\, p(e^{-\kappa s}) &\ge C_3 \min\{1,\kappa\sigma\} = C_3\kappa \min\{\kappa^{-1},\sigma\}\\ &\ge C_3\kappa \min\{\kappa^{-1},1\}\min\{1,\sigma\} \ge C_3\kappa\min\{\kappa_{\max}^{-1},1\}\underline{\sigma}.\end{aligned}$$

This finishes the proof. □

Proposition 4.8.5. *With the above hypotheses, any zero of p in the boundary of the unit disk is simple.*

Proof. See [53, Chapter 5, Theorem 1.5]. The result is obviously related to the fact that the unit ball is mapped to $\mathbb{C}_+$. A double zero would then make part of the unit disk to be mapped outside $\mathbb{C}_+$. □

4.9 Exercises

1. (Section 4.3) Prove Proposition 4.3.3.
2. (Section 4.5) The CQ method can also be described by means of a discrete differentiation operator

   $$s_\kappa := \kappa^{-1} p(e^{-s\kappa})$$

   that is used to define discrete versions of all symbols $\mathrm{F}_\kappa(s) := \mathrm{F}(s_\kappa)$. Show that Proposition 4.5.1 can be written in the alternative form

   $$|s_\kappa| \le C_1|s|, \qquad |s_\kappa - s| \le C_2\kappa^q|s|^{q+1}, \qquad \mathrm{Re}\, s_\kappa \ge C_3\underline{\sigma} \qquad \forall s \in \mathbb{C}_+.$$

3. (Section 4.5) Consider the discrete differentiation operator $\partial_\kappa u := \mathcal{L}^{-1}\{s_\kappa \mathrm{U}(s)\}$, where $s_\kappa := \kappa^{-1}p(e^{-s\kappa})$. Use the previous exercise and Proposition 3.2.2 to provide direct bounds for

   $$\|(\partial_\kappa u)(t)\| \quad \text{and} \quad \|u'(t) - (\partial_\kappa u)(t)\|.$$

4. (Section 4.5) Prove that if F is holomorphic in a convex set, then for all s_1, s_2 in that set

   $$\|\mathrm{F}(s_1) - \mathrm{F}(s_2)\| \le \sqrt{2}|s_1 - s_2| \max_{z\in[s_1,s_2]} \|\mathrm{F}'(z)\|,$$

   where $[s_1, s_2] = \{\alpha s_1 + (1-\alpha)s_2 \ : \ \alpha \in [0,1]\}$. (**Hint.** Use the Mean Value Theorem for bivariate functions and the Cauchy-Riemann equations.)

5. (Section 4.6) Show that if f is a causal distribution and for a sufficiently smooth causal g it holds that

$$\|(f * g)(t)\| \leq C(t) \int_0^\infty \|g^{(k)}(\tau)\| \mathrm{d}\tau,$$

then

$$\|(f * g)(t)\| \leq C(t) \int_0^t \|g^{(k)}(\tau)\| \mathrm{d}\tau.$$

(**Hint.** You can use the strategy of the proof of Proposition 3.2.2. The idea can be simplified to showing the result for $k = 0$ and then extending it for general k using differentiation and antidifferentiation operators.)

6. (Section 4.6) Prove part (b) of Proposition 4.6.1. (**Hint.** Use the same inequalities that are used to prove part (a). Assume that $\log \kappa < -1$ and take

$$\varepsilon := -\frac{r}{\log \kappa} < r, \qquad r = k - \mu - \tfrac{1}{2} = \beta$$

in Proposition 4.5.6. Note also that $t^{-r/\log \kappa} \leq \max\{1, t^r\}$.)

7. (Section 4.8) Show that in the hypotheses of Section 4.8, the function

$$h(s) := \frac{p(e^{-s}) - s}{s^{q+1}}$$

is holomorphic in a neighborhood of zero.

Chapter 5
The discrete layer potentials

In this chapter we explore fully discrete CQ-BEM methods (Convolution Quadrature in time, Galerkin Boundary Elements in space) for the exterior Dirichlet and Neumann problems associated with the wave equation. For the first one we will use a retarded single layer potential representation, and for the second one, a retarded double layer potential representation. We will detail the analysis of Galerkin semidiscretization error and some stability estimates that are needed to apply the results on Convolution Quadrature to the fully discrete scheme. It has already been mentioned, but let us repeat it here: the analysis of full discretizations for time domain integral equations in acoustics seems to be restricted to equations of the first kind, since the analysis is based on variational (energy) estimates. Not much is known about the behavior of the integral equations of the second kind, especially once they are discretized. The theory of full Galerkin discretizations is based on the weak weighted coercivity estimates given in Section 3.7. In the way the method is used, space and time Galerkin can be understood as a Galerkin-in-time discretization of a semidiscrete Galerkin problem. From that point of view, all the results given here about semidiscretization in space could eventually be used for the fully discrete method.

5.1 Sound-soft scattering with CQ-BEM

The method. A fully discrete method for the problem

$$\mathcal{V} * \lambda = \beta, \qquad u = \mathcal{S} * \lambda$$

can be devised by first discretizing in space with a Galerkin scheme and then using convolution quadrature for the resulting system. In order to do this, we start by

F.-J. Sayas, *Retarded Potentials and Time Domain Boundary Integral Equations*, Springer Series in Computational Mathematics 50,
DOI 10.1007/978-3-319-26645-9_5

choosing a finite dimensional space $X_h \subset H^{-1/2}(\Gamma)$ (in practice, the space is always taken as $X_h \subset L^\infty(\Gamma)$). In the time domain, we look for $\lambda^h \in \mathrm{TD}(X_h)$ (where X_h is endowed with the $H^{-1/2}(\Gamma)$-norm) such that

$$\langle \mu^h, \mathcal{V} * \lambda^h \rangle_\Gamma = \langle \mu^h, \beta \rangle_\Gamma \qquad \forall \mu^h \in X_h. \tag{5.1}$$

It is not difficult to see that this is again a convolution equation: to do that, we take the (steady state) restriction operator $R^h : H^{1/2}(\Gamma) \to X_h'$

$$H^{1/2}(\Gamma) \ni \beta \longmapsto R^h \beta := \langle \,\cdot\,, \beta \rangle_\Gamma : X_h \to \mathbb{R}$$

and note that the equation (5.1) is equivalent to

$$(\delta_0 \otimes R^h) * \mathcal{V} * \lambda^h = R^h \beta.$$

Convolution quadrature is applied to operator $X_h \to X_h'$ defined by

$$\lambda^h \mapsto (\delta_0 \otimes R^h) * \mathcal{V} * \lambda^h = R^h(\mathcal{V} * \lambda^h).$$

In a postprocessing step, CQ is used again for the discretization of the potential

$$u^h := \mathcal{S} * \lambda^h. \tag{5.2}$$

A look at the equations. If $\{\Phi_1, \ldots, \Phi_N\}$ is a basis of X_h, then we can write

$$\lambda^h = \sum_{j=1}^N \lambda_j \otimes \Phi_j, \qquad \lambda_j \in \mathrm{TD}(\mathbb{R}).$$

Assuming that the time distributions λ_j are functions, we can write the semidiscrete system (4.17) in the following practical form:

$$\sum_{j=1}^N \int_\Gamma \int_\Gamma \frac{\Phi_i(\mathbf{x})\Phi_j(\mathbf{y})}{4\pi|\mathbf{x}-\mathbf{y}|} \lambda_j(t - c^{-1}|\mathbf{x}-\mathbf{y}|)\, \mathrm{d}\Gamma(\mathbf{x})\mathrm{d}\Gamma(\mathbf{y}) = \int_\Gamma \Phi_i(\mathbf{x})\, \beta(\mathbf{x}, t)\mathrm{d}\Gamma(\mathbf{x})$$

$$i = 1, \ldots, N, \qquad t > 0.$$

The Laplace domain form of the operator in this system corresponds to the s-dependent matrix

$$\mathrm{V}_{ij}(s) := \int_\Gamma \int_\Gamma \frac{e^{-s|\mathbf{x}-\mathbf{y}|}}{4\pi|\mathbf{x}-\mathbf{y}|}\, \Phi_i(\mathbf{x})\Phi_j(\mathbf{y})\mathrm{d}\Gamma(\mathbf{x})\mathrm{d}\Gamma(\mathbf{y}),$$

which is an X^h-based Galerkin discretization of the integral operator $\mathrm{V}(s)$. The convolution quadrature method then needs an expansion

$$\mathrm{V}_{ij}\big(\tfrac{p(\zeta)}{\kappa}\big) = \sum_{m=0}^{\infty} \mathrm{V}_{ij}^m(\kappa)\zeta^m.$$

Time-stepping is carried out by solving the systems

$$\sum_{m=0}^{n}\sum_{j=1}^{N} \mathrm{V}_{ij}^m \lambda_j^{n-m} = \beta_i^n \qquad i = 1,\dots,N, \qquad n \geq 0,$$

where

$$\beta_i^n := \int_\Gamma \Phi_i(\mathbf{x})\beta(\mathbf{x}, t_n)\mathrm{d}\Gamma(\mathbf{x}) \qquad i = 1,\dots,N, \qquad n \geq 0.$$

The unknowns allow us to reconstruct functions

$$X_h \ni \lambda_\kappa^h(t_n) := \sum_{j=1}^{N} \lambda_j^n \Phi_j \approx \lambda^h(t_n),$$

for which only the values at discrete times are computed. Every time step requires the solution of an $N \times N$ system

$$\sum_{j=1}^{N} \mathrm{V}_{ij}^0(\kappa)\lambda_j^n = \beta_i^n - \sum_{m=1}^{n}\sum_{j=1}^{N} \mathrm{V}_{ij}^m(\kappa)\lambda_j^{n-m}, \qquad i = 1,\dots,N. \tag{5.3}$$

The matrix is

$$\mathrm{V}_{ij}^0(\kappa) = \int_\Gamma\int_\Gamma \frac{e^{-\frac{p(0)}{\kappa}|\mathbf{x}-\mathbf{y}|}}{4\pi|\mathbf{x}-\mathbf{y}|}\Phi_i(\mathbf{x})\Phi_j(\mathbf{y})\mathrm{d}\Gamma(\mathbf{x})\mathrm{d}\Gamma(\mathbf{y}),$$

which is, once again, the result of using an X^h-based Galerkin discretization to the operator $\mathrm{V}(\frac{p(0)}{\kappa})$. Actually, if we directly apply a convolution quadrature method to $\mathcal{V} * \lambda = \beta$, we end up with a sequence of integral equations

$$\mathrm{V}^0(\kappa)\lambda^n = \beta(t^n) - \sum_{m=1}^{n} \mathrm{V}^m(\kappa)\lambda^{n-m}, \tag{5.4}$$

where

$$\mathrm{V}\big(\tfrac{p(\zeta)}{\kappa}\big) = \sum_{m=0}^{\infty} \mathrm{V}^m(\kappa)\zeta^m$$

is the sequence of weights associated with the CQ scheme. Using an X^h-based Galerkin discretization to each of the integral equations (5.4) is equivalent to the fully discrete method we had derived before, which is to say, Galerkin semidiscretization in space and CQ in time commute for this kind of problem.

Final words. The sequence of problems (5.3) (and the time-semidiscrete sequence (5.4)) presents the serious disadvantage of having an infinite tail. In other words, the passage through the Laplace domain introduces a regularization of the wave equation that eliminates the Huygens' principle that so clearly appears in the time domain retarded operators and potentials. Some estimates exist concerning the norm of the operators $\mathrm{V}^m(\kappa)$ for large enough m and the effect of doing away with them in the time-stepping process (see [51]). They provide the theoretical support for a trick that can be played safely for in practice. Practical evaluation of the discrete kernels V_{ij}^n and the discrete convolutions where they appear can be accomplished through a battery of techniques based on the FFT and recursion. Details on how this is done can be found in [22].

5.2 The Galerkin (semidiscrete) solver

In this section we study the semidiscrete-in-space problem (5.1)–(5.2). We will study the operator that associates $\beta \mapsto (\lambda^h, u^h)$, by referring to its properties in the Laplace domain. We thus look at the problem of finding

$$\lambda^h \in X_h \qquad \text{such that} \qquad \langle \mu^h, \mathrm{V}(s)\lambda^h\rangle_\Gamma = \langle \mu^h, \beta\rangle_\Gamma \quad \forall \mu^h \in X_h. \tag{5.5}$$

and then constructing the potential

$$u^h = \mathrm{S}(s)\lambda^h. \tag{5.6}$$

We will then write

$$\lambda^h = \mathrm{G}_\lambda^h(s)\beta, \qquad u^h = \mathrm{G}_u^h(s)\beta = \mathrm{S}(s)\mathrm{G}_\lambda^h(s)\beta.$$

What we mean with the expression 'for all *h*'. We want the analysis to be independent (as much as possible) of the particular choice of the discrete space X_h. We will use the convention of referring to a property being satisfied for all h to refer to the fact that the bound is completely independent of the choice of X_h. As we will see, all estimates will only use the fact that X_h is a closed subspace of $H^{-1/2}(\Gamma)$, but we will never use that it is finite dimensional. The results will therefore hold for $X_h = H^{-1/2}(\Gamma)$, in which case $\mathrm{G}_\lambda^h(s) = \mathrm{V}^{-1}(s)$.

Direct estimates. Using Proposition 2.6.1, namely

$$\mathrm{Re}\Big(e^{\imath\,\mathrm{Arg}\,s}\langle\overline{\lambda}^h, \mathrm{V}(s)\lambda^h\rangle_\Gamma\Big) \geq C_\Gamma \frac{\sigma\underline{\sigma}}{|s|^2}\|\lambda^h\|^2_{-1/2,\Gamma} \qquad \forall\lambda^h \in X_h, \quad \forall s \in \mathbb{C}_+,$$

we can bound

$$\|\mathrm{G}^h_\lambda(s)\|_{H^{1/2}(\Gamma)\to H^{-1/2}(\Gamma)} \leq C_\Gamma^{-1}\frac{|s|^2}{\sigma\underline{\sigma}} \qquad \forall s \in \mathbb{C}_+ \qquad \forall h. \tag{5.7}$$

Using the estimate for the norm of $\mathrm{S}(s)$ (Proposition 2.6.2) we can bound

$$\|\mathrm{G}^h_u(s)\|_{H^{1/2}(\Gamma)\to H^1(\mathbb{R}^d)} \leq \|\mathrm{S}(s)\|_{H^{-1/2}(\Gamma)\to H^1(\mathbb{R}^d)}\|\mathrm{G}^h_\lambda(s)\|_{H^{1/2}(\Gamma)\to H^{-1/2}(\Gamma)} \leq C_\Gamma\frac{|s|^3}{\sigma^2\underline{\sigma}^3}. \tag{5.8}$$

If we compare this result with the one for the continuous operator ($X_h = H^{-1/2}(\Gamma)$) given in Section 3.3, we can easily notice that the latter result is not optimal. We will next give a detailed proof of how to obtain a better estimate (that reproduces the one in Section 3.3) by using a very similar approach to the one used in the continuous case. We emphasize the importance of lowering the powers of $|s|$ and σ^{-1} in the bounds, both for mapping properties (Proposition 3.2.2) and for smoothness requirements to obtain convergence of the Convolution Quadrature method (Propositions 4.6.1 and 4.6.2).

Analysis based on exotic transmission problems. We first need to introduce some notation. The polar set or annihilator of the set X_h is

$$X_h^\circ := \{\xi \in H^{1/2}(\Gamma) \,:\, \langle\lambda^h, \xi\rangle_\Gamma = 0 \quad \forall\lambda^h \in X_h\}.$$

A first observation is the fact that equation (5.5) is equivalent to

$$\lambda^h \in X_h \qquad \text{such that} \qquad \mathrm{V}(s)\lambda^h - \beta \in X_h^\circ. \tag{5.9}$$

The second observation is the equivalence

$$\lambda \in X_h \qquad \Longleftrightarrow \qquad \langle\lambda, \xi\rangle_\Gamma = 0 \quad \forall\xi \in X_h^\circ. \tag{5.10}$$

These two observations trigger the forthcoming analysis. Following [59], we will divide the process in four steps:

1. Find a transmission problem that characterizes $u^h = \mathrm{G}^h_u(s)\beta$. This means that we move away from the boundary and focus on the potential solution.
2. Derive the variational formulation for the transmission problem.
3. Use a coercivity estimate in the energy norm to bound u^h, and use that to find an $H^1(\mathbb{R}^d)$ estimate for u^h.
4. Finally, find a bound for $\lambda^h = \mathrm{G}^h_\lambda(s)\beta$, by noticing that it can be recovered as the jump of the normal derivative of u^h.

Let us emphasize that, ever since the seminal work of Nédélec and Planchard [70], the analysis of integral equations of the first kind is based on coercivity estimates that are proved by considering the potential postprocessing and not the integral equation. We take this idea much further by also working at the discrete level and showing that the Galerkin discretization of integral equations of the first kind can be understood variationally.

Proposition 5.2.1 (Step #1: transmission problem). *Let $\beta \in H^{1/2}(\Gamma)$. Then $u^h = \mathrm{G}_u^h(s)\beta$ if and only if u^h is the solution of the following problem:*

$$u^h \in H^1(\mathbb{R}^d), \tag{5.11a}$$

$$\Delta u^h - s^2 u^h = 0 \quad \text{in } \mathbb{R}^d \setminus \Gamma, \tag{5.11b}$$

$$\gamma u^h - \beta \in X_h^\circ, \tag{5.11c}$$

$$[\![\partial_\nu u^h]\!] \in X_h. \tag{5.11d}$$

Proof. If $u^h = \mathrm{G}_u^h(s)\beta = \mathrm{S}(s)\mathrm{G}_\lambda^h(s)\beta$, then it is clear that u^h satisfies (5.11a) and (5.11b). Moreover, defining $\lambda^h := \mathrm{G}_\lambda^h(s)\beta$, we have

$$[\![\partial_\nu u^h]\!] = \lambda^h \in X_h \qquad \gamma u^h - \beta = \mathrm{V}(s)\lambda^h - \beta \in X_h^\circ,$$

by (5.5). This proves that u^h solves (5.11).

Conversely, if u^h satisfies (5.11) and $\lambda^h := [\![\partial_\nu u^h]\!] \in X_h$, then by the representation formula for the solutions of $\Delta u - s^2 u = 0$ in $\mathbb{R}^d \setminus \Gamma$, we can write

$$u^h = \mathrm{S}(s)[\![\partial_\nu u^h]\!] - \mathrm{D}(s)[\![\gamma u^h]\!] = \mathrm{S}(s)\lambda^h$$

and therefore

$$\mathrm{V}(s)\lambda^h - \beta = \gamma u^h - \beta \in X_h^\circ,$$

which proves that $\lambda^h = \mathrm{G}_\lambda^h(s)\beta$ and $u^h = \mathrm{S}(s)\lambda^h = \mathrm{G}_u^h(s)\beta$. □

Proposition 5.2.2 (Step #2: variational formulation). *Consider the space*

$$H_h := \{v^h \in H^1(\mathbb{R}^d) \,:\, \gamma v^h \in X_h^\circ\}.$$

The transmission problem (5.11) *is equivalent to the variational problem*

$$\left[\begin{array}{ll} u^h \in H^1(\mathbb{R}^d) & \gamma u^h - \beta \in X_h^\circ, \\ a_{s,\mathbb{R}^d}(u^h, v^h) = 0 & \forall v^h \in H_h. \end{array}\right. \tag{5.12}$$

Proof. If u^h solves (5.11), then for all $v^h \in H_h$

$$\begin{aligned} a_{s,\mathbb{R}^d}(u^h, v^h) &= (\nabla u^h, \nabla v^h)_{\mathbb{R}^d} + (\Delta u^h, v^h)_{\mathbb{R}^d\setminus\Gamma} && (\Delta u^h = s^2 u^h) \\ &= \langle [\![\partial_\nu u^h]\!], \gamma v^h\rangle_\Gamma && (\text{definition of } [\![\partial_\nu \cdot]\!]) \\ &= 0. && ([\![\partial_\nu u^h]\!] \in X_h, \quad \gamma v^h \in X_h^\circ) \end{aligned}$$

Conversely, if u^h is a solution to (5.12) it follows readily that $\Delta u^h - s^2 u^h = 0$ in $\mathbb{R}^d \setminus \Gamma$, since $\mathcal{D}(\mathbb{R}^d \setminus \Gamma) \subset H_h$. Now let $\xi \in X_h^\circ$ and construct $v^h \in H^1(\mathbb{R}^d)$ such that $\gamma v^h = \xi$. It follows then that $v^h \in H_h$. Therefore

$$\begin{aligned} \langle [\![\partial_\nu u^h]\!], \xi\rangle_\Gamma &= \langle [\![\partial_\nu u^h]\!], \gamma v^h\rangle_\Gamma \\ &= (\nabla u^h, \nabla v^h)_{\mathbb{R}^d} + (\Delta u^h, v^h)_{\mathbb{R}^d\setminus\Gamma} && (\text{definition of } [\![\partial_\nu \cdot]\!]) \\ &= a_{s,\mathbb{R}^d}(u^h, v^h) && (\Delta u^h = s^2 u^h) \\ &= 0, && (v^h \in H_h) \end{aligned}$$

and therefore $[\![\partial_\nu u^h]\!] \in X_h$ by (5.10). □

Proposition 5.2.3 (Step #3: energy/coercivity estimate). *There exists C_Γ such that, for all $\beta \in H^{1/2}(\Gamma)$,*

$$|\!|\!| u^h |\!|\!|_{|s|,\mathbb{R}^d} \le C_\Gamma \frac{|s|^{3/2}}{\sigma\, \underline{\sigma}^{1/2}} \|\beta\|_{1/2,\Gamma}, \qquad \forall s \in \mathbb{C}_+, \quad \forall h, \tag{5.13}$$

where u^h is the solution of (5.12). *Therefore*

$$\|\mathrm{G}_u^h(s)\|_{H^{1/2}(\Gamma)\to H^1(\mathbb{R}^d)} \le C_\Gamma \frac{|s|^{3/2}}{\sigma\, \underline{\sigma}^{3/2}}. \tag{5.14}$$

Proof. Recall first that the bilinear form $a_{s,\mathbb{R}^d}$ is coercive (see (2.13)) in the energy norm

$$\mathrm{Re}\big(e^{-\imath \operatorname{Arg} s} a_{s,\mathbb{R}^d}(v, \overline{v})\big) = \frac{\sigma}{|s|} |\!|\!| v |\!|\!|_{s,\mathbb{R}^d}^2 \qquad \forall v \in H^1(\mathbb{R}^d). \tag{5.15}$$

Using the lifting of Proposition 2.5.1, it is easy to construct liftings (right-inverses of the trace operator) $\gamma^+(s) : H^{1/2}(\Gamma) \to H^1(\mathbb{R}^d)$ such that

$$|\!|\!| \gamma^+(s)\beta |\!|\!|_{|s|,\mathbb{R}^d} \le C_1 \max\{1, |s|\}^{1/2} \|\beta\|_{1/2,\Gamma} \qquad \forall \beta \in H^{1/2}(\Gamma) \qquad \forall s \in \mathbb{C}_+. \tag{5.16}$$

Let us now test equations (5.12) with $v^h = u^h - u_\beta \in H_h$, where $u_\beta := \gamma^+(s)\beta$. Then

$$\begin{aligned}
\frac{\sigma}{|s|}|||u^h - u_\beta|||^2_{|s|,\mathbb{R}^d} &= \operatorname{Re}\big(e^{-\imath \operatorname{Arg} s} a_{s,\mathbb{R}^d}(u^h - u_\beta, \overline{u^h - u_\beta})\big) && \text{(by (5.15) or (2.13))}\\
&\le |a_{s,\mathbb{R}^d}(u_\beta, \overline{u^h - u_\beta})| && \text{(u^h solves (5.12))}\\
&\le |||u_\beta|||_{|s|,\mathbb{R}^d}|||u^h - u_\beta|||_{|s|,\mathbb{R}^d}. && \text{(by (2.12))}
\end{aligned}$$

Therefore,

$$\begin{aligned}
|||u^h|||_{|s|,\mathbb{R}^d} &\le |||u_\beta|||_{|s|,\mathbb{R}^d} + \frac{|s|}{\sigma}|||u_\beta|||_{|s|,\mathbb{R}^d} \le 2\frac{|s|}{\sigma}|||u_\beta|||_{|s|,\mathbb{R}^d} && \text{(see above)}\\
&\le 2C_1\frac{|s|}{\sigma}\max\{1,|s|\}^{1/2}\|\beta\|_{1/2,\Gamma} && \text{(by (5.16))}\\
&\le 2C_1\frac{|s|^{3/2}}{\sigma\,\underline{\sigma}^{1/2}}\|\beta\|_{1/2,\Gamma}. && \text{(by (2.15))}
\end{aligned}$$

This proves (5.13). To prove (5.14) use Propositions 5.2.1 and 5.2.2 and the inequality $\underline{\sigma}\|u\|_{1,\mathbb{R}^d} \le |||u|||_{|s|,\mathbb{R}^d}$. Note that this entire group of proofs (from Proposition 5.2.1 and 5.2.3) just repeats at the discrete level what was proved at the continuous level in Section 3.3. □

Proposition 5.2.4 (Step #4: boundary wrap-up). *There exists C_Γ such that*

$$\|\mathrm{G}^h_\lambda(s)\|_{H^{1/2}(\Gamma)\to H^{-1/2}(\Gamma)} \le C_\Gamma\frac{|s|^2}{\sigma\underline{\sigma}} \qquad \forall s \in \mathbb{C}_+ \quad \forall h.$$

Proof. Proceeding in this four step program, this final step uses the fact that $\mathrm{G}^h_\lambda(s)\beta = [\![\partial_\nu \mathrm{G}^h_u(s)\beta]\!]$, Proposition 5.2.3 and Proposition 2.5.2, i.e., the fact that we can bound

$$\|[\![\partial_\nu u^h]\!]\|_{-1/2,\Gamma} \le C\frac{|s|^{1/2}}{\underline{\sigma}^{1/2}}|||u^h|||_{|s|,\mathbb{R}^d}.$$

Note that we obtain again the direct estimate given by coercivity in (5.7). This should not be a surprise since coercivity played a key role in the above proofs. □

5.3 The Galerkin error operator

Consider the problem of solving

$$\lambda^h \in X_h \qquad \text{such that} \qquad \langle \mu^h, \mathrm{V}(s)\lambda^h\rangle_\Gamma = \langle \mu^h, \mathrm{V}(s)\lambda\rangle_\Gamma \quad \forall \mu^h \in X_h$$

and then defining $u^h = \mathrm{S}(s)\lambda^h$. We want to study the following operators

$$\mathrm{E}_\lambda^h(s)\lambda := \lambda^h - \lambda, \qquad \mathrm{E}_u^h(s)\lambda := u^h - u = \mathrm{S}(s)\mathrm{E}_\lambda^h(s)\lambda.$$

It is clear that

$$\mathrm{E}_\lambda^h(s) = \mathrm{G}_\lambda^h(s)\mathrm{V}(s) - \mathrm{I}, \qquad \mathrm{E}_u^h(s) = \mathrm{G}_u^h(s)\mathrm{V}(s) - \mathrm{S}(s) = \mathrm{S}(s)(\mathrm{G}_\lambda^h(s)\mathrm{V}(s) - \mathrm{I})$$

and that

$$\mathrm{E}_\lambda^h(s) = [\![\partial_\nu \cdot]\!]\mathrm{E}_u^h(s).$$

The process is very similar to the one developed in Section 5.2, and we will only sketch the four-step program in the proof. Details are left to the reader. We will use again the space

$$H_h := \{v^h \in H^1(\mathbb{R}^d) \, : \, \gamma v^h \in X_h^\circ\}.$$

Proposition 5.3.1 (Step #1: transmission problem). *Let* $\lambda \in H^{-1/2}(\Gamma)$. *Then* $\varepsilon_u^h = \mathrm{E}_u^h(s)\lambda$ *if and only if*

$$\begin{aligned} \varepsilon_u^h &\in H^1(\mathbb{R}^d), && \text{(5.17a)}\\ \Delta\varepsilon_u^h - s^2\varepsilon_u^h &= 0 \quad \textit{in } \mathbb{R}^d \setminus \Gamma, && \text{(5.17b)}\\ \gamma\varepsilon_u^h &\in X_h^\circ, && \text{(5.17c)}\\ [\![\partial_\nu\varepsilon_u^h]\!] + \lambda &\in X_h. && \text{(5.17d)} \end{aligned}$$

Note the parallelism of the transmission problems (5.11) and (5.17). They are based on the same set of exotic transmission conditions. While the study of the Galerkin solver leads to a nonhomogeneous essential transmission condition, the Galerkin error operator takes us to a nonhomogeneous natural transmission condition.

Proposition 5.3.2 (Step #2: variational formulation). *The transmission problem* (5.17) *is equivalent to the variational problem*

$$\left[\begin{array}{l} \varepsilon_u^h \in H_h, \\ a_{s,\mathbb{R}^d}(\varepsilon_u^h, v^h) = -\langle \lambda, \gamma v^h\rangle_\Gamma \qquad \forall v^h \in H_h. \end{array}\right. \tag{5.18}$$

Proposition 5.3.3 (Step #3: energy/coercivity estimate). *There exists* C_Γ *such that, for all* $\lambda \in H^{-1/2}(\Gamma)$,

$$|\!|\!|\varepsilon_u^h|\!|\!|_{|s|,\mathbb{R}^d} \le C_\Gamma \frac{|s|}{\sigma\,\underline{\sigma}} \|\lambda\|_{1/2,\Gamma}, \qquad \forall s \in \mathbb{C}_+, \quad \forall h,$$

where ε_u^h is the solution of (5.18). *Therefore*

$$\|\mathrm{E}_u^h(s)\|_{H^{-1/2}(\Gamma)\to H^1(\mathbb{R}^d)} \le C_\Gamma \frac{|s|}{\sigma\,\underline{\sigma}^2}. \tag{5.19}$$

Proposition 5.3.4 (Step #4: boundary wrap-up). *There exists C_Γ such that*

$$\|\mathrm{E}_\lambda^h(s)\|_{H^{-1/2}(\Gamma)\to H^{-1/2}(\Gamma)} \le C_\Gamma \frac{|s|^{3/2}}{\sigma\underline{\sigma}^{3/2}} \qquad \forall s\in\mathbb{C}_+ \quad \forall h.$$

Galerkin semidiscretization error. Let finally $\Pi^h : H^{-1/2}(\Gamma) \to X_h$ be the orthogonal projection operator onto X_h. Note that

$$\mathrm{E}_\lambda^h(s)\Pi^h = 0,$$

as a consequence of $\lambda \in X_h$ implying $\lambda^h = \lambda$. In other words, this reflects the fact that Π^h and $\mathrm{E}_\lambda^h(s)$ are complementary projections. We will use this property in the following equivalent form

$$\mathrm{E}_\lambda^h(s) = \mathrm{E}_\lambda^h(s)(\mathrm{I}-\Pi^h), \qquad \mathrm{E}_u^h(s) = \mathrm{E}_u^h(s)(\mathrm{I}-\Pi^h),$$

which is a highly compacted form of a Céa estimate.

5.4 Error estimates

Review. Let us briefly recall the fully discrete method destined to approximate the solution of

$$\mathcal{V} * \lambda = \beta, \qquad u = \mathcal{S} * \lambda. \tag{5.20}$$

The Galerkin solver, defined in the Laplace domain in Section 5.2, corresponds to convolution with the operator

$$\mathcal{G}_\lambda^h := \mathcal{L}^{-1}\{\mathrm{G}_\lambda^h\}.$$

The semidiscrete system is then

$$\lambda^h = \mathcal{G}_\lambda^h * \beta, \qquad u^h = \mathcal{S} * \lambda^h, \tag{5.21}$$

while the full discrete CQ-BEM scheme consists of

$$\lambda_\kappa^h = \mathcal{G}_{\lambda,\kappa}^h * \beta, \qquad u_\kappa^h = \mathcal{S}_\kappa * \lambda_\kappa^h. \tag{5.22}$$

Note that CQ has been applied to both convolution processes in (5.21), i.e., to the semidiscrete version of the equation $\mathcal{V} * \lambda = g$ and to the potential postprocessing $u = \mathcal{S} * \lambda^h$. The Laplace domain version of (5.22) is

$$\Lambda_\kappa^h(s) = \mathrm{G}_{\lambda,\kappa}^h(s)\mathrm{B}(s), \qquad \mathrm{U}_\kappa^h(s) = \mathrm{S}_\kappa(s)\Lambda_\kappa^h(s) = \mathrm{G}_{u,\kappa}^h(s)\mathrm{B}(s),$$

or equivalently,

$$\Lambda_\kappa^h(s) = \mathrm{G}_\lambda^h(s_\kappa)\mathrm{B}(s), \qquad \mathrm{U}_\kappa^h(s) = \mathrm{G}_u^h(s_\kappa)\mathrm{B}(s), \quad \text{where} \quad s_\kappa = \kappa^{-1}p(e^{-s\kappa}).$$

Galerkin semidiscretization error. Finally let

$$\mathcal{E}_\lambda^h := \mathcal{L}^{-1}\{\mathrm{E}_\lambda^h\}, \qquad \mathcal{E}_u^h := \mathcal{L}^{-1}\{\mathrm{E}_u^h\}.$$

Then

$$\lambda^h - \lambda = \mathcal{E}_\lambda^h * \lambda = \mathcal{E}_\lambda^h * (\lambda - \Pi^h\lambda)$$

and

$$u^h - u = \mathcal{S} * (\lambda^h - \lambda) = \mathcal{E}_u^h * \lambda = \mathcal{E}_u^h * (\lambda - \Pi^h\lambda).$$

Proposition 3.2.2 applied to the convolution operators with symbols E_λ^h and E_u^h (studied in Propositions 5.3.3 and 5.3.4) gives a complete analysis of the semidiscretization error in terms of approximation errors. Here are the results that we obtain when we combine Proposition 3.2.2 with the Laplace domain analysis of the Galerkin semidiscretization error operator. For the potential we use the bound of Proposition 5.3.3 and Proposition 3.2.2 with $\mu = 1$, (therefore, $k = 3$ and $\varepsilon = 1$) and $C(\sigma) = 1/(\sigma \min\{1, \sigma^2\})$:

$$\begin{aligned}\|u(t) - u^h(t)\|_{1,\mathbb{R}^d} &\le C\frac{t^2}{1+t}\max\{1,t\}\int_0^t \|\mathcal{P}_3(\lambda - \Pi^h\lambda)(\tau)\|_{-1/2,\Gamma}d\tau \\ &\le 3C\frac{t^2}{1+t}\max\{1,t\}\sum_{\ell=0}^3\int_0^t \|\lambda^{(\ell)}(\tau) - \Pi^h\lambda^{(\ell)}(\tau)\|_{-1/2,\Gamma}d\tau.\end{aligned}$$

For the density, the Laplace domain bound is given in Proposition 5.3.4, which allows us to use Proposition 3.2.2 with $\mu = 3/2$ (and then $k = 3$, $\varepsilon = 1/2$) and $C(\sigma) = 1/(\sigma \min\{1, \sigma^{3/2}\})$:

$$\begin{aligned}\|\lambda(t) - \lambda^h(t)\|_{-1/2,\Gamma} &\le C\left(\frac{t}{1+t}\right)^{1/2} t\max\{1,t^{3/2}\}\int_0^t \|\mathcal{P}_3(\lambda - \Pi^h\lambda)(\tau)\|_{-1/2,\Gamma}d\tau \\ &\le 3C\left(\frac{t}{1+t}\right)^{1/2} t\max\{1,t^{3/2}\}\sum_{\ell=0}^3\int_0^t \|\lambda^{(\ell)}(\tau) - \Pi^h\lambda^{(\ell)}(\tau)\|_{-1/2,\Gamma}d\tau.\end{aligned}$$

Note that the bound for $u(t) - u^h(t)$ is given for a problem that is extended to the interior of the scatterer, where the Dirichlet problem with the same boundary data is solved: this is part of the single layer potential representation of the solution to the problem.

Convolution Quadrature error. Since we have been careful in showing how the constants that bound the Galerkin solvers in the Laplace domain are independent of h (see Propositions 5.2.3 and 5.2.4), the analysis of the time discretization is now just a case of using the Convolution Quadrature estimates (Proposition 4.6.1 and 4.6.2) to bound

$$\|(\lambda^h - \lambda^h_\kappa)(t)\|_{-1/2,\Gamma} = \|(\mathcal{G}^h_\lambda * \beta - \mathcal{G}^h_{\lambda,\kappa} * \beta)(t)\|_{-1/2,\Gamma}$$

and

$$\|(u^h - u^h_\kappa)(t)\|_{1,\mathbb{R}^d} = \|(\mathcal{G}^h_u * \beta - \mathcal{G}^h_{u,\kappa} * \beta)(t)\|_{1/2,\Gamma}.$$

To bound the difference between semidiscrete and fully discrete error for the potential we use the Laplace domain bound in Proposition 5.2.3 as the particular instance to use the Proposition 4.6.1. For BDF2 ($q = 2$), we have $\mu = 3/2$ and we need $k = 6$ for full convergence:

$$\|u^h(t) - u^h_\kappa(t)\|_{1,\mathbb{R}^d} \leq C\kappa^2 t^{3/2}(1 + t^{7-3/2}) \int_0^t \|\beta^{(6)}(\tau)\|_{1/2,\Gamma} \mathrm{d}\tau.$$

The behavior for small and large t is presented in a slightly simplified way: if the function behaves like t^{m_1} for small t and it behaves like t^{m_2} for large t and $m_2 > m_1$ (it will always be the case), we bound by $t^{m_1} + t^{m_2} = t^{m_1}(1 + t^{m_2 - m_1})$. The error for the density follows from the application of the bounds of Proposition 5.2.4 to the framework of Proposition 4.6.1: since $\mu = 2$ for full convergence of BDF2, we need $k = 7$ to obtain

$$\|\lambda^h(t) - \lambda^h_\kappa(t)\|_{-1/2,\Gamma} \leq C\kappa^2 t^2(1 + t^5) \int_0^t \|\beta^{(7)}(\tau)\|_{1/2,\Gamma} \mathrm{d}\tau.$$

Regularity requirements for full order of convergence of the density and of the potential are different. We will come back and improve all these results when we approach the analysis of potentials (continuous and semidiscrete) and of the fully discrete methods with time domain techniques. If we compare these results with those of Section 4.7 (CQ applied to the same problem without space discretization), we will see that we obtain the same type of bounds (identical regularity of data and behavior with respect to t). This is not a coincidence. The use of variational techniques to analyze the semidiscrete solvers allows us to obtain the same Laplace domain bounds for the continuous and the discrete inverses.

5.5 Sound-hard scattering with CQ-BEM

The method. We are interested in the exterior wave equation with Neumann boundary conditions. The Neumann data is $\alpha \in \mathrm{TD}(H^{-1/2}(\Gamma))$ and we look for $u \in \mathrm{TD}(H^1_\Delta(\Omega^+))$ satisfying

$$\ddot{u} = \Delta u \qquad \text{and} \qquad \partial_\nu u = -\alpha.$$

(The negative sign in the data is introduced to simplify the integral formulation.) We then look for $\varphi \in \mathrm{TD}(H^{1/2}(\Gamma))$ such that

$$\mathcal{W} * \varphi = \alpha$$

and recover the wave field with the double layer representation

$$u = \mathcal{D} * \varphi.$$

A boundary condition $\partial_\nu u = \alpha$ and an integral representation of the form $u = -\mathcal{D} * \varphi$ would also lead to the integral equation $\mathcal{W} * \varphi = \alpha$. However, this would force us to use a negative sign in every occurrence of the double layer potential.

The Galerkin solver. We choose a closed subspace $Y_h \subset H^{1/2}(\Gamma)$ and first think of the following Galerkin solver: for causal $\alpha \in \mathrm{TD}(H^{-1/2}(\Gamma))$ find a causal Y_h-valued distribution φ^h such that

$$\langle \mathcal{W} * \varphi^h, \psi^h \rangle_\Gamma = \langle \alpha, \psi^h \rangle_\Gamma \qquad \forall \psi^h \in Y_h \tag{5.23}$$

and then define

$$u^h := \mathcal{D} * \varphi^h. \tag{5.24}$$

Note that (5.23) can be written in our usual shorthand notation

$$\mathcal{W} * \varphi^h - \alpha \in Y_h^\circ := \{\mu \in H^{-1/2}(\Gamma) \, : \, \langle \mu, \psi^h \rangle_\Gamma = 0 \quad \forall \psi^h \in Y_h\}. \tag{5.25}$$

Note also that from the point of view of u^h, we have

$$[\![\gamma u^h]\!] \in Y_h, \qquad [\![\partial_\nu u^h]\!] = 0, \qquad \partial_\nu u^h + \alpha \in Y_h^\circ, \tag{5.26}$$

and that the density is recovered with the jump of the trace $\varphi^h := -[\![\gamma u^h]\!]$. We will write

$$\varphi^h = \mathcal{G}_\varphi^h * \alpha \tag{5.27}$$

and avoid any symbol for the operator

$$\alpha \mapsto \mathcal{D} * \mathcal{G}_\varphi^h * \alpha$$

to avoid confusion with the Galerkin solver of Section 5.2.

Full discretization. A fully discrete scheme follows from using convolution quadrature for the semidiscrete equation (5.23) and for the potential postprocessing (5.24).

The Galerkin error operator. As usual, we think of (5.23) as a discretization of the equation from the point of view of an exact solution. We thus start with $\varphi \in \mathrm{TD}(H^{1/2}(\Gamma))$ and look for a causal Y_h-valued distribution φ^h such that

$$\langle \mathcal{W} * (\varphi^h - \varphi), \psi^h \rangle_\Gamma = 0 \qquad \forall \psi^h \in Y_h. \tag{5.28}$$

What we care about is the error operator

$$\mathcal{E}_\varphi^h * \varphi = \varphi^h - \varphi = \mathcal{G}_\varphi^h * \mathcal{W} * \varphi - \varphi, \tag{5.29}$$

and its associated potential

$$\varepsilon_u^h := \mathcal{D} * \mathcal{E}_\varphi^h * \varphi = \mathcal{D} * \varphi^h - \mathcal{D} * \varphi. \tag{5.30}$$

Once more, from the point of view of the potential, we can write

$$[\![\gamma \varepsilon_u^h]\!] + \varphi \in Y_h, \qquad [\![\partial_\nu \varepsilon_u^h]\!] = 0, \qquad \partial_\nu \varepsilon_u^h \in Y_h^\circ. \tag{5.31}$$

These equations are to be compared with (5.26). At this time, the reader should be able to recognize how in this case the Galerkin solver is going to be simpler to analyze than the error operator, just by looking at which transmission condition is nonhomogeneous.

The corresponding symbols for the continuous problem. Let us first refresh our memory with the continuous operators. For

$$\mathrm{D} := \mathcal{L}\{\mathcal{D}\}, \qquad \mathrm{W} := \mathcal{L}\{\mathcal{W}\} = -\partial_\nu \mathrm{D}, \qquad \mathrm{K} := \mathcal{L}\{\mathcal{K}\} := \{\!\{\gamma \mathrm{D}\}\!\},$$

we have proved (see Proposition 3.4.1) for all $s \in \mathbb{C}_+$

$$\|\mathrm{W}(s)\|_{H^{1/2}(\Gamma)\to H^{-1/2}(\Gamma)} \le C \frac{|s|^2}{\sigma \underline{\sigma}}, \tag{5.32}$$

$$\|\mathrm{D}(s)\|_{H^{1/2}(\Gamma)\to H^1(\mathbb{R}^d\setminus\Gamma)} \le C \frac{|s|^{3/2}}{\sigma \underline{\sigma}^{3/2}}, \tag{5.33}$$

$$\|\mathrm{K}(s)\|_{H^{1/2}(\Gamma)\to H^1(\mathbb{R}^d\setminus\Gamma)} \le C \frac{|s|^{3/2}}{\sigma \underline{\sigma}^{3/2}}. \tag{5.34}$$

Since for each value of $s \in \mathbb{C}_+$, $\mathrm{K}(s)$ is the real transpose (not the adjoint) for $\mathrm{K}^t(s) : H^{-1/2}(\Gamma) \to H^{1/2}(\Gamma)$ and $\mathrm{K}^t(s) = \{\partial_\nu \mathrm{S}(s)\}$, the bound (5.34) is also a consequence of the bound we got way back in Proposition 2.6.2. We finally recall the coercivity estimate of Proposition 3.4.1

$$\mathrm{Re}\left(e^{-\imath \mathrm{Arg}\, s} \langle \mathrm{W}(s)\varphi, \overline{\varphi}\rangle_\Gamma\right) \ge C \frac{\sigma \underline{\sigma}^2}{|s|} \|\varphi\|^2_{1/2,\Gamma} \qquad \forall \varphi \in H^{1/2}(\Gamma) \quad \forall s \in \mathbb{C}_+. \tag{5.35}$$

Symbols for the semidiscrete operators. We now consider

$$\mathrm{G}^h_\varphi := \mathcal{L}\{\mathcal{G}^h_\varphi\} \quad \text{and} \quad \mathrm{E}^h_\varphi := \mathcal{L}\{\mathcal{E}^h_\varphi\}$$

and note that

$$\mathrm{E}^h_\varphi(s) = \mathrm{G}^h_\varphi(s)\mathrm{W}(s) - \mathrm{I}.$$

It follows from the coercivity estimate (5.35) that

$$\|\mathrm{G}^h_\varphi(s)\|_{H^{-1/2}(\Gamma)\to H^{1/2}(\Gamma)} \le C \frac{|s|}{\sigma\, \underline{\sigma}^2} \qquad \forall s \in \mathbb{C}_+.$$

We will need to study this operator again though, given our interest in the postprocessed operator $\mathrm{D}(s)\mathrm{G}^h_\varphi(s)$.

5.6 The discrete double layer potential

A useful space. The proofs of all the assertions in this paragraph are proposed as exercises. We first consider the space

$$H_h := \{u \in H^1(\mathbb{R}^d \setminus \Gamma) \,:\, [\![\gamma u]\!] \in Y_h\},$$

note that the range of operator that maps $H_h \ni u \mapsto (\gamma^- u, [\![\gamma u]\!])$ is $H^{1/2}(\Gamma) \times Y_h$, and then use this fact to prove this weak (but equivalent) formulation of the natural boundary conditions that we will be using.

Lemma 5.6.1. *Let $u \in H^1_\Delta(\mathbb{R}^d \setminus \Gamma)$. Then the conditions*

$$[\![\partial_\nu u]\!] = 0, \qquad \partial_\nu u + \alpha \in Y_h^\circ$$

are equivalent to

$$(\nabla u, \nabla v)_{\mathbb{R}^d\setminus\Gamma} + (\Delta u, v)_{\mathbb{R}^d\setminus\Gamma} = -\langle \alpha, [\![\gamma v]\!]\rangle_\Gamma \quad \forall v \in H_h.$$

Analysis of the Galerkin solver. We will follow the four-step program of Sections 5.2 and 5.3. Details are left to the reader. We start with $\alpha \in H^{-1/2}(\Gamma)$ and want to study

$$\varphi^h = \mathrm{G}^h_\varphi(s)\alpha \qquad \text{and} \qquad u^h = \mathrm{D}(s)\mathrm{G}^h_\varphi(s)\alpha = \mathrm{D}(s)\varphi^h$$

noticing that $\varphi^h = -[\![\gamma u^h]\!]$.

- Step # 1 (Transmission problem). Given $\alpha \in H^{-1/2}(\Gamma)$, $u^h = \mathrm{D}(s)\mathrm{G}^h_\varphi(s)\alpha$ if and only if

$$u^h \in H^1(\mathbb{R}^d \setminus \Gamma), \tag{5.36a}$$

$$\Delta u^h - s^2 u^h = 0 \qquad \text{in } \mathbb{R}^d \setminus \Gamma, \tag{5.36b}$$

$$[\![\gamma u^h]\!] \in Y_h, \tag{5.36c}$$

$$[\![\partial_\nu u^h]\!] = 0, \tag{5.36d}$$

$$\partial_\nu u^h + \alpha \in Y_h^\circ. \tag{5.36e}$$

  Note that, compared to the transmission problems related to the single layer potential (Propositions 5.2.1 and 5.3.1), there is one more transmission condition. In fact, in those problems, the condition $[\![\gamma \cdot]\!] = 0$ was implicit to the fact that solutions were elements of $H^1(\mathbb{R}^d)$.
- Step #2 (Variational formulation) Problem (5.36) is equivalent to

$$\left[\begin{array}{l} u^h \in H_h, \\ a_{s,\mathbb{R}^d\setminus\Gamma}(u^h, v^h) = -\langle \alpha, [\![\gamma v^h]\!]\rangle_\Gamma \qquad \forall v^h \in H_h. \end{array}\right. \tag{5.37}$$

- Step # 3 (Energy estimate) The solution of (5.37) satisfies

$$|\!|\!| u^h |\!|\!|_{|s|,\mathbb{R}^d\setminus\Gamma} \le C \frac{|s|}{\sigma \underline{\sigma}} \|\alpha\|_{-1/2,\Gamma},$$

  and therefore

$$\|\mathrm{D}(s)\mathrm{G}^h_\varphi(s)\|_{H^{-1/2}(\Gamma)\to H^1(\mathbb{R}^d\setminus\Gamma)} \le C \frac{|s|}{\sigma \underline{\sigma}^2} \qquad \forall s \in \mathbb{C}_+. \tag{5.38}$$

- Step # 4 (Boundary wrap-up). Taking the jump of the trace in (5.38), we obtain

$$\|\mathrm{G}^h_\varphi(s)\|_{H^{-1/2}(\Gamma)\to H^{1/2}(\Gamma)} \le C \frac{|s|}{\sigma \underline{\sigma}^2} \qquad \forall s \in \mathbb{C}_+, \tag{5.39}$$

  which was already known using a direct coercivity estimate.

Analysis of the Galerkin error operator. We now focus on the operators

$$\mathrm{E}^h_\varphi(s) = \mathrm{G}^h_\varphi(s)\mathrm{W}(s) - \mathrm{I} \qquad \text{and} \qquad \mathrm{D}(s)\mathrm{E}^h_\varphi(s).$$

The four-step process parallels the one above, with the only difference that we will have to come up with a proper lifting of the essential transmission condition (as we did in Proposition 5.2.3 for the Single Layer potential) at the time of obtaining the energy estimate. This being an error analysis, we start with $\varphi \in H^{1/2}(\Gamma)$ and proceed from here.

- Step # 1 (Transmission problem). Given $\varphi \in H^{1/2}(\Gamma)$, $\varepsilon^h_u = \mathrm{D}(s)\mathrm{E}^h_\varphi(s)\varphi$ if and only if

$$\varepsilon^h_u \in H^1(\mathbb{R}^d \setminus \Gamma), \tag{5.40a}$$

$$\Delta\varepsilon^h_u - s^2\varepsilon^h_u = 0 \qquad \text{in } \mathbb{R}^d \setminus \Gamma, \tag{5.40b}$$

$$[\![\gamma\varepsilon^h_u]\!] + \varphi \in Y_h, \tag{5.40c}$$

$$[\![\partial_\nu\varepsilon^h_u]\!] = 0, \tag{5.40d}$$

$$\partial_\nu\varepsilon^h_u \in Y^\circ_h. \tag{5.40e}$$

- Step #2 (Variational formulation) Problem (5.40) is equivalent to

$$\left[\begin{array}{ll} \varepsilon^h_u \in H^1(\mathbb{R}^d \setminus \Gamma) & [\![\gamma\varepsilon^h_u]\!] + \varphi \in Y_h, \\ a_{s,\mathbb{R}^d\setminus\Gamma}(\varepsilon^h_u, v^h) = 0 & \forall v^h \in H_h. \end{array}\right. \tag{5.41}$$

- Step # 3 (Energy estimate) Using the lifting of Proposition 2.5.1 in the interior domain Ω^-, extended by zero to the exterior domain, we can build

$$u_\varphi(s) \in H^1(\mathbb{R}^d\setminus\Gamma), \qquad [\![\gamma u_\varphi(s)]\!] = \varphi, \qquad |\!|\!| u_\varphi(s) |\!|\!|_{|s|,\mathbb{R}^d\setminus\Gamma} \le C\frac{|s|^{1/2}}{\underline{\sigma}^{1/2}}\|\varphi\|_{1/2,\Gamma}.$$

  Then, $\varepsilon^h_u + u_\varphi(s) \in H_h$ can be used as a test in (5.41) and therefore the solution of (5.41) satisfies

$$|\!|\!|\varepsilon^h_u|\!|\!|_{|s|,\mathbb{R}^d\setminus\Gamma} \le |\!|\!|\varepsilon^h_u + u_\varphi(s)|\!|\!|_{|s|,\mathbb{R}^d\setminus\Gamma} + |\!|\!|u_\varphi(s)|\!|\!|_{|s|,\mathbb{R}^d\setminus\Gamma} \le \left(\frac{|s|}{\sigma}+1\right)|\!|\!|u_\varphi(s)|\!|\!|_{|s|,\mathbb{R}^d\setminus\Gamma}.$$

  From these inequalities we easily derive the bound

$$\|\mathrm{D}(s)\mathrm{E}^h_\varphi(s)\|_{H^{1/2}(\Gamma)\to H^1(\mathbb{R}^d\setminus\Gamma)} \le C\frac{|s|^{3/2}}{\sigma\underline{\sigma}^{3/2}} \qquad \forall s \in \mathbb{C}_+. \tag{5.42}$$

- Step # 4 (Boundary wrap-up). Taking the jump of the trace in (5.42), we finally prove

$$\|\mathrm{E}^h_\varphi(s)\|_{H^{1/2}(\Gamma)\to H^{1/2}(\Gamma)} \le C\frac{|s|^{3/2}}{\sigma\underline{\sigma}^{3/2}} \qquad \forall s \in \mathbb{C}_+. \tag{5.43}$$

Error estimates. Error estimates for semidiscretization can be obtained using the same ideas developed in Section 5.4. From the estimates for $\mathrm{E}^h_\varphi(s)$ and $\mathrm{D}(s)\mathrm{E}^h_\varphi(s)$ and from the convolution theorem (Proposition 3.2.2) we can derive bounds for the error committed by Galerkin semidiscretization. These bounds will be written in terms of cumulative integrals for derivatives of $\alpha - \Pi^h\alpha$, where $\Pi^h : H^{1/2}(\Gamma) \to Y_h$ is the orthogonal projection onto Y_h. From the estimates for $\mathrm{G}^h_\varphi(s)$ and $\mathrm{D}(s)\mathrm{G}^h_\varphi(s)$, and using the convolution quadrature estimates of Proposition 4.6.1, we can provide estimates for the difference between the semidiscrete solution and the fully discrete approximation.

5.7 Exercises

1. (Section 5.2) Prove (5.10). You can actually prove the same statement in this more general setting: X_h is a closed subspace of a Hilbert space X and $X_h^\circ \subset X'$ is its polar set. (**Hint.** You can decompose $X = X_h \oplus X_h^\perp$ and relate $X_h^\perp \subset X$ with $X_h^\circ \subset X'$ with the Riesz representation map.)
2. (Section 5.3) Prove all the results of this section.
3. (Section 5.3) Show that the Galerkin error operator $\mathrm{E}^h_\lambda(s)$ is a projection. Give a characterization of its range. (**Hint.** The range depends on s.)
4. (Section 5.6) For the space $H_h := \{u \in H^1(\mathbb{R}^d \setminus \Gamma) \;:\; [\![\gamma u]\!] \in Y_h\}$, prove that the operator $H_h \to H^{1/2}(\Gamma) \times Y_h$ given by

$$u \mapsto (\gamma^- u, [\![\gamma u]\!])$$

   is surjective.
5. (Section 5.6) Prove Lemma 5.6.1 (**Hint.** Use the identity

$$(\nabla u, \nabla v)_{\mathbb{R}^d\setminus\Gamma} + (\Delta u, v)_{\mathbb{R}^d\setminus\Gamma} = \langle [\![\partial_\nu u]\!], \gamma^- v\rangle_\Gamma + \langle \partial_\nu^+ u, [\![\gamma v]\!]\rangle_\Gamma$$

   valid for all $u \in H^1_\Delta(\mathbb{R}^d \setminus \Gamma)$ and $v \in H^1(\mathbb{R}^d \setminus \Gamma)$, and apply the result of the previous exercise.)
6. (Section 5.6) Prove (5.38).
7. (Section 5.6) Prove (5.42).
8. (Section 5.6) Compare the bound (5.43) with the result of estimating

$$\|\mathrm{E}^h_\varphi(s)\|_{H^{1/2}(\Gamma)\to H^{1/2}(\Gamma)} \le 1 + \|\mathrm{G}^h_\varphi(s)\|_{H^{1/2}(\Gamma)\to H^{-1/2}(\Gamma)}\|\mathrm{W}(s)\|_{H^{1/2}(\Gamma)\to H^{-1/2}(\Gamma)}.$$

9. (Section 5.6) The Laplace domain bounds of Section 5.6 provide time domain bounds for the Galerkin solver and error operator associated with the semidiscretization of the equation $\mathcal{W} * \varphi = \alpha$. Write them down. In the case of the error operator, write them in terms of $\alpha - \Pi^h \alpha$, where $\Pi^h : H^{1/2}(\Gamma) \to Y_h$ is the orthogonal projector.
10. (Section 5.6) Write error estimates for the BDF2-CQ method applied to (5.23) and (5.24).

Chapter 6
A general class of second order differential equations

In this short chapter we introduce some elementary tools of evolution equations which will be applied to the analysis of time domain potentials and integral operators. Most of these results in the abstract treatment of evolution equations can be obtained with elementary tools of the theory of strongly continuous semigroups of operators (of groups of isometries actually). An introduction to semigroup techniques applied to partial differential equations can be found in [58, Chapter 4] or [44], with a very general treatment given in [71]. The theory of semigroups of operators is rarely part of the mathematical toolbox for users of boundary integral equations. This is the reason why we include a self-contained approach in Appendix B using a simple point of view, namely the method of separation of variables.

6.1 The basic setup

Two spaces to start with. We consider two real Hilbert spaces

$$V \subset H,$$

with **continuous, compact, and dense injection.** The norm of H, $\|\cdot\|_H$, will be the one we will use to measure kinetic energy. We also consider a finite dimensional space

$$M \subset V,$$

that we will call the space of **rigid motions,** as well as the orthogonal projection $P : H \to M$ and the closed subspace

F.-J. Sayas, *Retarded Potentials and Time Domain Boundary Integral Equations*, Springer Series in Computational Mathematics 50,
DOI 10.1007/978-3-319-26645-9_6

$$H_0 := \{u \in H \,:\, (u, m)_H = 0 \quad \forall m \in M\} = M^\perp.$$

On V we assume that the norm is given in the form

$$\|v\|_V^2 := [v, v] + (Pv, v)_H = [v, v] + \|Pv\|_H^2,$$

where $[\,\cdot\,,\,\cdot\,]$ is a symmetric positive semidefinite bilinear form such that

$$[\,\cdot\,, m] = 0 \quad \forall m \in M.$$

This property is equivalent to

$$[m, m] = 0 \quad \forall m \in M.$$

Since $\|\,\cdot\,\|_V$ is a norm, it is easy to prove that

$$[u, v] = 0 \quad \forall v \in V \qquad \Longleftrightarrow \qquad u \in M.$$

For an element of the space

$$V_0 := \{u \in V \,:\, Pu = 0\} = V \cap H_0,$$

the norm is just $\|u\|_V = [u, u]^{1/2}$. This seminorm in V will be the one measuring **potential energy**. The continuity of the embedding of V into H can be expressed with the (Poincaré type) inequality

$$\|u\|_H \le C_\circ \|u\|_V \qquad \forall u \in V. \tag{6.1}$$

For reasons that will be understood in the following chapters, it is important to keep track of all occurrences of this constant $C_\circ$, although, at the very end of each application, we will make an effort in eliminating $C_\circ$ from all the estimates.

A third space and an operator. We assume the existence of a third space

$$D(A) \subset V \subset H \qquad \text{with} \qquad M \subset D(A)$$

and a linear operator $A : D(A) \to H$ such that the norm

$$\|u\|_{D(A)}^2 := \|Au\|_H^2 + \|u\|_V^2$$

makes $D(A)$ a Hilbert space. The final elements for this abstract construction are:

(a) an **abstract Green's Identity**

$$(Au, v)_H + [u, v] = 0 \qquad \forall u \in D(A) \quad \forall v \in V, \tag{6.2}$$

(b) and a **surjectivity property**:

$$A - I : D(A) \to H \quad \text{is onto.}$$

In the theory of evolution equations it is customary to understand $A : D(A) \subset H \to H$ as an unbounded operator and then endow $D(A)$ with the graph norm to make it bounded. Our approach is equivalent, and it will be enough for our needs.

Proposition 6.1.1. *Assuming all the hypotheses above, it follows that:*

(a) $\ker A = M$,
(b) $R(A) := \{Au \,:\, u \in D(A)\} \subset H_0$,
(c) $A - I : D(A) \to H$ *is invertible,*
(d) $A - I : M \to M$ *is just the minus identity operator,*
(e) $A - I : D(A) \cap H_0 \to H_0$ *is invertible.*

Proof. To prove (a) note first that if $Au = 0$, then by (6.2) it follows that $[u, u] = 0$ and therefore $u \in M$. Conversely, if $u \in M$, then

$$0 = (Au, v)_H + [u, v] = (Au, v)_H \quad \forall v \in V.$$

This implies that $Au = 0$ since V is dense in H. To prove (b) note that if $f = -Au$, then

$$(f, m)_H = (-Au, m)_H = [u, m] = 0 \quad \forall m \in M.$$

To prove (c) we only need to show that $A - I$ is injective. If $-Au + u = 0$, then, using (6.2),

$$0 = -(Au, u)_H + (u, u)_H = [u, u] + (u, u)_H$$

and therefore $u = 0$. Properties (d) and (e) are straightforward. □

Removing density as a hypothesis. The density of the inclusion of V in H is only used in proving that $M \subset \ker A$ (we will use it later on). Assuming that $M \subset \ker A$, we can actually get to prove that V is dense in H as a byproduct of a more demanding theoretical development.

A fast list of examples. It is useful to keep in mind many of the following examples to see how this abstract framework will help to describe several problems related to wave propagation. In all the examples Ω is a bounded set with Lipschitz boundary and the norm of $L^2(\Omega)$ is the standard one.

(1) Take

$$H = L^2(\Omega), \quad V = H_0^1(\Omega), \quad D(A) = \{u \in H_0^1(\Omega) \,:\, \Delta u \in L^2(\Omega)\},$$

with the Dirichlet form as the bilinear form

$$[u, v] := (\nabla u, \nabla v)_\Omega,$$

and $A := \Delta$. In this case $M = \{0\}$ and the constant $C_\circ$ in (6.1) is that of the Poincaré-Friedrichs inequality.

(2) We can also take

$$H = L^2(\Omega), \quad V = H^1(\Omega), \quad D(A) := \{u \in H^1(\Omega) \, : \, \Delta u \in L^2(\Omega), \quad \partial_\nu u = 0\}$$

with the same $[\,\cdot\,,\,\cdot\,]$, $A := \Delta$ again, and $M = \mathbb{P}_0(\Omega)$, the space of constant functions. The abstract Green's Identity in these two first examples is just a consequence of Green's identity after substituting the corresponding homogeneous boundary condition.

(3) Related to the biharmonic operator we can take

$$H = L^2(\Omega), \quad V = H^2_0(\Omega), \quad D(A) := \{u \in H^2_0(\Omega) \, : \, \Delta^2 u \in L^2(\Omega)\}$$

with the bilinear form

$$[u, v] := (\Delta u, \Delta v)_\Omega,$$

the space of rigid motions $M := \mathbb{P}_1(\Omega)$, including all plane displacements, and $A := -\Delta^2$. The abstract Green's Identity is now a weak formulation of the Rayleigh-Green theorem. The bilinear form can also be written as follows

$$[u, v] := \nu(\Delta u, \Delta v)_\Omega + (1 - \nu)(\mathrm{H}\, u, \mathrm{H}\, v)_\Omega, \tag{6.3}$$

where $\mathrm{H}\, u$ is the Hessian matrix of u and $\nu \in [0, 1]$. The fact that these two bilinear forms coincide in $H^2_0(\Omega)$ follows from a simple density argument.

(4) The bilinear form (6.3) and the operator $A := -\Delta^2$ can be used for three choices of the space V

$$H^2(\Omega) \cap H^1_0(\Omega), \qquad H^2(\Omega), \quad \text{or} \quad \{u \in H^2(\Omega) : \partial_\nu u = 0\}.$$

In the case of smooth boundary $\partial\Omega$, the corresponding space $D(A)$ can be described using two higher order boundary conditions. (In the general case, unexpected difficulties are met because of paradoxes related to the impossibility of cleanly separating the boundary conditions γu and $\partial_\nu u$ for elements of $H^2(\Omega)$ in some domains.)

(5) The linear elasticity system fits in this frame by using the bilinear form

$$[\mathbf{u}, \mathbf{v}] := \big(\mathcal{C}\boldsymbol{\varepsilon}(\mathbf{u}), \boldsymbol{\varepsilon}(\mathbf{v})\big)_\Omega \qquad \boldsymbol{\varepsilon}(\mathbf{u}) := \tfrac{1}{2}(\nabla\mathbf{u} + (\nabla\mathbf{u})^\top)$$

where $\mathcal{C}$ is a linear operator (a tensor) allowing for the bilinear form to be symmetric and the coercivity condition

$$[\mathbf{u}, \mathbf{u}] \geq C_{\mathcal{C}} \|\boldsymbol{\varepsilon}(\mathbf{u})\|^2_\Omega \qquad \forall \mathbf{u} \in H^1(\Omega)^d$$

to hold. With the spaces

$$H := L^2(\Omega)^d, \qquad V := H^1_0(\Omega)^d, \qquad D(A) := \{\mathbf{u} \in V \, : \, \mathrm{div}\,(\mathcal{C}\boldsymbol{\varepsilon}(\mathbf{u})) \in L^2(\Omega)^d\},$$

$M = \{0\}$, and $A := \mathrm{div}\,(\mathcal{C}\boldsymbol{\varepsilon}(\mathbf{u}))$ we can describe problems where displacement is given on the boundary. The weak form of the normal stress can be defined as an element of $H^{-1/2}(\Gamma)^d$ with Betti's formula (Green's formula for the elasticity system):

$$\langle \boldsymbol{\sigma}_\nu(\mathbf{u}), \gamma\mathbf{v}\rangle_\Gamma := (\mathrm{div}\,(\mathcal{C}\boldsymbol{\varepsilon}(\mathbf{u})), \mathbf{v})_\Omega + (\mathcal{C}\boldsymbol{\varepsilon}(\mathbf{u}), \boldsymbol{\varepsilon}(\mathbf{v}))_\Omega$$

On the other hand, we can start with the same H and define

$$V := H^1(\Omega)^d, \qquad D(A) := \{\mathbf{u} \in V \,:\, \mathrm{div}\,(\mathcal{C}\boldsymbol{\varepsilon}(\mathbf{u})) \in L^2(\Omega)^d, \quad \boldsymbol{\sigma}_\nu(\mathbf{u}) = \mathbf{0}\},$$

to describe problems with free boundary conditions. In this case, there is a space of rigid motions related to the problem: in two dimensions, M contains the functions

$$\begin{bmatrix} a_1 \\ a_2 \end{bmatrix} + \begin{bmatrix} 0 & -a_3 \\ a_3 & 0 \end{bmatrix} \begin{bmatrix} x_1 \\ x_2 \end{bmatrix},$$

whereas in three dimensions, it contains the functions

$$\begin{bmatrix} a_1 \\ a_2 \\ a_3 \end{bmatrix} + \begin{bmatrix} 0 & a_4 & -a_5 \\ -a_4 & 0 & a_6 \\ a_5 & -a_6 & 0 \end{bmatrix} \begin{bmatrix} x_1 \\ x_2 \\ x_3 \end{bmatrix}.$$

6.2 The main results

Fast forward. The first result, dealing with nonhomogeneous initial conditions, will not be used in our forthcoming strong theory on layer potentials and is only given for the sake of completeness. The theory about the Cauchy problem is standard in the theory of C_0-groups of isometries generated by some unbounded operators. The theory for nonhomogeneous problems with the given regularity for f is not so standard.

Theorem 6.2.1 (The Cauchy problem). *If $u_0 \in D(A)$ and $v_0 \in V$, then the problem*

$$\ddot{u}(t) = Au(t) \quad t \geq 0, \qquad u(0) = u_0, \qquad \dot{u}(0) = v_0$$

has a unique solution in the space

$$\mathcal{C}^2([0,\infty); H) \cap \mathcal{C}^1([0,\infty); V) \cap \mathcal{C}([0,\infty); D(A)).$$

Proof. Existence of solution is proved in Proposition B.2.1. Uniqueness is proved in Proposition B.2.2. □

Strong solutions of nonhomogeneous problems. We now consider the problem

$$\ddot{u}(t) = Au(t) + f(t) \quad t \geq 0, \qquad u(0) = \dot{u}(0) = 0. \tag{6.4}$$

We will see that a continuous $f : [0, \infty) \to V_0$ is enough to guarantee existence of solution to this problem. We will need more regularity when f takes values in H_0. The case of right-hand sides taking values in the space of rigid motions is treated separately in Section 6.3.

Theorem 6.2.2. *If $f : [0, \infty) \to V_0$ is continuous, problem* (6.4) *has a unique solution in the space*

$$\mathcal{C}^2([0,\infty); H) \cap \mathcal{C}^1([0,\infty); V) \cap \mathcal{C}([0,\infty); D(A)).$$

Moreover, for all $t \geq 0$

$$C_\circ^{-1}\|u(t)\|_H \leq \|u(t)\|_V \leq \int_0^t \|f(\tau)\|_H \mathrm{d}\tau, \qquad \|Au(t)\|_H \leq \int_0^t \|f(\tau)\|_V \mathrm{d}\tau,$$

and

$$\|\dot{u}(t)\|_H \leq \int_0^t \|f(\tau)\|_H \mathrm{d}\tau, \qquad \|\dot{u}(t)\|_V \leq \int_0^t \|f(\tau)\|_V \mathrm{d}\tau.$$

Proof. Existence of solution is shown in Proposition B.3.2. The bounds are given in Proposition B.3.3. □

Theorem 6.2.3. *If $f \in \mathcal{C}^1([0,\infty); H_0)$ with $f(0) = 0$, problem* (6.4) *has a unique solution in the space*

$$\mathcal{C}^2([0,\infty); H) \cap \mathcal{C}^1([0,\infty); V) \cap \mathcal{C}([0,\infty); D(A)),$$

satisfying

$$C_\circ^{-1}\|u(t)\|_H \leq \|u(t)\|_V \leq \int_0^t \|f(\tau)\|_H \mathrm{d}\tau, \qquad \|\dot{u}(t)\|_H \leq \int_0^t \|f(\tau)\|_H \mathrm{d}\tau,$$

$$\|\dot{u}(t)\|_V \leq \int_0^t \|\dot{f}(\tau)\|_H \mathrm{d}\tau, \qquad \text{and} \qquad \|Au(t)\|_H \leq 2\int_0^t \|\dot{f}(\tau)\|_H \mathrm{d}\tau.$$

Proof. See Proposition B.4.2. □

Weak solutions. If V' is the representation of the dual of V that follows from identifying H with H' (that is, we are considering the Courant triad $V \subset H \cong H' \subset V'$), then

$$\|v\|_{V'} \leq C_\circ \|v\|_H \qquad \forall v \in H.$$

Moreover, A can be extended to a bounded operator $A : V \to V'$ by using the Green's Identity

$$\langle Au, v\rangle_{V'\times V} := -[u, v] \qquad \forall u, v \in V.$$

Hence, any solution of (6.4) is also a solution of

$$\langle \ddot{u}(t), v\rangle_{V'\times V} + [u(t), v] = (f(t), v)_H \quad \forall v \in V, \quad \forall t \geq 0, \qquad u(0) = \dot{u}(0) = 0, \tag{6.5}$$

and it therefore makes sense to study problems (6.4) and (6.5) when $f : [0, \infty) \to H_0$.

Theorem 6.2.4. *If $f : [0, \infty) \to H_0$ is continuous, problem* (6.5) *has a unique solution in the space*

$$\mathcal{C}^2([0,\infty); V') \cap \mathcal{C}^1([0,\infty); H) \cap \mathcal{C}([0,\infty); V).$$

Moreover, for all $t \geq 0$

$$C_\circ^{-1}\|u(t)\|_H \leq \|u(t)\|_V \leq \int_0^t \|f(\tau)\|_H \mathrm{d}\tau \quad \text{and} \quad \|\dot{u}(t)\|_H \leq \int_0^t \|f(\tau)\|_H \mathrm{d}\tau.$$

Finally, the function

$$w(t) := \int_0^t u(\tau)\mathrm{d}\tau$$

is in the space $\mathcal{C}([0,\infty), D(A))$.

Proof. See Proposition B.4.1. □

6.3 Purely kinetic motion

For some of the arguments below it will be convenient to make use of an orthonormal basis of the finite dimensional space of rigid motions, M, so that

$$Pu = \sum_{n=1}^{K} (u, m_n)_H m_n \qquad \forall u \in H.$$

Proposition 6.3.1. *For any continuous $f : [0, \infty) \to H$, the function*

$$m_f(t) := \int_0^t (t-\tau)Pf(\tau)\mathrm{d}\tau = \sum_{n=1}^{K} \Big(\int_0^t (t-\tau)(f(\tau), m_n)_H \mathrm{d}\tau \Big) m_n, \tag{6.6}$$

is the unique solution of

$$\begin{bmatrix} \ddot{u}(t) = Au(t) + Pf(t) & t \in [0, \infty), \\ u(0) = 0, & \\ \dot{u}(0) = 0, & \end{bmatrix}$$

Moreover $m_f \in \mathcal{C}^2([0, \infty); M)$ *(with any norm in the finite dimensional space* M*).*

Proof. Note that

$$\dot{m}_f(t) = \int_0^t Pf(\tau)\mathrm{d}\tau \qquad \ddot{m}_f(t) = Pf(t) = Am_f(t) + Pf(t).$$

Initial conditions are straightforward to verify as well. Note also that since

$$\langle \ddot{m}_f(t), v\rangle_{V'\times V} + [m_f(t), v] = \langle \ddot{m}_f(t), v\rangle_{V'\times V} = (\ddot{m}_f(t), v)_H = (Pf(t), v)_H \qquad \forall v \in V,$$

then m_f is also a weak solution of the equations. □

Final remark. For problems

$$\begin{bmatrix} \ddot{u}(t) = Au(t) + f(t) & t \in [0, \infty), \\ u(0) = 0, & \\ \dot{u}(0) = 0, & \end{bmatrix}$$

with $f : [0, \infty) \to V$ or

$$\begin{bmatrix} \langle \ddot{u}(t), v\rangle_{V'\times V} + [u(t), v] = (f(t), v)_H \quad \forall v \in V & t \in [0, \infty), \\ u(0) = 0, & \\ \dot{u}(0) = 0, & \end{bmatrix}$$

with $f : [0, \infty) \to H$, we can decompose the solution as the sum of: (a) the rigid motion (6.6), which satisfies

$$\|m_f(t)\|_V = \|m_f(t)\|_H \qquad \text{and} \qquad \|Am_f(t)\|_H = 0,$$

plus (b) the solution of the problem with $f(t) - Pf(t)$ as right-hand side. For the bounds of Theorems 6.2.2 and 6.2.3, note that

$$\int_0^t \|f(\tau) - Pf(\tau)\|_H \mathrm{d}\tau \le \int_0^t \|f(\tau)\|_H \mathrm{d}\tau$$

and

$$\int_0^t \|f(\tau) - Pf(\tau)\|_V \mathrm{d}\tau = \int_0^t |f(\tau)|_V \mathrm{d}\tau, \qquad |f(\tau)|_V^2 = [f(\tau), f(\tau)].$$

Also $|u(t)|_V = |u(t) - m_f(t)|_V = \|u(t) - m_f(t)\|_V$.

Chapter 7
Time domain analysis of the single layer potential

In this chapter we develop a systematic approach to a pure time domain analysis of the single layer potential and operator, the Galerkin semidiscrete error operator and the Galerkin solver. Namely, we want to obtain estimates of

$$\mathcal{S} * \lambda, \qquad \mathcal{V} * \lambda, \qquad \mathcal{K}^t * \lambda,$$

of its semidiscrete inverses

$$\mathcal{G}_\lambda^h * \beta, \qquad \mathcal{G}_u^h * \beta = \mathcal{S} * \mathcal{G}_\lambda^h * \beta$$

and of the error operators

$$\mathcal{E}_\lambda^h * \lambda = \mathcal{G}_\lambda^h * \mathcal{V} * \lambda - \lambda, \qquad \mathcal{E}_u^h * \lambda = \mathcal{S} * \mathcal{E}_\lambda^h * \lambda.$$

The analysis of the semidiscrete operators will be developed in the usual 'for all h' spirit of Chapter 5, meaning that we will look for bounds independent of the discrete space $X_h \subset H^{-1/2}(\Gamma)$. We will only use that the space X_h is closed. Given the fact that when $X_h = H^{-1/2}(\Gamma)$, $\mathcal{G}_\lambda^h = \mathcal{V}^{-1}$, as a byproduct we will obtain an analysis of the operators $\mathcal{V}^{-1} * \beta$ and $\mathcal{S} * \mathcal{V}^{-1} * \beta$.

A remark. As a side reference, let me mention that, to the best of my knowledge, only Brian Rynne [77] had approached the analysis of time domain integral equations using time domain theory. His results are based on the theory of hyperbolic equations and need smoothness of the scatterer.

F.-J. Sayas, *Retarded Potentials and Time Domain Boundary Integral Equations*, Springer Series in Computational Mathematics 50,
DOI 10.1007/978-3-319-26645-9_7

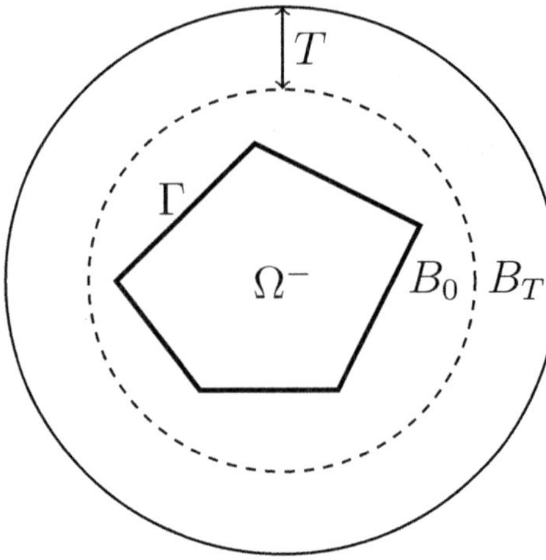

Fig. 7.1 A graphic representation of the cut-off process. The distance between the scatterer and ∂B_0 is denoted δ.

7.1 The cut-off process

Much of the technical effort of this chapter is related on the way of fitting transmission problems for the wave equation in the frame of the second order problems of Chapter 6. We are going to use finite speed of propagation of waves generated by potentials to cut-off the domain at a given distance from the scatterer. In this way, the potentials will coincide with the solution of wave propagation problems in bounded domains in a time interval $[0, T]$. Taking larger values of T, and following carefully all constants depending on the cut-off domain, we will be able to provide a full analysis for all times.

The cut-off geometry. Let $R > 0$ be such that

$$\overline{\Omega^-} \subset B(\mathbf{0}; R) = \{\mathbf{y} \in \mathbb{R}^d \ : \ |\mathbf{y}| < R\} =: B_0$$

and let

$$B_T := B(\mathbf{0}; R + T).$$

Consider also the distance of the boundary of B_0 to Γ (Figure 7.1):

$$\delta := \operatorname{dist}(\partial B_0, \Gamma) := \min\{|\mathbf{x} - \mathbf{y}| \ : \ \mathbf{x} \in \Gamma, |\mathbf{y}| = R\} > 0.$$

The trace operator and normal derivative operators on ∂B_T will have their own notation, making T explicit

$$\begin{aligned} \gamma_T &: H^1(B_T \setminus \overline{\Omega^-}) \to H^{1/2}(\partial B_T), \\ \partial_T^\nu &: H^1_\Delta(B_T \setminus \overline{\Omega^-}) \to H^{-1/2}(\partial B_T). \end{aligned}$$

Proposition 7.1.1 (Finite speed of propagation of potentials). *Let* $\lambda \in \mathrm{TD}(H^{-1/2}(\Gamma))$, $\varphi \in \mathrm{TD}(H^{1/2}(\Gamma))$ *and* $u := \mathcal{S} * \lambda + \mathcal{D} * \varphi$. *Then*

$$\operatorname{supp} \gamma_T u \subset [T + \delta, \infty) \qquad \operatorname{supp} \partial_T^\nu u \subset [T + \delta, \infty).$$

Proof. Note that $\gamma_T u \in \mathrm{TD}(H^{1/2}(\partial B_T))$ and $\partial_T^\nu u \in \mathrm{TD}(H^{-1/2}(\partial B_T))$. The result follows from the causality analysis of Section 3.6, in particular from Proposition 3.6.2. □

Some inequalities. It will be convenient to keep track of several constants related to classical inequalities.

(a) The Poincaré-Friedrichs inequality in B_T:

$$\|v\|_{B_T} \le C_T \|\nabla v\|_{B_T} \qquad \forall v \in H_0^1(B_T). \tag{7.1}$$

Note that we can take

$$C_T = (1 + T/R) C_0, \tag{7.2}$$

where C_0 is the constant in (7.1) for $T = 0$.

(b) Boundedness of the trace operators:

$$\|\gamma^\pm u\|_{1/2,\Gamma} \le C_\gamma \|u\|_{1,B_0\setminus\Gamma} \qquad \forall u \in H^1(B_0 \setminus \Gamma).$$

(c) Boundedness of the normal derivative operator:

$$\|\partial_\nu^\pm u\|_{-1/2,\Gamma} \le C_\nu \Big(\|\nabla u\|^2_{B_0\cap\Omega_\pm} + \|\Delta u\|^2_{B_0\cap\Omega_\pm} \Big)^{1/2} \qquad \forall u \in H^1_\Delta(B_0 \setminus \Gamma). \tag{7.3}$$

(d) A fixed two-sided lifting of the trace. We consider an operator $\gamma^\dagger : H^{1/2}(\Gamma) \to H^1_0(B_0)$ and a constant $C_\gamma^\dagger > 0$ such that

$$\gamma\gamma^\dagger \beta = \beta, \qquad \|\gamma^\dagger \beta\|_{1,B_0} \le C_\gamma^\dagger \|\beta\|_{1/2,\Gamma} \qquad \forall \beta \in H^{1/2}(\Gamma). \tag{7.4}$$

The space

$$\mathbb{X}_T := H^1_0(B_T) \cap H^1_\Delta(B_T \setminus \Gamma),$$

endowed with the norm

$$\|u\|^2_{\mathbb{X}_T} := \|\nabla u\|^2_{B_T} + \|\Delta u\|^2_{B_T\setminus\Gamma},$$

will play a prominent role in all what follows.

The general plan. Following a several step process which mimics in the time domain what we did in the Laplace domain (Sections 5.2 and 5.3), we will work in a systematic way. In this chapter, we will do it for three problems related to the single layer potential.

- Recognize the transmission problem for the wave equation that we want to analyze, then cut the domain off with the boundary ∂B_T and impose a Dirichlet boundary condition.
- Identify the underlying dynamical system framework $D(A) \subset V \subset H$ and verify all needed properties to fit in the frame of Chapter 6 (see below). Transmission conditions that appear in the definition of V will be called *essential*, while those that are imposed in the definition of $D(A)$ will be called *natural*.
- Find a lifting operator related to the nonhomogeneous transmission conditions.
- Prove existence of strong solutions to the wave propagation problem in bounded domain.
  - When transmission conditions are natural this will be doable in a single step (using strong solutions of nonhomogeneous second order equations after lifting the boundary condition).
  - When transmission conditions are essential, the lifting will lead to a weak solution of a nonhomogeneous second order equation, and a correction step (using higher data regularity) will then be applied to obtain bounds for all desired quantities.
- Show finite speed of propagation for the strong solution of the wave propagation problem in bounded domain. This will be phrased as a waiting time property.
- Identify potentials and strong solutions in the interval $[0, T]$ and use the previous bounds to estimate the potentials at time $t = T$.
- Finally use density arguments and the convolutional character of potentials to refine the continuity requirements of data. In a first approximation, we will not carry out this refining step. Instead, we will collect this final step for all three problems in Section 7.5.

The dynamical system framework (a checklist). In all examples below, we will need to identify the spaces

$$D(A) \subset V \subset H,$$

and their norms and inner products. (There will not be rigid motions here.) Recall that, in the absence of rigid motions, we denote by $[\,\cdot\,,\,\cdot\,]$ the inner product of V. We also have to define the operator A, which is related to the definition of $D(A)$. The checklist consists of the following points:

- Verify that the injection $V \subset H$ is compact and dense.
- Find the constant related to the (generalized Poincaré) inequality

$$\|u\|_H \leq C_\circ \|u\|_V \qquad \forall u \in V.$$

- Verify the associated Green identity:

$$[u, v] + (Au, v)_H = 0 \qquad \forall u \in D(A), v \in V.$$

- Verify the surjectivity of $I - A$, i.e., given $f \in H$ find

$$u \in D(A) \qquad \text{such that} \qquad -Au + u = f.$$

A word on causal functions and $[0, \infty)$. The dynamical system approach will work with data $[0, \infty) \to X$. In particular we will use the spaces

$$\mathcal{C}_0^k([0, \infty); X) := \{u \in \mathcal{C}^k([0, \infty); X) \,:\, u^{(\ell)}(0) = 0 \quad \ell \leq k-1\}.$$

Tempered or polynomial behavior at infinity will be often assumed to simplify arguments (mainly to have Laplace transforms or to be in the time domain class $\mathrm{TD}(X)$). However, given the fact that all operators involved in this analysis are causal convolution operators, the behavior of data functions at infinity will not be relevant. Unlike what we did in Chapter 2 (when we were still trying to distinguish derivatives), the dot symbol will be used for strong classical differentiation of functions $[0, \infty) \to X$ and for distributional differentiation of X-valued causal distributions.

Extension by zero to negative values of the time variable will be done using the following notation

$$u \in \mathcal{C}([0, \infty); X) \quad \longmapsto \quad Eu(t) := \begin{cases} u(t), & t \geq 0, \\ 0, & t < 0. \end{cases}$$

For ease of reference we now gather some simple properties of the extension operator E.

Lemma 7.1.2. *Let $u \in \mathcal{C}([0, \infty); X)$.*

(a) *Eu is a causal X-valued distribution.*
(b) *If $\|u(t)\| \leq Ct^m$ for all $t \geq 1$ with some positive m, then $Eu \in \mathrm{TD}(X)$.*
(c) *If $A \in \mathcal{B}(X, Y)$, then*

$$AEu = EAu,$$

where EAu is the extension of the continuous function $Au : [0, \infty) \to Y$ to a Y-valued distribution.
(d) *If $u \in \mathcal{C}^1([0, \infty); X)$, then*

$$\tfrac{\mathrm{d}}{\mathrm{d}t}(Eu) = E\dot{u} + \delta_0 \otimes u(0).$$

In particular, if $u \in \mathcal{C}_0^1([0, \infty); X)$, then

$$\tfrac{\mathrm{d}}{\mathrm{d}t}(Eu) = E\dot{u}.$$

7.2 Potentials and operators

In order to study $u = \mathcal{S} * \lambda$, its trace $\mathcal{V} * \lambda = \gamma(\mathcal{S} * \lambda)$ and the average of its normal derivatives $\mathcal{K}^t * \lambda = \{\!\{\partial_\nu(\mathcal{S} * \lambda)\}\!\}$, we consider the following cut-off problem: we look for $u_T : [0, \infty) \to H^1(B_T) \cap H^1_\Delta(B_T \setminus \Gamma)$ satisfying

$$\ddot{u}_T(t) = \Delta u_T(t) \qquad \forall t \geq 0, \tag{7.5a}$$

$$\gamma_T u_T(t) = 0 \qquad \forall t \geq 0, \tag{7.5b}$$

$$[\![\partial_\nu u_T]\!](t) = \lambda(t) \qquad \forall t \geq 0, \tag{7.5c}$$

$$u_T(0) = \dot{u}_T(0) = 0. \tag{7.5d}$$

The Laplace operator in (7.5a) is in the sense of distributions in $B_T \setminus \Gamma$ and this equality holds in the space $L^2(B_T) \equiv L^2(B_T \setminus \Gamma)$. The equation (7.5b) is imposed in $H^{1/2}(\partial B_T)$ and will be substituted by the implicit demand that

$$u_T : [0, \infty) \to \mathbb{X}_T = H^1_0(B_T) \cap H^1_\Delta(B_T \setminus \Gamma).$$

The jump condition (7.5c) will be imposed in $H^{-1/2}(\Gamma)$.

The dynamical system framework. Consider the spaces

$$\begin{aligned} H &:= L^2(B_T), \\ V &:= H^1_0(B_T), \\ D(A) &:= H^1_0(B_T) \cap H^2(B_T) = H^1_0(B_T) \cap H^1_\Delta(B_T) \\ &= \{u \in V \,:\, \Delta u \in L^2(B_T \setminus \Gamma), \quad [\![\partial_\nu u]\!] = 0\}, \end{aligned}$$

with

$$\|u\|_H := \|u\|_{B_T}, \qquad [u, v] := (\nabla u, \nabla v)_{B_T}$$

and where A is the distributional Laplacian in $B_T \setminus \Gamma$. Note that since functions in $D(A)$ do not have jumps for the trace and normal derivative across Γ, we can equally take A to be the Laplace operator in B_T. We now proceed with the checklist:

- Since $\mathcal{D}(B_T) \subset V$, the density of V in H follows. Compactness of the inclusion follows from Rellich's Compactness Theorem.
- The generalized Poincaré inequality is just the Poincaré inequality (7.1) and therefore $C_\circ = C_T$.
- The associated Green identity is just Green's Identity

$$(\nabla u, \nabla v)_{B_T} + (\Delta u, v)_{B_T} = 0 \qquad \forall u \in D(A) \subset H^1_\Delta(B_T) \qquad \forall v \in V = H^1_0(B_T).$$

- Given $f \in H = L^2(B_T)$ we can easily find $u \in H^1(B_T)$ satisfying

$$-\Delta u + u = f \quad \text{in } B_T \qquad \gamma_T u = 0.$$

Therefore $u \in D(A)$ and $-Au + u = f$.

The lifting operator. Consider the operator $L : H^{-1/2}(\Gamma) \to V = H_0^1(B_T)$ given by the solution of the variational problem

$$\left[\begin{array}{l} u_0 \in H_0^1(B_T), \\ (\nabla u_0, \nabla v)_{B_T} + (u_0, v)_{B_T} = \langle \lambda, \gamma v \rangle_\Gamma \qquad \forall v \in H_0^1(B_T), \end{array}\right.$$

or equivalently, the transmission problem

$$\begin{aligned} u_0 &\in H_0^1(B_T), \\ -\Delta u_0 + u_0 &= 0 \qquad \text{in } B_T \setminus \Gamma, \\ [\![\partial_\nu u_0]\!] &= \lambda. \end{aligned}$$

(Note that the boundary condition $\gamma_T u_0 = 0$ is implicit in the fact that $u_0 \in H_0^1(B_T)$.) Since we can easily bound

$$\|u_0\|_{\mathbb{X}_T} = \|u_0\|_{1,B_T} \le C_\gamma \|\lambda\|_{-1/2,\Gamma},$$

it follows that

$$\|L\|_{H^{-1/2}(\Gamma)\to Z} \le C_\gamma \qquad \text{where} \qquad Z \in \{H, V, H^1(B_T), \mathbb{X}_T\}. \tag{7.6}$$

Proposition 7.2.1 (Strong solutions in bounded domain). *If $\lambda \in \mathcal{C}_0^2([0,\infty); H^{-1/2}(\Gamma))$, then there exists a unique*

$$u_T : [0,\infty) \to \mathbb{X}_T = H_0^1(B_T) \cap H_\Delta^1(B_T \setminus \Gamma),$$

satisfying

$$\ddot{u}_T(t) = \Delta u_T(t) \qquad \forall t \ge 0, \tag{7.7a}$$

$$[\![\partial_\nu u_T]\!](t) = \lambda(t), \qquad \forall t \ge 0, \tag{7.7b}$$

$$u_T(0) = \dot{u}_T(0) = 0, \tag{7.7c}$$

with regularity

$$u_T \in \mathcal{C}^2([0,\infty); L^2(B_T)) \cap \mathcal{C}^1([0,\infty); H_0^1(B_T)) \cap \mathcal{C}([0,\infty); \mathbb{X}_T). \tag{7.8}$$

Moreover, for all $t \ge 0$,

$$\|\nabla u_T(t)\|_{B_T} \leq C_\gamma \big(\|\lambda(t)\|_{-1/2,\Gamma} + B(\lambda, t) \big), \tag{7.9a}$$

$$\|\Delta u_T(t)\|_{B_T \setminus \Gamma} \leq C_\gamma \big(\|\lambda(t)\|_{-1/2,\Gamma} + B(\lambda, t) \big), \tag{7.9b}$$

$$\|u_T(t)\|_{1,B_T} \leq C_\gamma \big(\|\lambda(t)\|_{-1/2,\Gamma} + c(T)\, B(\lambda, t) \big), \tag{7.9c}$$

$$\|\dot{u}_T(t)\|_{1,B_T} \leq C_\gamma \big(\|\dot{\lambda}(t)\|_{-1/2,\Gamma} + \sqrt{2}\, B(\lambda, t) \big), \tag{7.9d}$$

$$\|\{\partial_\nu u_T\}(t)\|_{-1/2,\Gamma} \leq \sqrt{2} C_\gamma C_\nu \big(\|\lambda(t)\|_{-1/2,\Gamma} + B(\lambda, t) \big), \tag{7.9e}$$

where

$$c(T) := \sqrt{1 + C_T^2} \quad \textit{and} \quad B(\lambda, t) := \int_0^t \|\lambda(\tau) - \ddot{\lambda}(\tau)\|_{-1/2,\Gamma} \mathrm{d}\tau.$$

Proof. Lifting and decomposition. Consider the functions

$$u_0(t) := L\lambda(t) \qquad \text{and} \qquad f(t) := L(\lambda(t) - \ddot{\lambda}(t)) = u_0(t) - \ddot{u}_0(t)$$

and note that $f : [0, \infty) \to V$ is continuous and $u_0 \in \mathcal{C}_0^2([0, \infty); \mathbb{X}_T)$ by (7.6). Also, by (7.6), it follows that

$$\int_0^t \|f(\tau)\|_{B_T} \mathrm{d}\tau \leq C_\gamma B(\lambda, t) \qquad \text{and} \qquad \int_0^t \|\nabla f(\tau)\|_{B_T} \mathrm{d}\tau \leq C_\gamma B(\lambda, t). \tag{7.10}$$

The key now is to work with the equation satisfied by $v_0 = u_T - u_0$. Thus, we look for $v_0 : [0, \infty) \to D(A)$ such that

$$\ddot{v}_0(t) = A v_0(t) + f(t) \quad \forall t \geq 0, \qquad v_0(0) = \dot{v}_0(0) = 0. \tag{7.11}$$

By Theorem 6.2.2, equation (7.11) has a solution

$$v_0 \in \mathcal{C}^2([0, \infty); L^2(B_T)) \cap \mathcal{C}^1([0, \infty); H_0^1(B_T)) \cap \mathcal{C}([0, \infty); H_\Delta^1(B_T)). \tag{7.12}$$

Moreover, by Theorem 6.2.2 and (7.10), we can bound

$$C_T^{-1} \|v_0(t)\|_{B_T} \leq \|\nabla v_0(t)\|_{B_T} \leq C_\gamma B(\lambda, t), \tag{7.13a}$$

$$\|\nabla \dot{v}_0(t)\|_{B_T} \leq C_\gamma B(\lambda, t), \tag{7.13b}$$

$$\|\dot{v}_0(t)\|_{B_T} \leq C_\gamma B(\lambda, t), \tag{7.13c}$$

$$\|\Delta v_0(t)\|_{B_T} \leq C_\gamma B(\lambda, t). \tag{7.13d}$$

Recomposition. Consider now the function $u_T(t) := u_0(t) + v_0(t)$. Note that u_T takes values in $\mathbb{X}_T$ for all t and that u_T satisfies (7.8); this follows from (7.12) and the fact that $u_0 \in \mathcal{C}^2([0, \infty); \mathbb{X}_T)$. Also, notice that $u_0(t) = \Delta u_0(t)$ for all t (see the definition of the lifting) and therefore

$$
\begin{aligned}
\ddot{u}_T(t) &= \ddot{u}_0(t) + \ddot{v}_0(t) \\
&= \ddot{u}_0(t) - u_0(t) + \Delta u_0(t) + \Delta v_0(t) + f(t) \\
&= \Delta(u_0(t) + v_0(t)) + \ddot{u}_0(t) - u_0(t) + u_0(t) - \ddot{u}_0(t) = \Delta u_T(t), \qquad \forall t \geq 0.
\end{aligned}
$$

Finally,

$$[\![\partial_\nu u_T]\!](t) = [\![\partial_\nu u_0]\!](t) = \lambda(t) \qquad \forall t \geq 0$$

and

$$u_T(0) = u_0(0) + v_0(0) = L\lambda(0) = 0 \qquad \dot{u}_T(0) = \dot{u}_0(0) + \dot{v}_0(0) = L\dot{\lambda}(0) = 0.$$

This proves that $u_T : [0,\infty) \to \mathbb{X}_T$ satisfies the problem (7.7). (Uniqueness of solution of this problem follows from uniqueness for the associated initial value problem, which was proved in Proposition B.2.2 in the abstract framework of Chapter 6.)

Bounds. The bounds for u_T follows from those for $v_0(t)$, i.e. (7.13), from the boundedness properties of the lifting L (7.6) and from the fact that

$$\|\{\partial_\nu u_T\}(t)\|_{-1/2,\Gamma} \leq C_\nu \|u_T(t)\|_{\mathbb{X}_T}$$

(see (7.3)). □

A note. From the bounds of Proposition 7.2.1 it is clear that polynomially bounded (in time) densities lead to polynomially bounded potentials. The solution of (7.7) will exist no matter what the growth of λ is as a function of t. More importantly, it is clear that the value of $u_T(t)$ depends only on the past values of $\lambda(t)$ (see the bounds of Proposition 7.2.1), which means that in the arguments below we will be able to artificially introduce a hypothesis on the possibility of controlling the norms of $\lambda(t)$ by a polynomial in t.

Reconciling solutions. The next couple of pages deal with some fine details that are needed to make sure that the solution we obtained in the cut-off setting is the same (for a preset time interval) that we had in the sense of causal distributions. This is going to be the process:

- We first show that the normal derivative of the solution of (7.5) on ∂B_T is zero for small enough values of time. This is just a reflection of the finite speed of propagation of the solution of (7.5), which is being created by a density on Γ. This might look like a silly thing to prove. However, the information on the finite speed of propagation from Γ towards ∂B_T is missing in view of (7.5) as an instance of an abstract second order equation. (This is similar to how the separation of variables solution to the wave equation on an interval does not show the finite speed of propagation.)

- We next show that we can extend u_T by zero to $\mathbb{R}^d \setminus B_T$ and obtain a function that satisfies

$$\ddot{u}(t) = \Delta u(t) \qquad 0 \le t \le T + \delta.$$

This extension to free space is exactly what is needed to compare u_T with a single layer potential.

This reconciliation process will be slightly more complicated in the study of boundary integral equations, where some quantity on the boundary Γ is being computed.

Proposition 7.2.2 (Waiting Time on ∂B_T). *If $u_T : [0,\infty) \to \mathbb{X}_T$ is the solution of (7.7) for $\lambda \in \mathcal{C}_0^2([0,\infty); H^{-1/2}(\Gamma))$, then*

$$\partial_T^\nu u_T(t) = 0 \qquad 0 \le t \le T + \delta.$$

Proof. A simple uniqueness argument shows that the value of $\lambda(t)$ for $t > T + \delta$ is not relevant. We thus assume that λ is polynomially bounded at infinity, so that $E\lambda \in \mathrm{TD}(H^{-1/2}(\Gamma))$ by Lemma 7.1.2.

Also by Lemma 7.1.2, Eu_T is an $\mathbb{X}_T$-valued causal distribution satisfying

$$\frac{\mathrm{d}^2}{\mathrm{d}t^2}(Eu_T) = E\ddot{u}_T = E\Delta u_T = \Delta E u_T$$

and

$$[\![\partial_\nu E u_T]\!] = E[\![\partial_\nu u_T]\!] = E\lambda.$$

These equations hold in the sense of $L^2(B_T \setminus \Gamma)$ and $H^{-1/2}(\Gamma)$-valued distributions respectively. Moreover, the bounds of Proposition 7.2.1 show that u_T grows polynomially in the $\mathbb{X}_T$ norm and therefore $Eu_T \in \mathrm{TD}(\mathbb{X}_T)$.

Consider now the distributions

$$w := Eu_T - (\mathcal{S} * E\lambda)|_{B_T} \qquad \text{and} \qquad \xi := \gamma_T w = -\gamma_T(\mathcal{S} * E\lambda).$$

Note that w is a causal solution of the problem

$$w \in \mathrm{TD}(H^1(B_T) \cap H^1_\Delta(B_T \setminus \Gamma)), \qquad \ddot{w} = \Delta w, \qquad [\![\partial_\nu w]\!] = 0, \qquad \gamma_T w = \xi.$$

(The condition $[\![\gamma w]\!] = 0$ is implicit in the space $H^1(B_T) \cap H^1_\Delta(B_T \setminus \Gamma)$.) We can now use a Laplace transform argument (see exercises) to show that

$$w = \mathcal{M} * \xi \qquad \text{where} \qquad \mathcal{M} \in \mathrm{TD}(\mathcal{B}(H^{1/2}(\partial B_T), H^1(B_T) \cap H^1_\Delta(B_T \setminus \Gamma)).$$

Since the speed of propagation of potentials is finite (Proposition 7.1.1), it follows that $\operatorname{supp}\xi \subset [T+\delta,\infty)$, but then, preservation of causality of causal convolution operators (Proposition 3.2.1) proves that $\operatorname{supp} w \subset [T+\delta,\infty)$. This shows that

$$\operatorname{supp} \partial_T^\nu w \subset [T+\delta,\infty),$$

which, combined with the fact that $\operatorname{supp}\partial_T^\nu(\mathcal{S} * E\lambda) \subset [T+\delta,\infty)$ (again by Proposition 7.1.1), implies that

$$\operatorname{supp} \partial_T^\nu E u_T \subset [T+\delta,\infty). \tag{7.14}$$

However, $\partial_T^\nu u_T : [0,\infty) \to H^{-1/2}(\partial B_T)$ is a continuous function as u_T is a continuous $\mathbb{X}_T$-valued function and $\partial_T^\nu : \mathbb{X}_T \to H^{-1/2}(\partial B_T)$ is bounded, and thus (7.14) implies the result. □

Proposition 7.2.3 (Extension to free space). *Let $\lambda \in \mathcal{C}_0^2([0,\infty); H^{-1/2}(\Gamma))$ and u_T be the solution of (7.7). The extension*

$$\underline{u}_T : [0,\infty) \to H^1_\Delta(\mathbb{R}^d \setminus (\Gamma \cup \partial B_T))$$

given by

$$\underline{u}_T(t)|_{B_T} = u_T(t) \qquad \text{and} \qquad \underline{u}_T(t)|_{\mathbb{R}^d\setminus B_T} \equiv 0$$

takes values in $H^1_\Delta(\mathbb{R}^d\setminus\Gamma)$ for $t \le T+\delta$. Furthermore, $\underline{u}_T : [0,T+\delta] \to H^1_\Delta(\mathbb{R}^d\setminus\Gamma)$ is continuous and

$$\underline{u}_T(t) = (\mathcal{S} * E\lambda)(t) \qquad 0 \le t < T+\delta. \tag{7.15}$$

Proof. It is clear that $\underline{u}_T : [0,\infty) \to H^1(\mathbb{R}^d) \cap H^1_\Delta(\mathbb{R}^d \setminus (\Gamma \cup \partial B_T))$ is continuous and that $\underline{u}_T$ satisfies the equations:

$$\ddot{\underline{u}}_T(t) = \Delta \underline{u}_T(t) \qquad \forall t \ge 0, \tag{7.16a}$$

$$[\![\partial_\nu \underline{u}_T]\!](t) = \lambda(t) \qquad \forall t \ge 0, \tag{7.16b}$$

$$[\![\partial_T^\nu \underline{u}_T]\!](t) = \partial_T^\nu u_T(t) \qquad \forall t \ge 0, \tag{7.16c}$$

$$\underline{u}_T(0) = \dot{\underline{u}}_T(0) = 0. \tag{7.16d}$$

The Laplace operator in (7.16a) is the distributional Laplacian in $\mathbb{R}^d \setminus (\Gamma \cup \partial B_T)$, and the jump operator in (7.16c) corresponds to the jump of normal derivative across ∂B_T. By Lemma 7.1.2 (see the argument in the proof of Proposition 7.2.2), the extension $E\underline{u}_T$ is a causal solution of the wave equation

$$E\underline{u}_T \in \mathrm{TD}(H^1_\Delta(\mathbb{R}^d \setminus (\Gamma \cup \partial B_T)) \qquad \tfrac{\mathrm{d}^2}{\mathrm{d}t^2}(E\underline{u}_T) = \Delta E\underline{u}_T$$

and therefore, by Kirchhoff's formula (Proposition 3.5.1), we can represent

$$E\underline{u}_T = \mathcal{S} * [\![\partial_\nu E\underline{u}_T]\!] + \mathcal{S}_{\partial B_T} * [\![\partial_T^\nu E\underline{u}_T]\!] = \mathcal{S} * E\lambda + \mathcal{S}_{\partial B_T} * E\partial_T^\nu u_T,$$

where $\mathcal{S}_{\partial B_T}$ is the distribution that defines the single layer operator stemming from the boundary ∂B_T. By Proposition 7.2.2 and the causality of (causal) convolution operators, it follows that

$$\operatorname{supp}(E\underline{u}_T - \mathcal{S} * E\lambda) = \operatorname{supp}(\mathcal{S}_{\partial B_T} * \partial_T^\nu u_T) \subset [T+\delta, \infty).$$

This proves (7.15). Finally,

$$H^1(\mathbb{R}^d) \cap H^1_\Delta(\mathbb{R}^d \setminus \Gamma) = \{u \in H^1(\mathbb{R}^d) \cap H^1_\Delta(\mathbb{R}^d \setminus (\Gamma \cup \partial B_T)) \,:\, [\![\partial_T^\nu u]\!] = 0\}$$

and the different Laplace operators in these two sets (Laplacian in $\mathbb{R}^d \setminus \Gamma$ and in $\mathbb{R}^d \setminus (\Gamma \cup \partial B_T)$ respectively) coincide. The fact that $\underline{u}_T$ satisfies the equations (7.16) and Proposition 7.2.2 prove then that $\underline{u}_T(t) \in H^1(\mathbb{R}^d) \cap H^1_\Delta(\mathbb{R}^d \setminus \Gamma)$ for all $t \le T+\delta$ and that $\Delta \underline{u}_T(t) = \underline{\Delta u_T}(t)$ for $t \le T+\delta$. □

The results so far. Let us now start with $\lambda \in \mathrm{TD}(H^{-1/2}(\Gamma))$ such that the restriction $\lambda|_{(0,\infty)}$ is in $\mathcal{C}^2_0([0,\infty); H^{-1/2}(\Gamma))$. In order to observe all potentials and operators at time $t = T$, we use Proposition 7.2.3 with the cut-off at T, and evaluate all functions at $t = T$. We then use the bounds of Proposition 7.2.1 to obtain:

$$\begin{aligned}
\|\nabla(\mathcal{S} * \lambda)(T)\|_{\mathbb{R}^d} &= \|\nabla \underline{u}_T(T)\|_{\mathbb{R}^d} && \text{(Proposition 7.2.3)}\\
&= \|\nabla u_T(T)\|_{B_T} && \text{(definition of } \underline{u}_T)\\
&\le C_\gamma(\|\lambda(T)\|_{-1/2,\Gamma} + B(\lambda, T)). && \text{(Proposition 7.2.1)}
\end{aligned}$$

Similarly

$$\begin{aligned}
\|(\mathcal{V} * \lambda)(T)\|_{1/2,\Gamma} &\le C_\gamma \|(\mathcal{S} * \lambda)(T)\|_{1,\mathbb{R}^d} && \text{(trace operator)}\\
&= C_\gamma \|u_T(T)\|_{1,B_T} && \text{(Proposition 7.2.3)}\\
&\le C_\gamma^2(\|\lambda(T)\|_{-1/2,\Gamma} + c(T)\, B(\lambda, T)), && \text{(Proposition 7.2.1)}
\end{aligned}$$

and

$$\begin{aligned}
\|(\mathcal{K}^t * \lambda)(T)\|_{-1/2,\Gamma} &= \|[\![\partial_\nu u_T]\!](T)\|_{-1/2,\Gamma} && \text{(Proposition 7.2.3)}\\
&\le \sqrt{2}\, C_\nu C_\gamma(\|\lambda(T)\|_{-1/2,\Gamma} + B(\lambda, T)). && \text{(Proposition 7.2.1)}
\end{aligned}$$

Using the regularity part of Proposition 7.2.1 and the identification of the potential $\mathcal{S} * \lambda$ with the extensions $\underline{u}_T$, it is also clear that we have proved that

$$\mathcal{S} * \lambda \in \mathcal{C}^2(\mathbb{R}; L^2(\mathbb{R}^d)) \cap \mathcal{C}^1(\mathbb{R}; H^1(\mathbb{R}^d)) \cap \mathcal{C}(\mathbb{R}; H^1_\Delta(\mathbb{R}^d \setminus \Gamma)), \qquad (7.17)$$

and therefore

$$\mathcal{V} * \lambda \in \mathcal{C}^1(\mathbb{R}; H^{1/2}(\Gamma))$$

and

$$\mathcal{K}^t * \lambda \in \mathcal{C}(\mathbb{R}; H^{-1/2}(\Gamma)).$$

Note that (7.17) includes the information about the fact that $\ddot{\underline{u}}_T(0)=\Delta\underline{u}_T(0)=0$, which allows us to show that the potential is actually in the spaces that appear in (7.17).

Concluding remark. We will come back to these estimates in Section 7.5, where we will use some density and shift arguments to reduce the continuity requirements on the side of the density λ. We will also get rid of the quantity $c(T)$ by changing the way we accumulate information from the data function λ.

7.3 The Galerkin error operator

Recall that the Galerkin (semidiscrete) equations, seen as an approximation process, look for a causal X_h-valued density λ^h such that

$$\langle \mu^h, \mathcal{V} * (\lambda^h - \lambda)\rangle_\Gamma = 0 \qquad \forall \mu^h \in X_h \tag{7.18}$$

and then build $u^h = \mathcal{S} * \lambda^h$. The error operators map λ to

$$\mathcal{E}^h_\lambda * \lambda = \lambda^h - \lambda \qquad \text{and} \qquad \mathcal{E}^h_u * \lambda = u^h - u = \mathcal{S} * \mathcal{E}^h_\lambda * \lambda.$$

These operators have been studied in the Laplace domain in Section 5.3. As in that section, the analysis is carried out through the potential component ($\lambda \mapsto \mathcal{E}^h_u * \lambda$) and then conclusions are drawn using the fact that $\mathcal{E}^h_\lambda * \lambda = [\![\partial_\nu(\mathcal{E}^h_u * \lambda)]\!]$. Note that $\varepsilon_u := \mathcal{E}^h_u * \lambda$ can be characterized by the uniquely solvable problem

$$\varepsilon_u \in \mathrm{TD}(H^1(\mathbb{R}^d) \cap H^1_\Delta(\mathbb{R}^d \setminus \Gamma)), \qquad \ddot{\varepsilon}_u = \Delta\varepsilon_u, \qquad \gamma\varepsilon_u \in X_h^\circ, \qquad [\![\partial_\nu \varepsilon_u]\!] + \lambda \in X_h.$$

We cut off with a boundary ∂B_T and look for $\varepsilon_T : [0,\infty) \to H^1(B_T) \cap H^1_\Delta(B_T \setminus \Gamma)$ satisfying

$$\ddot{\varepsilon}_T(t) = \Delta\varepsilon_T(t) \qquad \forall t \ge 0, \tag{7.19a}$$

$$\gamma_T\varepsilon_T(t) = 0 \qquad \forall t \ge 0, \tag{7.19b}$$

$$\gamma\varepsilon_T(t) \in X_h^\circ \qquad \forall t \ge 0, \tag{7.19c}$$

$$[\![\partial_\nu\varepsilon_T]\!](t) + \lambda(t) \in X_h \qquad \forall t \ge 0, \tag{7.19d}$$

$$\varepsilon_T(0) = \dot{\varepsilon}_T(0) = 0. \tag{7.19e}$$

The first condition on Γ, (7.19c), encodes the discrete Galerkin equation (7.18) from the point of view of $u^h - u$. At the same time, (7.19d) is used to impose that the semidiscrete unknown in (7.18) takes values in X_h. As we did in (7.5), the boundary condition on ∂B_T will be directly imposed as part of the definition of the space, by considering

$$\varepsilon_T : [0,\infty) \to \mathbb{X}_T = H_0^1(B_T) \cap H^1_\Delta(B_T \setminus \Gamma).$$

The associated dynamical system. The spaces for the second order equations are suggested by the transmission and boundary conditions in (7.19). We thus define

$$\begin{aligned} H &:= L^2(B_T), \\ V &:= \{u \in H_0^1(B_T) \,:\, \gamma u \in X_h^\circ\}, \\ D(A) &:= \{u \in V \,:\, \Delta u \in L^2(B_T \setminus \Gamma),\ [\![\partial_\nu u]\!] \in X_h\}, \end{aligned}$$

with

$$\|u\|_H := \|u\|_{B_T}, \qquad [u,v] := (\nabla u, \nabla v)_{B_T}.$$

We take A to be the distributional Laplacian in $B_T \setminus \Gamma$. Note that if $u \in H^1_\Delta(B_T \setminus \Gamma)$, then

$$[\![\partial_\nu u]\!] \in X_h \qquad \Longleftrightarrow \qquad (\nabla u, \nabla v)_{B_T\setminus\Gamma} + (\Delta u, v)_{B_T\setminus\Gamma} = 0 \quad \forall v \in V. \tag{7.20}$$

We now verify all the hypotheses in our checklist:

- The fact that $\mathcal{D}(B_T \setminus \Gamma) \subset V \subset H^1(B_T) \subset H$ implies the density and the compactness of the embedding of V in H.
- The constant to relate the norms of H and V is the one of the Poincaré inequality (7.1) and thus $C_\circ = C_T$, with C_T given by (7.1).
- The Green Identity is just another way of phrasing (7.20).
- Finally, if $f \in H = L^2(B_T)$, we can solve the coercive problem

$$\left[\begin{array}{l} u \in V, \\ (\nabla u, \nabla v)_{B_T} + (u,v)_{B_T} = (f,v)_{B_T} \qquad \forall v \in V, \end{array}\right.$$

  and then verify that $-\Delta u + u = f$ in $B_T \setminus \Gamma$ (by testing with a general element of $\mathcal{D}(B_T \setminus \Gamma)$) and then

$$(\nabla u, \nabla v)_{B_T} + (\Delta u, v)_{B_T\setminus\Gamma} = (\nabla u, \nabla v)_{B_T} + (u,v)_{B_T} - (f,v)_{B_T} = 0 \qquad \forall v \in V,$$

  which proves that $u \in D(A)$.

The lifting. We now define $L : H^{-1/2}(\Gamma) \to V$ through the solution of the problem

$$\left[\begin{array}{ll} u_0 \in V, & \\ (\nabla u_0, \nabla v)_{B_T} + (u_0, v)_{B_T} = \langle \lambda, \gamma v\rangle_\Gamma & \forall v \in V, \end{array}\right. \tag{7.21}$$

which is the variational formulation of

$$u_0 \in H_0^1(B_T), \tag{7.22a}$$

$$-\Delta u_0 + u_0 = 0 \qquad \text{in } B_T \setminus \Gamma, \tag{7.22b}$$

$$\gamma u_0 \in X_h^\circ, \tag{7.22c}$$

$$[\![\partial_\nu u_0]\!] - \lambda \in X_h. \tag{7.22d}$$

Testing equations (7.21) with $v = u_0$ we can bound

$$\|u_0\|_{\mathbb{X}_T} = \|u_0\|_{1,B_T} \leq C_\gamma \|\lambda\|_{-1/2,\Gamma}$$

which yields the estimates

$$\|L\|_{H^{-1/2}(\Gamma)\to Z} \leq C_\gamma \qquad \text{where} \qquad Z \in \{H, V, H^1(B_T), \mathbb{X}_T\}. \tag{7.23}$$

Proposition 7.3.1 (Strong solutions in bounded domain). *If* $\lambda \in \mathcal{C}_0^2([0,\infty); H^{-1/2}(\Gamma))$, *then there exists a unique*

$$\varepsilon_T : [0,\infty) \to \mathbb{X}_T = H_0^1(B_T) \cap H_\Delta^1(B_T \setminus \Gamma),$$

satisfying

$$\ddot{\varepsilon}_T(t) = \Delta \varepsilon_T(t) \qquad \forall t \geq 0, \tag{7.24a}$$

$$\gamma \varepsilon_T(t) \in X_h^\circ \qquad \forall t \geq 0, \tag{7.24b}$$

$$[\![\partial_\nu \varepsilon_T]\!](t) + \lambda(t) \in X_h \qquad \forall t \geq 0, \tag{7.24c}$$

$$\varepsilon_T(0) = \dot{\varepsilon}_T(0) = 0, \tag{7.24d}$$

with regularity

$$\varepsilon_T \in \mathcal{C}^2([0,\infty); L^2(B_T)) \cap \mathcal{C}^1([0,\infty); H_0^1(B_T)) \cap \mathcal{C}([0,\infty); \mathbb{X}_T). \tag{7.25}$$

Moreover, for all $t \geq 0$,

$$\|\nabla \varepsilon_T(t)\|_{B_T} \leq C_\gamma \big(\|\lambda(t)\|_{-1/2,\Gamma} + B(\lambda, t)\big), \tag{7.26a}$$

$$\|\Delta \varepsilon_T(t)\|_{B_T \setminus \Gamma} \leq C_\gamma \big(\|\lambda(t)\|_{-1/2,\Gamma} + B(\lambda, t)\big), \tag{7.26b}$$

$$\|\varepsilon_T(t)\|_{1,B_T} \leq C_\gamma\big(\|\lambda(t)\|_{-1/2,\Gamma} + c(T)\,B(\lambda,t)\big), \tag{7.26c}$$

$$\|\dot\varepsilon_T(t)\|_{1,B_T} \leq C_\gamma\big(\|\dot\lambda(t)\|_{-1/2,\Gamma} + \sqrt{2}\,B(\lambda,t)\big), \tag{7.26d}$$

$$\|[\![\partial_\nu\varepsilon_T]\!](t)\|_{-1/2,\Gamma} \leq \sqrt{2}C_\gamma C_\nu\big(\|\lambda(t)\|_{-1/2,\Gamma} + B(\lambda,t)\big), \tag{7.26e}$$

where

$$c(T) := \sqrt{1+C_T^2} \quad \textit{and} \quad B(\lambda,t) := \int_0^t \|\lambda(\tau) - \ddot\lambda(\tau)\|_{-1/2,\Gamma}\mathrm{d}\tau.$$

Proof. The proof is very similar to that of Proposition 7.2.1. We first define

$$u_0(t) := -L\lambda(t) \qquad \text{and} \qquad f(t) := u_0(t) - \ddot u_0(t) = L(\ddot\lambda(t) - \lambda(t)),$$

consider next the solution $v_0 : [0,\infty) \to D(A)$ of the problem

$$\ddot v_0(t) = Av_0(t) + f(t) \quad \forall t \geq 0, \qquad v_0(0) = \dot v_0(0) = 0, \tag{7.27}$$

and finally propose $\varepsilon_T := u_0 + v_0$. The key facts are the continuity of $f : [0,\infty) \to V$ (which means that (7.27) has strong solutions in the sense of Section 6.2) and the equality $\Delta u_0(t) = u_0(t)$ for all t.

Since $L\lambda(t) \in V$ and $v_0(t) \in D(A) \subset V$ for all $t \geq 0$, it is clear that $\varepsilon_T(t) \in V$ for all t. To verify the other transmission condition note that

$$[\![\partial_\nu\varepsilon_T]\!](t) + \lambda(t) = \underbrace{-[\![\partial_\nu L\lambda(t)]\!] + \lambda(t)}_{\text{(see (7.22))}} + \underbrace{[\![\partial_\nu v_0(t)]\!]}_{(v_0(t)\in D(A))} \in X_h.$$

With this we show that ε_T satisfies equations (7.24). Using the fact that $u_0 \in \mathcal{C}^2([0,\infty);\mathbb{X}_T)$ and Theorem 6.2.2 we can prove the desired regularity (7.25).

We can finally use Theorem 6.2.2 to bound v_0 in terms of f, as well as (7.23) to bound f in terms of λ. This leads to the estimates

$$\begin{aligned} C_T^{-1}\|v_0(t)\|_{B_T} \leq \|\nabla v_0(t)\|_{B_T} &\leq C_\gamma B(\lambda,t),\\ \|\nabla \dot v_0(t)\|_{B_T} &\leq C_\gamma B(\lambda,t),\\ \|\dot v_0(t)\|_{B_T} &\leq C_\gamma B(\lambda,t),\\ \|\Delta v_0(t)\|_{B_T\setminus\Gamma} &\leq C_\gamma B(\lambda,t). \end{aligned}$$

Finally, (7.23) is used again to bound u_0 and this finishes the proof. □

Reconciling solutions. The process to recognize that the zero extension of ε_T (to $\mathbb{R}^d \setminus B_T$ and to negative t) is $\mathcal{E}_u^h * \lambda$ is going to be divided into several steps. This proof is more complicated than the one we used for the single layer potential since we will need to recover the density on the boundary. This is a rough sketch of the process as a sequence of steps:

- Show that the normal derivative of ε_T on ∂B_T vanishes for $t \in [0, T+\delta]$.
- Show that $[\![\partial_\nu \varepsilon_T]\!](t)$ does not depend on T for $t \in [0, T+\delta]$. This means that the family of equations (7.19) —parametrized in T— does not depend on T for small enough t.
- Show that the extension by zero to $\mathbb{R}^d \setminus B_T$ satisfies

$$\ddot{\varepsilon}(t) = \Delta\varepsilon(t) \qquad 0 \le t \le T+\delta,$$

and use this fact to write the extension as a single layer retarded potential.
- Finally, use the previous results to relate the strong solution on the cut-off domain with $\mathcal{E}_u^h * \lambda$ and $\mathcal{E}_\lambda^h * \lambda$.

This reconciliation process will be part of the analysis of all other problems related to the single and double layer potentials. While the proofs require a certain amount of careful mathematics, they follow a repeated pattern that the reader will be able to grasp easily.

Proposition 7.3.2 (Waiting Time on ∂B_T). *If $\varepsilon_T : [0,\infty) \to \mathbb{X}_T$ be the solution of* (7.24) *for $\lambda \in \mathcal{C}_0^2([0,\infty); H^{-1/2}(\Gamma))$, then*

$$\partial_T^\nu \varepsilon_T(t) = 0 \qquad \forall t \le T+\delta.$$

Proof. We can restrict our attention to densities λ with polynomial growth in the time variable, which implies that ε_T has also polynomial growth (see the bounds in Proposition 7.3.1). We then define the continuous function $[\![\partial_\nu \varepsilon_T]\!] : [0,\infty) \to H^{-1/2}(\Gamma)$. What is left of the proof consists of a straightforward adaption of the proof of Proposition 7.2.2: we first show that

$$\operatorname{supp}\big(E\varepsilon_T - (\mathcal{S} * E[\![\partial_\nu \varepsilon_T]\!])|_{B_T}\big) \subset [T+\delta, \infty)$$

via a causality argument; then we use finite speed of propagation of potentials (Proposition 7.1.1) to prove that

$$\operatorname{supp} \partial_T^\nu E\varepsilon_T \subset [T+\delta,\infty).$$

Details are left to the reader. □

Proposition 7.3.3 (Recovery of the density). *Let $\varepsilon_T : [0,\infty) \to \mathbb{X}_T$ be the solution of* (7.24) *for $\lambda \in \mathcal{C}_0^2([0,\infty); H^{-1/2}(\Gamma))$. The function*

$$\varepsilon_\lambda : [0,\infty) \to H^{-1/2}(\Gamma), \qquad \varepsilon_\lambda(T) := [\![\partial_\nu \varepsilon_T]\!](T) \qquad \forall T \ge 0,$$

is continuous and satisfies

$$\varepsilon_\lambda(t) = [\![\partial_\nu \varepsilon_T]\!](t) \qquad t \le T+\delta.$$

Proof. A new causality argument can be invoked (see exercises) to show that

$$\varepsilon_T(t) = \varepsilon_{T+M}(t)|_{B_T} \qquad \forall t \le T+\delta \qquad M \ge 0.$$

Therefore

$$[\![\partial_\nu \varepsilon_T]\!](T) = [\![\partial_\nu \varepsilon_{T+M}]\!](T) \qquad \forall M \ge 0.$$

Finally, this implies that

$$\varepsilon_\lambda(t) = [\![\partial_\nu \varepsilon_t]\!](t) = [\![\partial_\nu \varepsilon_T]\!](t) \qquad \forall t \in [0,T].$$

Since $\varepsilon_T : [0,\infty) \to \mathbb{X}_T$ is continuous (Proposition 7.3.1) and $[\![\partial_\nu \cdot]\!] : \mathbb{X}_T \to H^{-1/2}(\Gamma)$ is bounded, it follows that $\varepsilon_\lambda : [0,T] \to H^{-1/2}(\Gamma)$ is continuous for all T. □

Proposition 7.3.4 (Extension to free space). *Let $\lambda \in \mathcal{C}^2_0([0,\infty);X)$, ε_T be the solution of* (7.24) *and ε_λ be defined by Proposition 7.3.3. Consider the extension*

$$\underline{\varepsilon}_T : [0,\infty) \to H^1_\Delta(\mathbb{R}^d \setminus (\Gamma \cup \partial B_T)) \qquad \underline{\varepsilon}_T(t)|_{B_T} = \varepsilon_T(t), \quad \underline{\varepsilon}_T(t)|_{\mathbb{R}^d\setminus B_T} \equiv 0$$

The function $\underline{\varepsilon}_T$ is continuous from $[0, T+\delta]$ to $H^1_\Delta(\mathbb{R}^d \setminus \Gamma)$ and

$$\underline{\varepsilon}_T(t) = (\mathcal{S} * E\varepsilon_\lambda)(t) \qquad 0 \le t < T+\delta. \tag{7.28}$$

Proof. It is clear that $\underline{\varepsilon}_T : [0,\infty) \to H^1(\mathbb{R}^d) \cap H^1_\Delta(\mathbb{R}^d \setminus (\Gamma \cup \partial B_T))$ is continuous. Let then

$$\mu := [\![\partial_\nu \varepsilon_T]\!] : [0,\infty) \to H^{-1/2}(\Gamma), \qquad \mu_T := \partial^\nu_T \varepsilon_T : [0,\infty) \to H^{-1/2}(\partial B_T),$$

and note that by Proposition 7.3.3, $\mu(t) = \varepsilon_\lambda(t)$ for $t \le T+\delta$, and that $\underline{u}_T$ satisfies the equations:

$$\ddot{\underline{\varepsilon}}_T(t) = \Delta \underline{\varepsilon}_T(t) \qquad \forall t \ge 0, \tag{7.29a}$$

$$[\![\partial_\nu \underline{\varepsilon}_T]\!](t) = \mu(t) \qquad \forall t \ge 0, \tag{7.29b}$$

$$[\![\partial^\nu_T \underline{\varepsilon}_T]\!](t) = \mu_T(t) \qquad \forall t \ge 0, \tag{7.29c}$$

$$\underline{\varepsilon}_T(0) = \dot{\underline{\varepsilon}}_T(0) = 0. \tag{7.29d}$$

Proceeding as in the proof of Proposition 7.2.3, we can write

$$E\underline{\varepsilon}_T = \mathcal{S} * E\mu + \mathcal{S}_{\partial B_T} * E\mu_T = \mathcal{S} * E\varepsilon_\lambda + \mathcal{S} * E(\mu - \varepsilon_\lambda) + \mathcal{S}_{\partial B_T} * E\mu_T.$$

Since $\operatorname{supp} E\mu_T \subset [T+\delta,\infty)$ (by the Waiting Time Property – Proposition 7.3.2) and $\operatorname{supp} E(\mu - \varepsilon_\lambda) \subset [T+\delta,\infty)$, the result follows readily. □

Corollary 7.3.1 (Identification of the Galerkin error operators). *In the conditions and notation of Proposition 7.3.4, it follows that*

$$E\varepsilon_\lambda = \mathcal{E}^h_\lambda * E\lambda,$$

and therefore

$$\varepsilon_\lambda(t) = (\mathcal{E}^h_\lambda * E\lambda)(t) \qquad \forall t \geq 0,$$

and

$$\underline{\varepsilon}_T(t) = (\mathcal{E}^h_u * E\lambda)(t) \qquad 0 \leq t < T + \delta.$$

Proof. By Proposition 7.3.3,

$$\varepsilon_\lambda(t) + \lambda(t) = [\![\partial_\nu \varepsilon_t]\!](t) + \lambda(t) \in X_h \qquad \forall t \geq 0.$$

On the other hand, since $\gamma \varepsilon_T(t) \in X_h^\circ$ for all t, by Proposition 7.3.4,

$$(\mathcal{V} * E\varepsilon_\lambda)(t) = \gamma(\mathcal{S} * E\varepsilon_\lambda)(t) = \gamma \underline{\varepsilon}_t(t) \in X_h^\circ.$$

This shows that $E\varepsilon_\lambda = \mathcal{E}^h_\lambda * E\lambda$. □

Conclusions. Like we did in Section 7.2, this collection of results can be gathered in some regularity statements plus some estimates. For a causal distribution λ such that $\lambda|_{(0,\infty)} \in \mathcal{C}^2_0(\mathbb{R}; H^{-1/2}(\Gamma))$, we thus prove that

$$\mathcal{E}^h_u * \lambda \in \mathcal{C}^2(\mathbb{R}; L^2(B_T)) \cap \mathcal{C}^1(\mathbb{R}; H^1(\mathbb{R}^d)) \cap \mathcal{C}(\mathbb{R}; H^1_\Delta(\mathbb{R}^d \setminus \Gamma)) \tag{7.30}$$

and (therefore)

$$\mathcal{E}^h_\lambda * \lambda \in \mathcal{C}(\mathbb{R}; H^{-1/2}(\Gamma)).$$

In addition to this, we get the estimates:

$$\|\nabla(\mathcal{E}^h_u * \lambda)(t)\|_{\mathbb{R}^d} \leq C_\gamma(\|\lambda(t)\|_{-1/2,\Gamma} + B(\lambda, t)), \tag{7.31a}$$

$$\|(\mathcal{E}^h_u * \lambda)(t)\|_{1,\mathbb{R}^d} \leq C_\gamma(\|\lambda(t)\|_{-1/2,\Gamma} + c(t)\, B(\lambda, t)), \tag{7.31b}$$

$$\|(\mathcal{E}^h_\lambda * \lambda)(t)\|_{-1/2,\Gamma} \leq \sqrt{2} C_\nu C_\gamma(\|\lambda(t)\|_{-1/2,\Gamma} + B(\lambda, t)). \tag{7.31c}$$

We will come back to these results in Section 7.5.

7.4 The Galerkin solver

The Galerkin equations, seen from the point of view of data, look for a causal X_h-valued density λ^h such that

$$\langle \mu^h, \mathcal{V} * \lambda^h \rangle_\Gamma = \langle \mu^h, \beta \rangle_\Gamma \qquad \forall \mu^h \in X_h,$$

and then build $u^h = \mathcal{S} * \lambda^h$. The corresponding operators are

$$\mathcal{G}^h_\lambda * \beta = \lambda^h \qquad \text{and} \qquad \mathcal{G}^h_u * \beta = u^h = \mathcal{S} * \mathcal{G}^h_\lambda \beta.$$

These operators were studied in the Laplace domain in Section 5.2. Once again, we will study the operator $\beta \mapsto \mathcal{G}^h_u * \beta$ and then use that $\mathcal{G}^h_\lambda * \beta = [\![\partial_\nu(\mathcal{G}^h_u * \beta)]\!]$. The distribution $u^h = \mathcal{G}^h_u * \beta$ is characterized as the solution of the evolution problem

$$u^h \in \mathrm{TD}(H^1(\mathbb{R}^d) \cap H^1_\Delta(\mathbb{R}^d \setminus \Gamma)), \qquad \ddot{u}^h = \Delta u^h, \qquad \gamma u^h - \beta \in X_h^\circ, \qquad [\![\partial_\nu u^h]\!] \in X_h.$$

The process is triggered again with a cut-off on ∂B_T, so that we look for $u^h_T : [0,\infty) \to H^1(B_T) \cap H^1_\Delta(B_T \setminus \Gamma)$

$$\ddot{u}^h_T(t) = \Delta u^h_T(t) \qquad \forall t \ge 0, \tag{7.32a}$$

$$\gamma_T u^h_T(t) = 0 \qquad \forall t \ge 0, \tag{7.32b}$$

$$\gamma u^h_T(t) - \beta(t) \in X_h^\circ \qquad \forall t \ge 0, \tag{7.32c}$$

$$[\![\partial_\nu u^h_T]\!](t) \in X_h \qquad \forall t \ge 0, \tag{7.32d}$$

$$u^h_T(0) = \dot{u}^h_T(0) = 0. \tag{7.32e}$$

This problem uses the same set of transmission conditions as the one satisfied by the Galerkin error operator (7.19), with the main difference that the essential transmission condition is nonhomogeneous while the natural one is homogeneous. As usual, the homogeneous boundary condition will be implicitly given by looking for $u^h_T : [0,\infty) \to \mathbb{X}_T$.

The associated dynamical system. It will be no surprise to the reader that, given the coincidences between (7.19) and (7.32), the functional setting for this section is exactly the same as that of Section 7.3.

The lifting. Let $L : H^{1/2}(\Gamma) \to H^1_0(B_T) \neq V$ be defined with the variational equations

$$\left[\begin{array}{l} u_0 \in H^1_0(B_T), \quad \gamma u_0 - \beta \in X_h^\circ, \\ (\nabla u_0, \nabla v)_{B_T} + (u_0, v)_{B_T} = 0 \quad \forall v \in V, \end{array}\right. \tag{7.33}$$

or equivalently, with the transmission problem

$$u_0 \in H_0^1(B_T), \tag{7.34a}$$

$$-\Delta u_0 + u_0 = 0 \qquad \text{in } B_T \setminus \Gamma, \tag{7.34b}$$

$$\gamma u_0 - \beta \in X_h^\circ, \tag{7.34c}$$

$$[\![\partial_\nu u_0]\!] \in X_h. \tag{7.34d}$$

Testing equations (7.33) with $v = u_0 - \gamma^\dagger \beta$ (recall the lifting of the trace $\gamma^\dagger$ in (7.4)), we can bound

$$\|u_0\|_{\mathbb{X}_T} = \|u_0\|_{1,B_T} \le \|\gamma^\dagger \beta\|_{1,B_T} \le C_\gamma^\dagger \|\beta\|_{1/2,\Gamma},$$

and thus

$$\|L\|_{H^{1/2}(\Gamma)\to Z} \le C_\gamma^\dagger \qquad \text{where} \qquad Z \in \{H, H^1(B_T), \mathbb{X}_T\}. \tag{7.35}$$

Proposition 7.4.1 (Weak solutions in bounded domain). *If* $\beta \in C_0^2([0,\infty); H^{1/2}(\Gamma))$., *then there exists a unique*

$$u_T^h : [0,\infty) \to H_0^1(B_T),$$

satisfying

$$\langle \ddot{u}_T^h(t), v\rangle_{V'\times V} + (\nabla u_T^h(t), \nabla v)_{B_T} = 0 \qquad \forall v \in V \qquad \forall t \ge 0, \tag{7.36a}$$

$$\gamma u_T^h(t) - \beta(t) \in X_h^\circ \qquad \forall t \ge 0, \tag{7.36b}$$

$$u_T^h(0) = \dot{u}_T^h(0) = 0, \tag{7.36c}$$

with regularity

$$u_T^h \in C^2([0,\infty); V') \cap C^1([0,\infty); L^2(B_T)) \cap C([0,\infty); H_0^1(B_T)). \tag{7.37}$$

Moreover, for all $t \ge 0$,

$$\|\nabla u_T^h(t)\|_{B_T} \le C_\gamma^\dagger \big(\|\beta(t)\|_{1/2,\Gamma} + B(\beta,t)\big),$$

$$\|u_T^h(t)\|_{1,B_T} \le C_\gamma^\dagger \big(\|\beta(t)\|_{1/2,\Gamma} + c(T)\, B(\beta,t)\big),$$

$$\|\dot{u}_T^h(t)\|_{B_T} \le C_\gamma^\dagger \big(\|\beta(t)\|_{1/2,\Gamma} + B(\beta,t)\big),$$

where

$$c(T) := \sqrt{1 + C_T^2} \quad \textit{and} \quad B(\beta,t) := \int_0^t \|\beta(\tau) - \ddot{\beta}(\tau)\|_{1/2,\Gamma} \mathrm{d}\tau.$$

Proof. Let

$$u_0(t) := L\beta(t), \qquad f(t) := u_0(t) - \ddot{u}_0(t) = L(\beta(t) - \ddot{\beta}(t))$$

and note that $u_0 \in \mathcal{C}_0^2([0,\infty)); \mathbb{X}_T)$ and $f \in \mathcal{C}([0,\infty); H)$. However, f does not take values in V. We then consider the solution $v_0 : [0,\infty) \to V$ to the problem

$$\langle \ddot{v}_0(t), v\rangle_{V'\times V} + (\nabla v_0(t), \nabla v)_{B_T} = (f(t), v)_{B_T} \quad \forall v \in V \qquad \forall t \geq 0, \tag{7.38a}$$

$$v_0(0) = \dot{v}_0(0) = 0. \tag{7.38b}$$

By Theorem 6.2.4,

$$v_0 \in \mathcal{C}^1([0,\infty); L^2(B_T)) \cap \mathcal{C}([0,\infty); H^1(B_T))$$

and

$$C_T^{-1}\|v_0(t)\|_{B_T} \leq \|\nabla v_0(t)\|_{B_T} \leq \int_0^t \|f(\tau)\|_{B_T} \mathrm{d}\tau \leq C_\gamma^\dagger B(\beta, t),$$

where we have used (7.35) in the last inequality. We can thus define $u_T^h := u_0 + v_0$. What is left is the proof that u_T^h satisfies equations (7.36), since everything else follows readily from what has been shown for v_0 and from the properties of the lifting L (7.35). To show that u_T^h satisfies (7.36) we first note that

$$\gamma u_T^h(t) - \beta(t) = \underbrace{\gamma L\beta(t) - \beta(t)}_{(\text{see } (7.33))} + \underbrace{\gamma v_0(t)}_{(v_0(t)\in V)} \in X_h^\circ.$$

Also, using the fact that $\langle\cdot,\cdot\rangle_{V'\times V}$ extends $(\cdot,\cdot)_H$, we prove that

$$\begin{aligned}
&\langle \ddot{u}_T^h(t), v\rangle_{V'\times V} + (\nabla u_T^h(t), \nabla v)_V \\
&\qquad = (\ddot{u}_0(t), v)_{B_T} + (\nabla u_0(t), \nabla v)_{B_T} + (f(t), v)_V && (\text{by } (7.38a)) \\
&\qquad = (u_0(t), v)_{B_T} + (\nabla u_0(t), \nabla v)_{B_T} && (\text{definition of } f) \\
&\qquad = 0 \qquad \forall v \in V. && (\text{by } (7.33),\ u_0 = L\beta)
\end{aligned}$$

This finishes the proof. □

Proposition 7.4.2 (Strong solutions in bounded domain). *If $\beta \in \mathcal{C}_0^3([0,\infty); H^{1/2}(\Gamma))$, then there exists a unique*

$$u_T^h : [0,\infty) \to \mathbb{X}_T = H_0^1(B_T) \cap H_\Delta^1(B_T \setminus \Gamma),$$

satisfying

$$\ddot{u}_T^h(t) = \Delta u_T^h(t) \qquad \forall t \geq 0, \tag{7.39a}$$

$$\gamma u_T^h(t) - \beta(t) \in X_h^\circ \qquad \forall t \geq 0, \tag{7.39b}$$

$$[\![\partial_\nu u_T^h]\!](t) \in X_h \qquad \forall t \geq 0, \tag{7.39c}$$

$$u_T^h(0) = \dot{u}_T^h(0) = 0, \tag{7.39d}$$

with regularity

$$u_T^h \in \mathcal{C}^2([0,\infty); L^2(B_T)) \cap \mathcal{C}^1([0,\infty); H_0^1(B_T)) \cap \mathcal{C}([0,\infty); \mathbb{X}_T). \tag{7.40}$$

Moreover, u_T^h *is also the solution of* (7.36) *and*

$$\|\Delta u_T^h(t)\|_{B_T\setminus\Gamma} \leq C_\gamma^\dagger(\|\beta(t)\|_{1/2,\Gamma} + 2B(\dot{\beta},t)),$$

$$\|\nabla \dot{u}_T^h(t)\|_{B_T\setminus\Gamma} \leq C_\gamma^\dagger(\|\dot{\beta}(t)\|_{1/2,\Gamma} + B(\dot{\beta},t)),$$

$$\|[\![\partial_\nu u_T^h]\!](t)\|_{-1/2,\Gamma} \leq \sqrt{2}C_\nu C_\gamma^\dagger(2\|\beta(t)\|_{1/2,\Gamma} + B(\beta,t) + 2B(\dot{\beta},t)).$$

Proof. We define $u_0 \in \mathcal{C}_0^3([0,\infty); \mathbb{X}_T)$ and $f \in \mathcal{C}_0^1([0,\infty); L^2(B_T))$ as in the proof of Proposition 7.4.1:

$$u_0(t) := L\beta(t), \qquad f(t) := u_0(t) - \ddot{u}_0(t).$$

Then we consider the solution $v_0 : [0,\infty) \to D(A)$ of the problem

$$\ddot{v}_0(t) = \Delta v_0(t) + f(t) \qquad t \geq 0, \qquad v_0(0) = \dot{v}_0(0) = 0,$$

and we construct $u_T^h := u_0 + v_0$. It is clear that $u_T^h(0) = \dot{u}_T^h(0) = 0$. Also,

$$\gamma u_T^h(t) - \beta(t) = \underbrace{\gamma L\beta(t) - \beta(t)}_{\text{(see (7.33))}} + \underbrace{\gamma v_0(t)}_{(v_0(t)\in D(A)\subset V)} \in X_h^\circ$$

and

$$[\![\partial_\nu u_T^h]\!](t) = \underbrace{[\![\partial_\nu u_0]\!](t)}_{\text{(see (7.34))}} + \underbrace{[\![\partial_\nu v_0]\!](t)}_{(v_0(t)\in D(A))} \in X_h.$$

The right-hand side of the problem satisfied by v_0 is designed so that u_T^h satisfies the differential equation

$$\ddot{u}_T^h(t) = \Delta u_T^h(t) \qquad t \geq 0,$$

and therefore u_T^h is the solution of (7.39). It follows from Theorem 6.2.3 that u_T^h satisfies (7.40). The bound for $\|\Delta u_T^h(t)\|_{B_T\setminus\Gamma}$ follows then from (7.35) and Theorem 6.2.3. This bound and the bound for $\|\nabla u_T^h(t)\|_{B_T}$ given in Proposition 7.4.1 provide the estimate for $\|[\![\partial_\nu u_T^h]\!](t)\|_{-1/2,\Gamma}$, which finishes the proof. □

The final steps. Identification of the Galerkin solver from extensions of the functions u_T^h follows exactly the same strategy as in Section 7.3, with very much the same proofs. Here is a simple sketch of the process.

- Comparing Eu_T^h with $(\mathcal{S} * E[\![\partial_\nu u_T^h]\!])|_{B_T}$ (see Propositions 7.2.2 and 7.3.2) we prove that

$$\partial_T^\nu u_T^h(t) = 0 \qquad 0 \leq t \leq T + \delta.$$

- Proceeding as in Proposition 7.3.3 we define

$$\lambda^h : [0, \infty) \to H^{-1/2}(\Gamma), \qquad \lambda^h(T) := [\![\partial_\nu u_T^h]\!](T) \quad T \geq 0 \tag{7.41}$$

  and show that $\lambda^h(t) = [\![\partial_\nu u_T^h]\!](t)$ for all $t \in [0, T + \delta]$. This implies that $\lambda^h \in \mathcal{C}([0, \infty); H^{-1/2}(\Gamma))$.
- With the same proof as in Proposition 7.3.4, we next show that

$$\underline{u}_T^h(t) = (\mathcal{S} * E\lambda^h)(t) \qquad 0 \leq t \leq T + \delta,$$

  and that $\underline{u}_T^h(t) \in H^1_\Delta(\mathbb{R}^d \setminus \Gamma)$ for all $t \leq T + \delta$. (As usual, the underlining makes reference to extension by zero to $\mathbb{R}^d \setminus B_T$.)

All of this leads to the final goal. We defer the collection of bounds for the Galerkin solvers to the next section.

Proposition 7.4.3 (Identification of the Galerkin solver). *If u_T^h are the functions given by Proposition 7.4.2 and λ^h is defined by* (7.41), *we can then identify*

$$E\lambda^h = \mathcal{G}_\lambda^h * E\beta$$

and therefore

$$\lambda^h(t) = (\mathcal{G}_\lambda^h * \beta)(t) \qquad \forall t \geq 0,$$

and

$$\underline{u}_T^h(t) = (\mathcal{G}_u^h * E\beta)(t) \qquad 0 \leq t < T + \delta.$$

7.5 Final version of all the results

In this section, we look carefully to everything we obtained in Sections 7.2, 7.3, and 7.4 and refine the results so that they hold for weaker densities or right-hand sides. Some ingredients first:

- We will use the spaces

$$\mathcal{W}_+^k(\mathbb{R};X) := \{\beta \in \mathcal{C}^{k-1}(\mathbb{R};X)\cap \mathrm{TD}(X) : \operatorname{supp}\beta \subset [0,\infty), \quad \beta^{(k)} \in L^1_{\mathrm{loc}}(\mathbb{R};X)\},$$
$$\mathcal{C}_+^k(\mathbb{R};X) := \{\beta \in \mathcal{C}^k(\mathbb{R};X) \,:\, \operatorname{supp}\beta \subset [0,\infty)\}.$$

- A generic constant $C = C(\Gamma)$ will be used to collect all constants C_γ, $C_\gamma^\dagger$, C_ν, etc. that appeared before. We will also consider a generic function $c(t) \le C_1 + C_2\, t$ to handle linear growth of constants at infinity.
- So far, dependence of operators with respect to data has been displayed using the growing norms

$$B(\psi,t) := \int_0^t \|\psi(\tau) - \ddot{\psi}(\tau)\| \mathrm{d}\tau,$$

where the norm that we are integrating was not made explicit. We will present the results with respect to the growing seminorms (note that they miss anything happening after time t)

$$H_k(\psi,t\,|\,X) := \sum_{\ell=0}^k \int_0^t \|\psi^{(\ell)}(\tau)\|_X \,\mathrm{d}\tau.$$

Note that if $\psi \in \mathcal{W}_+^1(\mathbb{R};X)$, then

$$\|\psi(t)\|_X = \left\| \int_0^t \dot{\psi}(\tau)\mathrm{d}\tau \right\|_X \le H_1(\psi,t\,|\,X).$$

- Finally we consider the operator

$$\partial^{-1}u(t) := \int_0^t u(\tau)\mathrm{d}\tau,$$

i.e., $\mathcal{L}\{\partial^{-1}u\} = s^{-1}\mathrm{U}(s)$.

Theorem 7.5.1 (Mapping properties for the single layer potential). *If $\lambda \in \mathcal{W}_+^1(\mathbb{R};H^{-1/2}(\Gamma))$, then*

$$\mathcal{S} * \lambda \in \mathcal{C}_+^1(\mathbb{R};L^2(\mathbb{R}^d)) \cap \mathcal{C}_+(\mathbb{R};H^1(\mathbb{R}^d)) \tag{7.42}$$

and therefore

$$\mathcal{V} * \lambda \in \mathcal{C}_+(\mathbb{R};H^{1/2}(\Gamma)).$$

Moreover, for all $t \geq 0$

$$\|(\mathcal{S} * \lambda)(t)\|_{1,\mathbb{R}^d} \leq C H_2(\partial^{-1}\lambda, t \,|\, H^{-1/2}(\Gamma)), \tag{7.43a}$$

$$\|(\mathcal{V} * \lambda)(t)\|_{1/2,\Gamma} \leq C H_2(\partial^{-1}\lambda, t \,|\, H^{-1/2}(\Gamma)). \tag{7.43b}$$

If additionally $\lambda \in \mathcal{W}^2_+(\mathbb{R}; H^{-1/2}(\Gamma))$, then

$$\mathcal{K}^t * \lambda \in \mathcal{C}_+(\mathbb{R}; H^{-1/2}(\Gamma)). \tag{7.44}$$

and for all $t \geq 0$

$$\|(\mathcal{K}^t * \lambda)(t)\|_{-1/2,\Gamma} \leq C H_2(\lambda, t \,|\, H^{-1/2}(\Gamma)). \tag{7.45}$$

Proof. For causal $\lambda \in \mathcal{C}^2(\mathbb{R}; H^{-1/2}(\Gamma))$, we let $u := \mathcal{S} * \lambda$. From the results of Section 7.2 (see specifically Propositions 7.2.1 and 7.2.3, as well as the conclusions drawn from the latter at the end of Section 7.2), it follows that

$$u \in \mathcal{C}^2_+(\mathbb{R}; L^2(\mathbb{R}^d)) \cap \mathcal{C}^1_+(\mathbb{R}; H^1(\mathbb{R}^d)) \cap \mathcal{C}_+(\mathbb{R}; H^1_\Delta(\mathbb{R}^d \setminus \Gamma))$$

and

$$\|u(t)\|_{1,\mathbb{R}^d} \leq C(\|\lambda(t)\|_{-1/2,\Gamma} + c(t)B(\lambda, t)), \tag{7.46a}$$

$$\|\dot{u}(t)\|_{1,\mathbb{R}^d} \leq C(\|\dot{\lambda}(t)\|_{-1/2,\Gamma} + B(\lambda, t)), \tag{7.46b}$$

$$\|\ddot{u}(t)\|_{\mathbb{R}^d\setminus\Gamma} + \|\{\partial_\nu u\}(t)\|_{-1/2,\Gamma} \leq C(\|\lambda(t)\|_{-1/2,\Gamma} + B(\lambda, t)). \tag{7.46c}$$

(Note that the equation $\ddot{u}(t) = \Delta u(t)$ provides the bound for the second derivative.) We now endow the space

$$X_{\text{aux}} := \{\lambda \in \mathcal{C}^2([0, T]; H^{-1/2}(\Gamma)) \, : \, \lambda(0) = \dot{\lambda}(0) = 0\},$$

with the norm

$$H_2(\lambda, T \,|\, H^{-1/2}(\Gamma)) = \sum_{\ell=0}^{2} \int_0^T \|\lambda^{(\ell)}(\tau)\|_{-1/2,\Gamma} \mathrm{d}\tau,$$

and consider the space

$$\begin{aligned} Y_{\text{aux}} := \{ & u \in \mathcal{C}^2([0, T]; L^2(\mathbb{R}^d)) \cap \mathcal{C}^1([0, T]; H^1(\mathbb{R}^d)) \cap \mathcal{C}([0, T]; H^1_\Delta(\mathbb{R}^d \setminus \Gamma)) \\ & : \, u(0) = \dot{u}(0) = \ddot{u}(0) = 0\}, \end{aligned}$$

(i.e., elements of Y_{aux} can be extended by zero to negative times so that they are still continuous in $(-\infty, T]$), endowed with the norm

$$\max_{0 \leq t \leq T} \|\ddot{u}(t)\|_{\mathbb{R}^d} + \max_{0 \leq t \leq T} \|\nabla \dot{u}(t)\|_{\mathbb{R}^d} + \max_{0 \leq t \leq T} \|\Delta u(t)\|_{\mathbb{R}^d\setminus\Gamma}.$$

Note that Y_{aux} is a Banach space. Looking at values in the interval $[0,T]$, the bounds (7.46) provide the continuity of the map $X_{\text{aux}} \ni \lambda \mapsto u \in Y_{\text{aux}}$. We can thus extend this map to the closure of X_{aux}

$$\{\lambda \in \mathcal{C}^1([0,T]; H^{-1/2}(\Gamma)) \,:\, \ddot{\lambda} \in L^1((0,T); H^{-1/2}(\Gamma)),\ \lambda(0) = \dot{\lambda}(0) = 0\}.$$

With this we have proved that the bounds (7.46) can be extended for $u = \mathcal{S} * \lambda$ with $\lambda \in \mathcal{W}^2_+(\mathbb{R}; H^{-1/2}(\Gamma))$. This proves the mapping property for the operator $\lambda \mapsto \mathcal{K}^t * \lambda$ in the statement, as well as the bound (7.45).

If $\lambda \in \mathcal{W}^1_+(\mathbb{R}; H^{-1/2}(\Gamma))$, then $\partial^{-1}\lambda \in \mathcal{W}^2_+(\mathbb{R}; H^{-1/2}(\Gamma))$. Considering then $u := \mathcal{S}*\partial^{-1}\lambda$, and noticing that $\dot{u} = \mathcal{S}*\lambda$, we have proved (7.42). Using (7.46b), we prove (7.43a). The continuity of $\mathcal{V}*\lambda$ and (7.43b) are straightforward consequences of the fact that $\mathcal{V} * \lambda = \gamma(\mathcal{S} * \lambda)$. □

Theorem 7.5.2 (Mapping properties of the Galerkin error operator for $\mathcal{V}$). *If $\lambda \in \mathcal{W}^1_+(\mathbb{R}; H^{-1/2}(\Gamma))$, then*

$$\mathcal{E}^h_u * \lambda \in \mathcal{C}^1_+(\mathbb{R}; L^2(\mathbb{R}^d)) \cap \mathcal{C}_+(\mathbb{R}; H^1(\mathbb{R}^d))$$

and for all $t \geq 0$

$$\|(\mathcal{E}^h_u * \lambda)(t)\|_{1,\mathbb{R}^d} \leq C\, H_2(\partial^{-1}\lambda, t \,|\, H^{-1/2}(\Gamma)).$$

If additionally $\lambda \in \mathcal{W}^2_+(\mathbb{R}; H^{-1/2}(\Gamma))$, then

$$\mathcal{E}^h_\lambda * \lambda \in \mathcal{C}_+(\mathbb{R}; H^{-1/2}(\Gamma))$$

and for all $t \geq 0$

$$\|(\mathcal{E}^h_\lambda * \lambda)(t)\|_{-1/2,\Gamma} \leq C\, H_2(\lambda, t \,|\, H^{-1/2}(\Gamma)).$$

All the bounds are independent of the choice of the discrete space X_h.

Proof. The argument is very similar to the one of Theorem 7.5.1. Given $\lambda \in \mathcal{C}^2_+(\mathbb{R}; H^{-1/2}(\Gamma))$, we define $\varepsilon^h_u := \mathcal{E}^h_u * \lambda$ and $\varepsilon^h_\lambda = [\![\partial_\nu \varepsilon^h_u]\!] = \mathcal{E}^h_\lambda * \lambda$. The results of Section 7.3 (see Propositions 7.3.1 and 7.3.4 and the conclusions at the end of the section) prove that

$$\varepsilon^h_u \in \mathcal{C}^2_+(\mathbb{R}; L^2(\mathbb{R}^d)) \cap \mathcal{C}^1_+(\mathbb{R}; H^1(\mathbb{R}^d)) \cap \mathcal{C}_+(\mathbb{R}; H^1_\Delta(\mathbb{R}^d \setminus \Gamma))$$

and

$$\begin{aligned}
\|\varepsilon^h_u(t)\|_{1,\mathbb{R}^d} &\leq C(\|\lambda(t)\|_{-1/2,\Gamma} + c(t)\, B(\lambda, t)),\\
\|\dot{\varepsilon}^h_u(t)\|_{1,\mathbb{R}^d} &\leq C(\|\dot{\lambda}(t)\|_{-1/2,\Gamma} + B(\lambda, t)),\\
\|\ddot{\varepsilon}^h_u(t)\|_{\mathbb{R}^d\setminus\Gamma} + \|[\![\partial_\nu \varepsilon^h_u]\!](t)\|_{-1/2,\Gamma} &\leq C(\|\lambda(t)\|_{-1/2,\Gamma} + B(\lambda, t)).
\end{aligned}$$

The remainder of the proof of Theorem 7.5.1 can be applied almost verbatim: first we apply a density argument to weaken λ to be in $\mathcal{W}^2_+(\mathbb{R}; H^{-1/2}(\Gamma))$ and then we use a shifting argument to work with $\mathcal{E}^h_\lambda * \lambda = \frac{\mathrm{d}}{\mathrm{d}t}(\mathcal{E}^h_\lambda * \partial^{-1}\lambda)$. □

Theorem 7.5.3 (Mapping properties of the Galerkin solver for $\mathcal{V}$). *If* $\beta \in \mathcal{W}^2_+(\mathbb{R}; H^{1/2}(\Gamma))$, *then*

$$\mathcal{G}^h_u * \beta \in \mathcal{C}^1_+(\mathbb{R}; L^2(\mathbb{R}^d)) \cap \mathcal{C}_+(\mathbb{R}; H^1(\mathbb{R}^d))$$

and for all $t \geq 0$

$$\|(\mathcal{G}^h_u * \beta)(t)\|_{1,\mathbb{R}^d} \leq C H_3(\partial^{-1}\beta, t \,|\, H^{1/2}(\Gamma)).$$

If additionally $\beta \in \mathcal{W}^3_+(\mathbb{R}; H^{1/2}(\Gamma))$, *then*

$$\mathcal{G}^h_\lambda * \beta \in \mathcal{C}_+(\mathbb{R}; H^{-1/2}(\Gamma))$$

and for all $t \geq 0$

$$\|(\mathcal{G}^h_\lambda * \beta)(t)\|_{-1/2,\Gamma} \leq C H_3(\beta, t \,|\, H^{1/2}(\Gamma)).$$

All the bounds are independent of the choice of the discrete space X_h *and therefore the above assertions hold for the case* $X_h = H^{-1/2}(\Gamma)$, *in which case* $\mathcal{G}^h_u = \mathcal{S} * \mathcal{V}^{-1}$ *and* $\mathcal{G}^h_\lambda = \mathcal{V}^{-1}$.

Proof. The argument is very similar to the proofs of Theorems 7.5.1 and 7.5.2. We start with $\beta \in \mathcal{C}^3_+(\mathbb{R}; H^{1/2}(\Gamma))$, define $u^h := \mathcal{S} * \mathcal{G}^h_\lambda * \beta = \mathcal{G}^h_u * \lambda$ and note that by the results of Section 7.4,

$$u^h \in \mathcal{C}^2_+(\mathbb{R}; L^2(\mathbb{R}^d)) \cap \mathcal{C}^1_+(\mathbb{R}; H^1(\mathbb{R}^d)) \cap \mathcal{C}_+(\mathbb{R}; H^1_\Delta(\mathbb{R}^d \setminus \Gamma))$$

and

$$\begin{aligned}
\|u^h(t)\|_{1,\mathbb{R}^d} &\leq C(\|\beta(t)\|_{1/2,\Gamma} + c(t)\, B(\beta, t)),\\
\|\dot{u}^h(t)\|_{1,\mathbb{R}^d} &\leq C(\|\beta(t)\|_{1/2,\Gamma} + \|\dot{\beta}(t)\|_{1/2,\Gamma} + B(\beta, t) + B(\dot{\beta}, t)),\\
\|\ddot{u}^h(t)\|_{\mathbb{R}^d\setminus\Gamma} + \|[\![\partial_\nu u^h]\!](t)\|_{-1/2,\Gamma} &\leq C(\|\beta(t)\|_{1/2,\Gamma} + \|\dot{\beta}(t)\|_{1/2,\Gamma} + B(\dot{\beta}, t)).
\end{aligned}$$

(These bounds come from the collection of bounds in Propositions 7.4.1 and 7.4.2.) The remainder of the proof follows the usual pattern of a density argument followed by a shifting argument. □

7.6 Exercises

1. (Section 7.1) Prove that if

$$\|v\|_{B_0} \leq C_0 \|\nabla v\|_{B_0} \qquad \forall v \in H_0^1(B_0),$$

then

$$\|v\|_{B_T} \leq C_0(1 + T/R)\|\nabla v\|_{B_T} \qquad \forall v \in H_0^1(B_T).$$

2. (Section 7.1) Prove Lemma 7.1.2.
3. (Section 7.2) Prove the bounds given in Proposition 7.2.1.
4. (Section 7.2) Let $\xi \in \mathrm{TD}(H^{1/2}(\partial B_T))$. Show that there exists

$$\mathcal{M} \in \mathrm{TD}(\mathcal{B}(H^{1/2}(\partial B_T), H^1(B_T) \cap H^1_\Delta(B_T \setminus \Gamma))$$

such that $w = \mathcal{M} * \xi$ is the unique solution in $\mathrm{TD}(H^1(B_T) \cap H^1_\Delta(B_T \setminus \Gamma))$ of

$$\ddot{w} = \Delta w \qquad [\![\partial_\nu w]\!] = 0 \qquad \gamma_T w = \xi.$$

(**Hint.** This can be done using the Laplace transform or writing w in terms of potentials and operators defined on the boundary ∂B_T.)

5. (Section 7.2) Let $\lambda \in \mathrm{TD}(H^{-1/2}(\Gamma))$ be such that $\lambda|_{(0,\infty)} \in \mathcal{C}_0^2([0,\infty); H^{-1/2}(\Gamma))$ and let $u := \mathcal{S} * \lambda$. Show that

$$u \in \mathcal{C}^2([0,\infty); L^2(\mathbb{R}^d)) \cap \mathcal{C}^1([0,\infty); H^1(\mathbb{R}^d)) \cap \mathcal{C}([0,\infty); H^1_\Delta(\mathbb{R}^d \setminus \Gamma)).$$

Use the initial values and the differential equation to show that this property can be extended through the origin $t = 0$ to

$$u \in \mathcal{C}^2(\mathbb{R}; L^2(\mathbb{R}^d)) \cap \mathcal{C}^1(\mathbb{R}; H^1(\mathbb{R}^d)) \cap \mathcal{C}(\mathbb{R}; H^1_\Delta(\mathbb{R}^d \setminus \Gamma)).$$

6. (Section 7.3) Let $X_h \subset H^{-1/2}(\Gamma)$ be closed and let

$$V := \{v \in H_0^1(B_T) \ : \ \gamma v \in X_h^\circ\}.$$

Show that if $u \in H^1_\Delta(B_T \setminus \Gamma)$, then

$$[\![\partial_\nu u]\!] \in X_h \qquad \Longleftrightarrow \qquad (\nabla u, \nabla v)_{B_T \setminus \Gamma} + (\Delta u, v)_{B_T \setminus \Gamma} = 0 \quad \forall v \in V.$$

7. (Section 7.3) Show that problems (7.21) and (7.22) are equivalent.
8. (Section 7.3) Let ε_T and ε_{T+M} be the solutions of (7.24) for cut-off domains B_T and B_{T+M} respectively, where $M > 0$.

(a) Use the Waiting Time property (as phrased in Proposition 7.3.2) to show that the function

$$\underline{\varepsilon}_T : [0,\infty) \to H_0^1(B_{T+M}) \cap H_\Delta^1(B_{T+M} \setminus (\Gamma \cup \partial B_T))$$

given by

$$\underline{\varepsilon}_T(t)(\mathbf{x}) := \begin{cases} \varepsilon_T(t)(\mathbf{x}), & \mathbf{x} \in B_T, \\ 0, & \mathbf{x} \in B_{T+M} \setminus B_T, \end{cases}$$

is a strong solution of the same problem as ε_{T+M} in the time interval $[0, T+\delta]$.

(b) Use an energy conservation argument (see the final arguments in the proof of Proposition B.2.2) to show that $\varepsilon_{T+M} = \underline{\varepsilon}_T$ in $[0, T+\delta]$.

9. (Section 7.3) Give all details leading to the proof of (7.30) and (7.31). (**Hint.** The arguments are the same as those used at the end of Section 7.2 to derive regularity and estimates for the single layer potential.)
10. (Section 7.4) Show that problems (7.33) and (7.34) are equivalent and uniquely solvable.
11. (Section 7.4) **Weak solutions are distributional solutions.** Consider the function $u_T^h : [0,\infty) \to H_0^1(B_T)$ defined by (7.36)–(7.37) and assume that β is polynomially bounded.

(a) Show that we can understand $Eu_T^h \in \mathrm{TD}(H_0^1(B_T))$ and also $Eu_T^h \in \mathrm{TD}(\mathbb{X}_T)$. (**Hint.** Use the last part of Theorem 6.2.4 to study the extension of $\int_0^t u_T^h$.)

(b) Show that Eu_T^h is the unique solution of the problem

$$w \in \mathrm{TD}(\mathbb{X}_T) \qquad \ddot{w} = \Delta w, \qquad \gamma w - E\beta \in X_h^\circ, \qquad [\![\partial_\nu w]\!] \in X_h.$$

(**Hint.** Define $\mathrm{U}(s) := \mathcal{L}\{Eu_T^h\} \in \mathbb{X}_T$ and show that

$$\gamma \mathrm{U}(s) - \mathcal{L}\{E\beta\} \in X_h^\circ \qquad s^2\langle \mathrm{U}(s), v\rangle_{V'\times V} + (\nabla \mathrm{U}(s), \nabla v)_{B_T} = 0 \qquad \forall v \in V.$$

This can be done in several different ways, one of which consists of rewriting (7.36a) as distributional equation in V'.)

The slightly surprising result of this exercise consists of the fact that even if the strong equations are satisfied in V' (because there is no regularity to have them satisfied in $L^2(B_T)$ and to separate the natural transmission condition), the distributional equation is actually satisfied in $L^2(B_T)$ and the natural transmission condition is satisfied in the sense of distributions. The reader should compare this with the paradox of the moving waves in the exercise list of Chapter 2.

12. (Section 7.5) Show that

$$H_k(\partial^{-1}\psi, t \,|\, X) \le t \int_0^t \|\psi(\tau)\|_X \mathrm{d}\tau + H_{k-1}(\psi, t \,|\, X).$$

Chapter 8
Time domain analysis of the double layer potential

This chapter is the double layer counterpart of Chapter 7. We will start by studying

$$\mathcal{D} * \varphi, \qquad \mathcal{K} * \varphi, \quad \text{and} \quad \mathcal{W} * \varphi.$$

We will next focus on the semidiscrete inversion of the operator $\varphi \mapsto \mathcal{W}*\varphi$, namely, given a closed subspace $Y_h \subset H^{1/2}(\Gamma)$ we will look for the causal solution φ^h of

$$\mathcal{W} * \varphi^h - \alpha \in Y_h^\circ$$

and input it in a double potential representation

$$u^h = \mathcal{D} * \varphi^h.$$

This process was studied in the Laplace domain in Sections 5.5 and 5.6.

8.1 The cut-off process

Geometry. The set up is the same as in Chapter 7. We write

$$\overline{\Omega^-} \subset B_0 := B(\mathbf{0}; R), \qquad B_T := B(\mathbf{0}; R+T), \qquad \delta := \operatorname{dist}(\partial B_0, \Gamma)$$

and consider the operators on ∂B_T

$$\gamma_T : H^1(B_T \setminus \overline{\Omega^-}) \to H^{1/2}(\partial B_T) \quad \text{and} \quad \partial_T^\nu : H^1_\Delta(B_T \setminus \overline{\Omega^-}) \to H^{-1/2}(\partial B_T).$$

F.-J. Sayas, *Retarded Potentials and Time Domain Boundary Integral Equations*, Springer Series in Computational Mathematics 50,
DOI 10.1007/978-3-319-26645-9_8

Inequalities. We now collect an extended version of the inequality toolbox of Section 7.1. Two new spaces will be relevant. The first one contains a Dirichlet boundary condition on ∂B_T but allows jumps across Γ:

$$H^1_{\partial B_T}(B_T \setminus \Gamma) := \{u \in H^1(B_T \setminus \Gamma) \, : \, \gamma_T u = 0\}.$$

It is clear that $u \mapsto \|\nabla u\|_{B_T \setminus \Gamma}$ does not define a norm in $H^1_{\partial B_T}(B_T \setminus \Gamma)$, but the simple correction

$$\|\nabla u\|^2_{B_T \setminus \Gamma} + |\jmath(u)|^2, \quad \text{where} \quad \jmath(u) := \frac{1}{|\Omega^-|^{1/2}} \int_{\Omega^-} u, \tag{8.1}$$

does. The second space involves the Laplacian

$$\mathbb{Y}_T := \{u \in H^1_{\partial B_T}(B_T \setminus \Gamma) \, : \, \Delta u \in L^2(B_T \setminus \Gamma)\} = \{u \in H^1_\Delta(B_T \setminus \Gamma) \, : \, \gamma_T u = 0\}, \tag{8.2}$$

and is endowed with the seminorm

$$|u|^2_{\mathbb{Y}_T} := \|\nabla u\|^2_{B_T \setminus \Gamma} + \|\Delta u\|^2_{B_T \setminus \Gamma}.$$

(a) In addition to the Poincaré-Friedrichs inequality

$$\|v\|_{B_T} \le C_T \|\nabla v\|_{B_T} \quad \forall v \in H^1_0(B_T), \qquad C_T = C_0(1 + T/R), \tag{8.3}$$

we will need a scale-dependent generalized Poincaré inequality

$$\|u\|_{B_T} \le \widehat{C}_T \Big(\|\nabla u\|^2_{B_T \setminus \Gamma} + |\jmath(u)|^2 \Big)^{1/2} \qquad \forall u \in H^1_{\partial B_T}(B_T \setminus \Gamma). \tag{8.4}$$

We could *conjecture* that $\widehat{C}_T \le C_1 + C_2 T$, as it happened in Chapter 7, but this will be not relevant in the sequel, since the constant $\widehat{C}_T$ will be eliminated from all expressions in the final version of the estimates.

(b) Boundedness of the trace operators: for simplicity we will use a joint constant for the jump and average of the trace on Γ

$$\|[\![\gamma u]\!]\|_{1/2,\Gamma} + \|\{\!\{\gamma u\}\!\}\|_{1/2,\Gamma} \le C_\gamma \|u\|_{1,B_0 \setminus \Gamma} \qquad \forall u \in H^1(B_0 \setminus \Gamma). \tag{8.5}$$

(c) Boundedness of the normal derivative operator:

$$\|\partial_\nu^\pm u\|_{-1/2,\Gamma} \le C_\nu \Big(\|\nabla u\|^2_{B_0 \cap \Omega^\pm} + \|\Delta u\|^2_{B_0 \cap \Omega^\pm} \Big)^{1/2} \qquad \forall u \in H^1(B_0 \setminus \Gamma). \tag{8.6}$$

(d) A one-sided lifting of the trace will be used as a lifting of the jump of the trace across Γ. We consider an operator $[\![\gamma]\!]^\dagger : H^{1/2}(\Gamma) \to H^1_{\partial B_0}(B_0 \setminus \Gamma)$ and a constant $C_\gamma^\dagger > 0$ such that

$$[\![[\![\gamma]\!]^\dagger \alpha]\!] = \alpha, \quad \|[\![\gamma]\!]^\dagger \alpha\|_{1,B_0\setminus\Gamma} = \|[\![\gamma]\!]^\dagger \alpha\|_{1,\Omega_-} \le C_\gamma^\dagger \|\alpha\|_{1/2,\Gamma} \quad \forall \alpha \in H^{1/2}(\Gamma). \tag{8.7}$$

8.2 Potentials and operators

We start with the operators

$$\mathcal{D} * \varphi, \qquad \mathcal{W} * \varphi = -\partial_\nu(\mathcal{D} * \varphi), \quad \text{and} \quad \mathcal{K} * \varphi = \{\gamma(\mathcal{D} * \varphi)\}.$$

In terms of the potential, the cut-off problem looks for $u_T : [0, \infty) \to \mathbb{Y}_T$ (the homogeneous Dirichlet boundary condition on ∂B_T is implied by u_T taking values in $\mathbb{Y}_T$) such that

$$\ddot{u}_T(t) = \Delta u_T(t) \qquad \forall t \ge 0, \tag{8.8a}$$

$$[\![\gamma u_T]\!](t) + \varphi(t) = 0 \qquad \forall t \ge 0, \tag{8.8b}$$

$$[\![\partial_\nu u_T]\!](t) = 0 \qquad \forall t \ge 0, \tag{8.8c}$$

$$u_T(0) = \dot{u}_T(0) = 0. \tag{8.8d}$$

Compare with (7.5), the corresponding cut-off problem for the single layer potential. The process will mimic that of the Galerkin solver for the single layer operator, using first a weak solution to get some bounds and then a strong solution to obtain what is left.

The dynamical system triplet. The spaces and norms are the same as those of Section 7.2, namely

$$\begin{aligned} H &:= L^2(B_T), \\ V &:= H^1_0(B_T), \\ D(A) &:= H^1_0(B_T) \cap H^2(B_T) = H^1_0(B_T) \cap H^1_\Delta(B_T) \\ &= \{u \in V : \Delta u \in L^2(B_T \setminus \Gamma), \quad [\![\partial_\nu u]\!] = 0\}, \end{aligned}$$

with

$$\|u\|_H := \|u\|_{B_T}, \qquad [u, v] := (\nabla u, \nabla v)_{B_T}.$$

The lifting. We consider the operator $L : H^{1/2}(\Gamma) \to \mathbb{Y}_T$ given by the solution of

$$\begin{aligned} u_0 \in H^1_{\partial B_T}(B_T \setminus \Gamma), \\ -\Delta u_0 + u_0 &= 0 \qquad \text{in } B_T \setminus \Gamma, \\ [\![\gamma u_0]\!] &= -\varphi, \\ [\![\partial_\nu u_0]\!] &= 0, \end{aligned}$$

that is,

$$\left[\begin{array}{ll} u_0 \in H^1_{\partial B_T}(B_T \setminus \Gamma), & [\![\gamma u_0]\!] = -\varphi, \\ (\nabla u_0, \nabla v)_{B_T\setminus\Gamma} + (u_0, v)_{B_T} = 0 & \forall v \in H^1_0(B_T). \end{array}\right.$$

Testing then with $v = u_0 + [\![\gamma]\!]^\dagger \varphi$, it is easy to prove that

$$\|L\|_{H^{1/2}(\Gamma)\to Z} \le C_\gamma^\dagger \qquad Z \in \{L^2(B_T), H^1(B_T \setminus \Gamma), \mathbb{Y}_T\}. \tag{8.9}$$

Proposition 8.2.1 (Weak solutions in bounded domain). *For $\varphi \in \mathcal{C}_0^2([0,\infty); H^{1/2}(\Gamma))$, there exists a unique*

$$u_T : [0,\infty) \to H^1_{\partial B_T}(B_T \setminus \Gamma)$$

such that

$$\langle \ddot u_T(t), v\rangle_{V'\times V} + (\nabla u_T(t), \nabla v)_{B_T\setminus\Gamma} = 0 \qquad \forall v \in V \qquad \forall t \ge 0, \tag{8.10a}$$

$$[\![\gamma u_T]\!](t) + \varphi(t) = 0 \qquad \forall t \ge 0, \tag{8.10b}$$

$$u_T(0) = \dot u_T(0) = 0, \tag{8.10c}$$

and

$$u_T \in \mathcal{C}^2([0,\infty); V') \cap \mathcal{C}^1([0,\infty); L^2(B_T)) \cap \mathcal{C}([0,\infty); H^1_{\partial B_T}(B_T \setminus \Gamma)). \tag{8.11}$$

Moreover, for all $t \ge 0$,

$$\begin{aligned} \|u_T(t)\|_{B_T} &\le C_\gamma^\dagger \big(\|\varphi(t)\|_{1/2,\Gamma} + C_T B(\varphi, t)\big), \\ \|\nabla u_T(t)\|_{B_T\setminus\Gamma} &\le C_\gamma^\dagger \big(\|\varphi(t)\|_{1/2,\Gamma} + B(\varphi, t)\big), \\ \|\dot u_T(t)\|_{B_T} &\le C_\gamma^\dagger \big(\|\dot\varphi(t)\|_{1/2,\Gamma} + B(\varphi, t)\big), \\ \|\{\gamma u_T\}(t)\|_{1/2,\Gamma} &\le C_\gamma C_\gamma^\dagger \big(\|\varphi(t)\|_{1/2,\Gamma} + c(T)\, B(\varphi, t)\big), \end{aligned}$$

where

$$c(T) := \sqrt{1 + C_T^2} \quad \text{and} \quad B(\varphi, t) := \int_0^t \|\varphi(\tau) - \ddot\varphi(\tau)\|_{1/2,\Gamma} \mathrm{d}\tau.$$

Proof. The entire proof is based on showing that if $u_0(t) := L\varphi(t), f(t) := u_0(t) - \ddot{u}_0(t)$, and if $v_0 : [0,\infty) \to H_0^1(B_T) = V$ is the solution of

$$\begin{aligned} \langle \ddot{v}_0(t), v\rangle_{V'\times V} + (\nabla v_0(t), \nabla v)_{B_T} &= (f(t), v)_{B_T} \quad \forall v \in V \qquad \forall t \geq 0, \\ v_0(0) = \dot{v}_0(0) &= 0, \end{aligned}$$

then $u_T := u_0 + v_0$ is the unique solution of (7.36). It is clear that $u_0 \in \mathcal{C}_0^2([0,\infty); \mathbb{Y}_T)$, while (see Theorem 6.2.4)

$$v_0 \in \mathcal{C}^2([0,\infty); V') \cap \mathcal{C}^1([0,\infty); L^2(B_T)) \cap \mathcal{C}([0,\infty); H_0^1(B_T)).$$

The bounds follow then from (8.9) (to estimate u_0) and Theorem 6.2.4 (to estimate v_0). To bound $\{\gamma u_T\}(t)$, use a bound for the $H^1(B_T)$ norm and (8.5). □

Proposition 8.2.2 (Strong solutions in bounded domain). *For $\varphi \in \mathcal{C}_0^3([0,\infty); H^{1/2}(\Gamma))$, there exists a unique*

$$u_T : [0,\infty) \to \mathbb{Y}_T,$$

satisfying

$$\ddot{u}_T(t) = \Delta u_T(t) \qquad \forall t \geq 0, \tag{8.13a}$$

$$[\![\gamma u_T(t)]\!] + \varphi(t) = 0 \qquad \forall t \geq 0, \tag{8.13b}$$

$$[\![\partial_\nu u_T]\!](t) = 0 \qquad \forall t \geq 0, \tag{8.13c}$$

$$u_T(0) = \dot{u}_T(0) = 0, \tag{8.13d}$$

and

$$u_T \in \mathcal{C}^2([0,\infty); L^2(B_T)) \cap \mathcal{C}^1([0,\infty); H^1_{\partial B_T}(B_T \setminus \Gamma)) \cap \mathcal{C}([0,\infty); \mathbb{Y}_T). \tag{8.14}$$

Moreover, u_T is also the solution of (8.10) *and*

$$\begin{aligned} \|\Delta u_T(t)\|_{B_T\setminus\Gamma} &\leq C_\gamma^\dagger(\|\varphi(t)\|_{1/2,\Gamma} + 2B(\dot{\varphi}, t)), \\ \|\nabla \dot{u}_T(t)\|_{B_T\setminus\Gamma} &\leq C_\gamma^\dagger(\|\dot{\varphi}(t)\|_{1/2,\Gamma} + B(\dot{\varphi}, t)), \\ \|\partial_\nu u_T(t)\|_{-1/2,\Gamma} &\leq C_\nu C_\gamma^\dagger(2\|\varphi(t)\|_{1/2,\Gamma} + B(\varphi, t) + 2B(\dot{\varphi}, t)). \end{aligned}$$

Proof. We now follow the steps of the proof of Proposition 7.4.2. To do that, we define $u_0(t) := L\varphi(t), f(t) := u_0(t) - \ddot{u}_0(t)$, as in the proof of Proposition 8.2.1. We then consider $v_0 : [0,\infty) \to D(A)$ as the solution of

$$\ddot{v}_0(t) = \Delta v_0(t) + f(t) \quad \forall t \geq 0, \qquad v_0(t) = \dot{v}_0(t) = 0.$$

Note that $u_0 \in \mathcal{C}^3_0([0,\infty);\mathbb{Y}_T)$ and

$$v_0 \in \mathcal{C}^2([0,\infty);L^2(B_T)) \cap \mathcal{C}^1([0,\infty);H^1_0(B_T)) \cap \mathcal{C}([0,\infty);D(A)).$$

It is then easy to show that $u_T := u_0 + v_0$ satisfies (8.13) and (8.14). It is also easy to show that a solution of (8.13) is a solution of (8.10).

To bound the Laplacian of u_T we first use (8.9) to obtain

$$\|\Delta u_0(t)\|_{B_T\setminus\Gamma} \le C^\dagger_\gamma \|\varphi(t)\|_{1/2,\Gamma}. \tag{8.15}$$

At the same time, by Theorem 6.2.3 and (8.9), it follows that

$$\|\Delta v_0(t)\|_{B_T} \le 2\int_0^t \|\dot f(\tau)\|_{B_T}\mathrm{d}\tau \le 2C^\dagger_\gamma B(\dot\varphi,t). \tag{8.16}$$

The bound for $\|\Delta u_T(t)\|_{B_T\setminus\Gamma}$ is just the collection of (8.15) and (8.16). The bound for $\|\partial_\nu u_T(t)\|_{-1/2,\Gamma}$ follows now from (8.6), the bound for $\|\Delta u_T(t)\|_{B_T\setminus\Gamma}$ and the estimate of $\|\nabla u_T(t)\|_{B_T\setminus\Gamma}$ in Proposition 8.2.1. □

Proposition 8.2.3 (Waiting Time on ∂B_T). *If $u_T : [0,\infty) \to \mathbb{Y}_T$ is the solution of* (8.13) *for $\varphi \in \mathcal{C}^3_0([0,\infty);H^{1/2}(\Gamma))$, then*

$$\partial^\nu_T u_T(t) = 0 \qquad 0 \le t \le T+\delta.$$

Proof. As usual with this kind of result, assume that φ is polynomially bounded as $t\to\infty$. The plan of the proof is similar to that of Proposition 7.2.2. Consider first

$$w := Eu_T - (\mathcal{D}*E\varphi)_{B_T} \in \mathrm{TD}(H^1_\Delta(B_T\setminus\Gamma)), \qquad \xi := -\gamma_T(\mathcal{D}*\varphi) \in \mathrm{TD}(H^{1/2}(\partial B_T)).$$

We can then prove

$$w \in \mathrm{TD}(H^1_\Delta(B_T\setminus\Gamma)), \quad \ddot w = \Delta w, \quad [\![\gamma w]\!] = 0, \quad [\![\partial_\nu w]\!] = 0, \qquad \gamma_T w = \xi,$$

and that this problem is uniquely solvable. Using the finite speed of propagation of the double layer potential (Proposition 7.1.1), we show that $\operatorname{supp}\xi \subset [T+\delta,\infty)$ and therefore

$$\operatorname{supp} w \subset [T+\delta,\infty).$$

Using again Proposition 7.1.1 to show that $\operatorname{supp}\partial^\nu_T Eu_T \subset [T+\delta,\infty)$, we finish the proof. □

Proposition 8.2.4 (Extension to free space). *Let $\varphi \in \mathcal{C}^3_0([0,\infty);H^{1/2}(\Gamma))$ and u_T be the solution of* (8.13). *Consider the extension*

$$\underline{u}_T : [0,\infty) \to H^1_\Delta(\mathbb{R}^d\setminus(\Gamma\cup\partial B_T))$$

given by

$$\underline{u}_T(t)|_{B_T} = u_T(t) \qquad \text{and} \qquad \underline{u}_T(t)|_{\mathbb{R}^d\setminus B_T} = 0.$$

The function $\underline{u}_T : [0, T + \delta] \to H^1_\Delta(\mathbb{R}^d \setminus \Gamma)$ *is continuous and*

$$\underline{u}_T(t) = (\mathcal{D} * E\varphi)(t) \qquad 0 \le t < T + \delta.$$

Proof. The proof is a slight variant of the proof of Proposition 7.2.3. We first prove that

$$E\underline{u}_T = \mathcal{D} * E\varphi + \mathcal{S}_{\partial B_T} * E\partial^\nu_T u_T$$

and then use Proposition 8.2.3 to show that supp $(E\underline{u}_T - \mathcal{D} * E\varphi) \subset [T + \delta, \infty)$. The fact that $\underline{u}_T(t) \in H^1_\Delta(\mathbb{R}^d \setminus \Gamma)$ for $t \le T + \delta$ is a direct consequence of Proposition 7.2.3. □

Theorem 8.2.5 (Mapping properties for the double layer potential). *If* $\varphi \in \mathcal{W}^2_+(\mathbb{R}; H^{1/2}(\Gamma))$, *then*

$$\mathcal{D} * \varphi \in \mathcal{C}^1_+(\mathbb{R}; L^2(\mathbb{R}^d)) \cap \mathcal{C}_+(\mathbb{R}; H^1(\mathbb{R}^d \setminus \Gamma))$$

and therefore

$$\mathcal{K} * \varphi \in \mathcal{C}_+(\mathbb{R}; H^{1/2}(\Gamma)).$$

Moreover, for all $t \ge 0$

$$\|(\mathcal{D} * \varphi)(t)\|_{1,\mathbb{R}^d\setminus\Gamma} \le C H_3(\partial^{-1}\varphi, t \,|\, H^{1/2}(\Gamma)),$$
$$\|(\mathcal{K} * \varphi)(t)\|_{1/2,\Gamma} \le C H_3(\partial^{-1}\varphi, t \,|\, H^{1/2}(\Gamma)).$$

If additionally $\varphi \in \mathcal{W}^3_+(\mathbb{R}; H^{1/2}(\Gamma))$, *then*

$$\mathcal{W} * \varphi \in \mathcal{C}_+(\mathbb{R}; H^{-1/2}(\Gamma)).$$

and for all $t \ge 0$

$$\|(\mathcal{W} * \varphi)(t)\|_{-1/2,\Gamma} \le C H_3(\varphi, t \,|\, H^{1/2}(\Gamma)).$$

Proof. For causal $\varphi \in \mathcal{C}^3(\mathbb{R}; H^{-1/2}(\Gamma))$, we let $u := \mathcal{D} * \varphi$. From the results above (Propositions 8.2.1, 8.2.2, and 8.2.4) it follows that

$$u \in \mathcal{C}^2_+(\mathbb{R}; L^2(\mathbb{R}^d)) \cap \mathcal{C}^1_+(\mathbb{R}; H^1(\mathbb{R}^d \setminus \Gamma)) \cap \mathcal{C}_+(\mathbb{R}; H^1_\Delta(\mathbb{R}^d \setminus \Gamma)).$$

Also, taking $T = t$ in Propositions 8.2.1 and 8.2.2, it follows that

$$\begin{aligned}
\|u(t)\|_{1,\mathbb{R}^d\setminus\Gamma} &\le C(\|\varphi(t)\|_{-1/2,\Gamma} + c(t)B(\varphi,t)),\\
\|\dot{u}(t)\|_{1,\mathbb{R}^d\setminus\Gamma} &\le C(\|\dot{\varphi}(t)\|_{-1/2,\Gamma} + B(\varphi,t) + B(\dot{\varphi},t)),\\
\|\ddot{u}(t)\|_{\mathbb{R}^d\setminus\Gamma} &\le C(\|\varphi(t)\|_{-1/2,\Gamma} + B(\dot{\varphi},t)),\\
\|(\partial_\nu u)(t)\|_{-1/2,\Gamma} &\le C(\|\varphi(t)\|_{-1/2,\Gamma} + B(\varphi,t) + B(\dot{\varphi},t)).
\end{aligned}$$

The remainder of the proof follows, line by line, the proof of Theorem 7.5.1. □

8.3 Galerkin solver for the hypersingular operator

Given a causal $H^{-1/2}(\Gamma)$-valued distribution α, we deal here with the discretization of the equation $\mathcal{W} * \varphi = \alpha$. This is carried out in a discrete space $Y_h \subset H^{1/2}(\Gamma)$. We will assume that

$$\mathbb{P}_0(\Gamma) \subset Y_h, \tag{8.17}$$

although, more precisely, what we need is $\gamma\mathbb{P}_0(\Omega_-) \subset Y_h$. We then look for a causal $H^{1/2}(\Gamma)$-valued distribution φ^h such that

$$\varphi^h \in Y_h, \qquad \mathcal{W} * \varphi^h - \alpha \in Y_h^\circ$$

and consider the associated potential

$$u^h = \mathcal{D} * \varphi^h.$$

In terms of the potential u^h, we are looking for

$$\begin{aligned}
u^h &\in \mathrm{TD}(H^1_\Delta(\mathbb{R}^d \setminus \Gamma)),\\
\ddot{u}^h &= \Delta u^h,\\
[\![\gamma u^h]\!] &\in Y_h,\\
[\![\partial_\nu u^h]\!] &= 0,\\
\partial_\nu u^h + \alpha &\in Y_h^\circ.
\end{aligned}$$

The cut-off at an external (eventually moving) boundary ∂B_T will be done as usual, leading to a dynamical system, which, as a novelty, will contain a rigid motion.

Dynamical system. We start with

$$H := L^2(B_T) \qquad \|u\|_H := \|u\|_{B_T}, \qquad M := \operatorname{span}\{\chi_{\Omega_-}\},$$

so that the projection onto M is

$$Pu = \Big(\frac{1}{|\Omega_-|}\int_{\Omega_-} u\Big)\chi_{\Omega_-} = \Big(\frac{1}{|\Omega_-|^{1/2}}\int_{\Omega_-} u\Big)\frac{1}{|\Omega_-|^{1/2}}\chi_{\Omega_-},$$

and (see (8.1))

$$\|Pu\|_{B_T} = \|Pu\|_{\Omega_-} = |\jmath(u)|.$$

We next consider

$$V := \{u \in H^1_{\partial B_T}(B_T \setminus \Gamma) \,:\, [\![\gamma u]\!] \in Y_h\} \qquad [u, v] := (\nabla u, \nabla v)_{B_T\setminus\Gamma},$$

as well as

$$D(A) := \{u \in V \,:\, \Delta u \in L^2(B_T \setminus \Gamma) \,:\, [\![\partial_\nu u]\!] = 0, \quad \partial_\nu u \in Y_h^\circ\},$$

with the operator A being the distributional Laplacian in $B_T\setminus\Gamma$. The condition (8.17) is used to show that $M \subset V$. Let us now check all the hypotheses:

- Since $\mathcal{D}(B_T \setminus \Gamma) \subset V \subset H^1(B_T \setminus \Gamma) \subset L^2(B_T)$, the compactness and density of the injection $V \subset H$ is straightforward to prove.
- The constant $C_\circ$ in $\|v\|_H \le C_\circ\|v\|_V$ is given by the constant of the generalized Poincaré inequality (8.4).
- To prove the associated Green's formula, we just need to use Lemma 5.6.1.
- Finally, given $f \in L^2(B_T)$, we can solve

$$\left[\begin{array}{l} u \in V, \\ (\nabla u, \nabla v)_{B_T\setminus\Gamma} + (u, v)_{B_T} = (f, v)_{B_T} \qquad \forall v \in V. \end{array}\right.$$

By Lemma 5.6.1, it follows that $u \in D(A)$ and that $-\Delta u + u = f$.

Lifting. Consider the operator $L : H^{1/2}(\Gamma) \to \mathbb{Y}_T$ that maps $\alpha \mapsto u_0$ by solving

$$\left[\begin{array}{l} u_0 \in V, \\ (\nabla u_0, \nabla v)_{B_T\setminus\Gamma} + (u_0, v)_{B_T} = -\langle\alpha, [\![\gamma v]\!]\rangle_\Gamma \qquad \forall v \in V, \end{array}\right. \tag{8.18}$$

or equivalently

$$\begin{aligned} u_0 &\in H^1_{\partial B_T}(B_T \setminus \Gamma), \\ -\Delta u_0 + u_0 &= 0 \qquad \text{in } B_T \setminus \Gamma, \\ [\![\gamma u_0]\!] &\in Y_h, \\ [\![\partial_\nu u_0]\!] &= 0, \\ \partial_\nu u_0 + \alpha &\in Y_h^\circ. \end{aligned}$$

It is easy to see that

$$\|L\alpha\|_{1,B_T} = |L\alpha|_{\mathbb{Y}_T} \le C_\gamma \|\alpha\|_{-1/2,\Gamma}. \tag{8.19}$$

Testing (8.18) with $v = |\Omega_-|^{-1/2}\chi_{\Omega_-}$, it follows that

$$PL\alpha = -\Big(\frac{1}{|\Omega_-|}\langle \alpha, 1\rangle_\Gamma\Big)\chi_{\Omega_-}. \tag{8.20}$$

Proposition 8.3.1. *For $\alpha \in \mathcal{C}_0^2([0,\infty); H^{-1/2}(\Gamma))$, there exists a unique*

$$u_T^h : [0,\infty) \to H^1_{\partial B_T}(B_T \setminus \Gamma),$$

such that

$$\begin{aligned} \ddot{u}_T^h(t) &= \Delta u_T^h(t) \qquad \forall t \ge 0, && (8.21a) \\ [\![\gamma u_T^h]\!](t) &\in Y_h \qquad \forall t \ge 0, && (8.21b) \\ [\![\partial_\nu u_T^h]\!](t) &= 0 \qquad \forall t \ge 0, && (8.21c) \\ \partial_\nu u_T^h(t) + \alpha(t) &\in Y_h^\circ, && (8.21d) \\ u_T^h(0) = \dot{u}_T^h(0) &= 0, && (8.21e) \end{aligned}$$

and

$$u_T^h \in \mathcal{C}^2([0,\infty); L^2(B_T)) \cap \mathcal{C}^1([0,\infty); H^1(B_T \setminus \Gamma)) \cap \mathcal{C}([0,\infty); \mathbb{Y}_T)). \tag{8.22}$$

Moreover, for all $t \ge 0$,

$$\begin{aligned} \|u_T^h(t)\|_{1,B_T\setminus\Gamma} &\le |\Omega_-|^{-1/2}|\rho(t)| + C_\gamma(\|\alpha(t)\|_{-1/2,\Gamma} + \hat{c}(T)B(\alpha,t)), \\ \|\dot{u}_T^h(t)\|_{1,B_T\setminus\Gamma} &\le |\Omega_-|^{-1/2}|\dot{\rho}(t)| + C_\gamma(\|\dot{\alpha}(t)\|_{-1/2,\Gamma} + \sqrt{2}\,B(\alpha,t)), \\ \|\Delta u_T^h(t)\|_{B_T\setminus\Gamma} &\le C_\gamma(\|\alpha(t)\|_{-1/2,\Gamma} + B(\alpha,t)), \end{aligned}$$

where

$$\hat{c}(T) := \sqrt{1+\widehat{C}_T^2}, \qquad B(\alpha,t) := \int_0^t \|\alpha(\tau)-\ddot{\alpha}(\tau)\|_{-1/2,\Gamma}\mathrm{d}\tau,$$

and

$$\rho(t) := \int_0^t (t-\tau)\langle\alpha(\tau),1\rangle_\Gamma \mathrm{d}\tau.$$

Proof. The proof of this result requires taking care of the rigid motion that is excited in the system. This will be done using the results of Section 6.3. We define

$$u_0(t) := L\alpha(t), \qquad f(t) := u_0(t)-\ddot{u}_0(t) = L(\alpha(t)-\ddot{\alpha}(t)),$$

consider the rigid motion (see Proposition 6.3.1 and (8.20))

$$\begin{aligned} m_\alpha(t) &:= Pu_0(t)+m_f(t) = Pu_0(t)+\int_0^t (t-\tau)Pf(\tau)\mathrm{d}\tau \\ &= PL\alpha(t)+\int_0^t (t-\tau)PL\alpha(\tau)\mathrm{d}\tau - \int_0^t (t-\tau)PL\ddot{\alpha}(\tau)\mathrm{d}\tau \\ &= \int_0^t (t-\tau)PL\alpha(\tau)\mathrm{d}\tau = -\frac{1}{|\Omega_-|}\Big(\int_0^t (t-\tau)\langle\alpha(\tau),1\rangle_\Gamma\mathrm{d}\tau\Big)\chi_{\Omega_-} = -\frac{\rho(t)}{|\Omega_-|}\chi_{\Omega_-}, \end{aligned}$$

and then consider the solution (see Theorem 6.2.2) $v_0 : [0,\infty) \to D(A)$ of the evolution equation

$$\ddot{v}_0(t) = \Delta v_0(t)+f(t)-Pf(t) \qquad t\ge 0, \qquad v_0(0)=\dot{v}_0(0)=0.$$

The solution of (8.21) is easily recomposed as the sum

$$u_T^h = (u_0-Pu_0)+m_\alpha+v_0 = u_0+m_f+v_0.$$

Note that

$$u_0-Pu_0 = (I-P)L\alpha \in \mathcal{C}_0^2([0,\infty);\mathbb{Y}_T), \qquad m_\alpha \in \mathcal{C}_0^4([0,\infty);M),$$

and

$$v_0 \in \mathcal{C}^2([0,\infty);H)\cap\mathcal{C}^1([0,\infty);V)\cap\mathcal{C}([0,\infty);D(A)),$$

which proves (8.22). The end of the proof is proposed as an exercise. □

Identification of u_T^h as the Galerkin solver. The next sequence of steps does not vary from what we have repeatedly done in previous examples:

- By comparison with $\mathcal{D} * [\![\gamma u_T^h]\!]|_{B_T}$, we show that
$$\partial_\nu^T u_T^h(t) = 0 \qquad 0 \le t \le T + \delta.$$
- Next we define
$$\varphi^h(T) := -[\![\gamma u_T^h]\!](T),$$
and prove that $\varphi^h \in \mathcal{C}^1([0,\infty); H^{1/2}(\Gamma))$ and
$$\varphi^h(t) = -[\![\gamma u_T^h]\!](t) \qquad t \le T + \delta.$$
- Finally, we consider the function $\underline{u}_T^h$ (extension by zero to $\mathbb{R}^d \setminus B_T$) and show that
$$\underline{u}_T^h(t) = (\mathcal{D} * E\varphi^h)(t) \qquad t < T + \delta,$$
which closes the loop.

Theorem 8.3.2 (Mapping properties of the Galerkin solver for $\mathcal{W}$). *If* $\alpha \in \mathcal{W}_+^1(\mathbb{R}; H^{-1/2}(\Gamma))$*, then*
$$\mathcal{D} * \mathcal{G}_\varphi^h * \alpha \in \mathcal{C}_+^1(\mathbb{R}; L^2(\mathbb{R}^d)) \cap \mathcal{C}_+(\mathbb{R}; H^1(\mathbb{R}^d \setminus \Gamma))$$
and therefore
$$\mathcal{G}_\varphi^h * \alpha \in \mathcal{C}_+(\mathbb{R}; H^{1/2}(\Gamma)).$$
Moreover, for all $t \ge 0$,
$$\|(\mathcal{D} * \mathcal{G}_\varphi^h * \alpha)(t)\|_{1,\mathbb{R}^d\setminus\Gamma} \le C\,(|\rho(t)| + H_2(\partial^{-1}\alpha, t\,|\,H^{-1/2}(\Gamma))),$$
where
$$\rho(t) := \int_0^t (t-\tau)\langle \alpha(\tau), 1\rangle_\Gamma \mathrm{d}\tau.$$

8.4 Galerkin error operator

In this section the input is a causal $H^{1/2}(\Gamma)$-valued distribution φ. We consider the causal solution of
$$\varphi^h \in Y_h \qquad \mathcal{W} * (\varphi^h - \varphi) \in Y_h^\circ,$$

and the associated potential

$$\varepsilon^h := \mathcal{D} * (\varphi^h - \varphi) = \mathcal{D} * \mathcal{E}^h_\varphi * \varphi,$$

which is characterized by the problem

$$\begin{aligned} \varepsilon^h &\in \mathrm{TD}(H^1_\Delta(\mathbb{R}^d \setminus \Gamma)), \\ \ddot{\varepsilon}^h &= \Delta \varepsilon^h, \\ [\![\gamma \varepsilon^h]\!] + \varphi &\in Y_h, \\ [\![\partial_\nu \varepsilon^h]\!] &= 0, \\ \partial_\nu \varepsilon^h &\in Y_h^\circ. \end{aligned}$$

The results of this section are given without proof, but it might benefit the reader to check them. After cut-off, the **dynamical system triplet** is the one of Section 8.3. The **lifting** $L : H^{1/2}(\Gamma) \to \mathbb{Y}_T$ is given by the solution of the variational problem:

$$\left[\begin{array}{ll} u_0 \in H^1_{\partial B_T}(B_T \setminus \Gamma), & [\![\gamma u_0]\!] + \varphi \in Y_h, \\ (\nabla u_0, \nabla v)_{B_T \setminus \Gamma} + (u_0, v)_{B_T} = 0 & \forall v \in V, \end{array}\right. \tag{8.23}$$

yielding the bound (see (8.7))

$$\|L\varphi\|_{1,B_T \setminus \Gamma} = |L\varphi|_{\mathbb{Y}_T} \le C^\dagger_\gamma \|\varphi\|_{1/2,\Gamma}.$$

It is important to note that

$$PL\varphi = 0$$

(take $v = \chi_{\Omega_-}$ in (8.23)), which will eliminate the need of taking care of rigid motions in the **associated evolution problem in a bounded domain**. Taking now $\varphi \in \mathcal{C}^3_0([0,\infty); H^{1/2}(\Gamma))$, defining

$$u_0(t) := L\varphi(t),$$

and letting $v_0 : [0,\infty) \to D(A)$ be the solution of

$$\ddot{v}_0(t) = \Delta v_0(t) + L(\varphi(t) - \ddot{\varphi}(t)) \qquad t \ge 0, \qquad v_0(0) = \dot{v}_0(0) = 0,$$

we can build the function

$$\varepsilon^h_T := u_0 + v_0 \in \mathcal{C}^2([0,\infty); L^2(B_T)) \cap \mathcal{C}^1([0,\infty); H^1(B_T \setminus \Gamma)) \cap \mathcal{C}([0,\infty); \mathbb{Y}_T))$$

that solves

$$\begin{aligned}
\ddot{\varepsilon}_T^h(t) &= \Delta \varepsilon_T^h(t) && \forall t \ge 0,\\
\gamma_T \varepsilon_T^h(t) &= 0 && \forall t \ge 0,\\
[\![\gamma \varepsilon_T^h]\!](t) + \varphi(t) &\in Y_h && \forall t \ge 0,\\
[\![\partial_\nu \varepsilon_T^h]\!](t) &= 0 && \forall t \ge 0,\\
\partial_\nu \varepsilon_T^h(t) &\in Y_h^\circ && \forall t \ge 0,\\
\varepsilon_T^h(0) = \dot{\varepsilon}_T^h(0) &= 0.
\end{aligned}$$

Here are the resulting bounds for $t \ge 0$:

$$\begin{aligned}
\|\varepsilon_T^h(t)\|_{1,B_T\setminus\Gamma} &\le C_\gamma^\dagger(\|\varphi(t)\|_{1/2,\Gamma} + \hat{c}(T)\, B(\varphi,t)),\\
\|\dot{\varepsilon}_T^h(t)\|_{1,B_T\setminus\Gamma} &\le C_\gamma^\dagger(\|\dot{\varphi}(t)\|_{1/2,\Gamma} + B(\varphi,t) + B(\dot{\varphi},t)),\\
\|\Delta\varepsilon_T^h(t)\|_{B_T\setminus\Gamma} &\le C_\gamma^\dagger(\|\varphi(t)\|_{1/2,\Gamma} + 2B(\dot{\varphi},t)).
\end{aligned}$$

What is left to complete the argument is following the usual pattern of identification of the Galerkin error operator as the extension $\underline{\varepsilon}_T^h$ for $t < T + \delta$ and then using a density argument to relax the continuity hypotheses required for φ. The following theorem is the result that comes out of this process.

Theorem 8.4.1 (Mapping properties of the Galerkin error operator for $\mathcal{W}$). *If $\varphi \in \mathcal{W}_+^2(\mathbb{R}; H^{1/2}(\Gamma))$, then*

$$\mathcal{D} * \mathcal{E}_\varphi^h * \varphi \in \mathcal{C}_+^1(\mathbb{R}; L^2(\mathbb{R}^d)) \cap \mathcal{C}_+(\mathbb{R}; H^1(\mathbb{R}^d \setminus \Gamma))$$

and therefore

$$\mathcal{E}_\varphi^h * \varphi \in \mathcal{C}_+(\mathbb{R}; H^{1/2}(\Gamma)).$$

Moreover, for all $t \ge 0$

$$\|(\mathcal{D} * \mathcal{E}_\varphi^h * \varphi)(t)\|_{1,\mathbb{R}^d\setminus\Gamma} \le C\, H_3(\partial^{-1}\varphi, t\,|\,H^{1/2}(\Gamma)).$$

Final remark. It is not difficult to show then that the bound of Theorem 8.4.1 is actually an error estimate for Galerkin semidiscretization. If $\Pi^h : H^{1/2}(\Gamma) \to Y_h$ is the best approximation operator on Y_h, then

$$\|u^h(t) - u(t)\|_{1,\mathbb{R}^d\setminus\Gamma} \le C\, H_3(\partial^{-1}(\varphi - \Pi^h\varphi), t\,|\,H^{1/2}(\Gamma)),$$

which follows from the fact that $\mathcal{E}_\varphi^h * (\varphi - \Pi^h\varphi) = 0$ for all φ.

8.5 Exercises

1. (Section 8.2) Prove the bounds (8.9).
2. (Section 8.3) Prove the bounds of Proposition 8.3.1.
3. (Section 8.3) Prove Theorem 8.3.2.
4. (Section 8.4) Prove the results of Section 8.4.

Chapter 9
Full discretization revisited

In this chapter we give some estimates for the fully discrete method (using Galerkin semidiscretization is space and CQ in time) for the single layer potential representation of the Dirichlet problem. Our goal is to prove all the bounds using time-domain techniques instead of estimates in the Laplace domain. The contents of this chapter follow (on a different example and with slightly different results) what appears in the last section of [13]. The techniques of this chapter mimic the traditional analysis of low order discretization methods for the wave equation. The context where these tools are used is, nevertheless, quite different from what is commonly seen in the literature.

9.1 Round and around Convolution Quadrature

Let us bring back our main example from Chapters 5 and 7: given a causal function $\beta : \mathbb{R} \to H^{1/2}(\Gamma)$, we look for a causal $H^{-1/2}(\Gamma)$-valued function λ^h satisfying (for all t)

$$\lambda^h \in X_h, \tag{9.1a}$$

$$\mathcal{V} * \lambda^h - \beta \in X_h^\circ, \tag{9.1b}$$

and we then define

$$u^h = \mathcal{S} * \lambda^h.$$

F.-J. Sayas, *Retarded Potentials and Time Domain Boundary Integral Equations*, Springer Series in Computational Mathematics 50,
DOI 10.1007/978-3-319-26645-9_9

Here X_h is a closed subspace of $H^{-1/2}(\Gamma)$. In the Laplace transformed domain we have (for all $s \in \mathbb{C}_+$)

$$\Lambda^h(s) \in X_h, \tag{9.2a}$$

$$\mathrm{V}(s)\Lambda^h(s) - \mathrm{B}(s) \in X_h^\circ, \tag{9.2b}$$

$$\mathrm{U}^h(s) = \mathrm{S}(s)\Lambda^h(s). \tag{9.2c}$$

Discrete causal derivatives. Consider now one of the following three choices of discrete symbol for the derivative

$$s_\kappa = \begin{cases} \frac{1}{\kappa}(1 - s^{-s\kappa}), & \text{(Backward Euler – BE)}, \\ \frac{1}{\kappa}(\frac{3}{2} - 2e^{-s\kappa} + \frac{1}{2}e^{-2s\kappa}), & \text{(BDF2)}, \\ \frac{2}{\kappa}\frac{1 - e^{-s\kappa}}{1 + e^{-s\kappa}}, & \text{(Trapezoidal rule – TR)}, \end{cases} \tag{9.3}$$

which can also be presented as

$$s_\kappa = \frac{1}{\kappa}p(e^{-s\kappa}), \qquad p(\zeta) = \begin{cases} 1 - \zeta, & \text{(BE)}, \\ \frac{3}{2} - 2\zeta + \frac{1}{2}\zeta^2, & \text{(BDF2)}, \\ \frac{1-\zeta}{1+\zeta}, & \text{(TR)}. \end{cases}$$

These symbols are associated with a discrete differentiation operator

$$\mathcal{L}\{\partial_\kappa f\} = s_\kappa \mathrm{F}(s).$$

In the case of BE and BDF2, this is plain backward (and therefore causal) differentiation

$$\partial_\kappa f = \frac{1}{\kappa}(f - f(\cdot - \kappa)), \qquad \text{or} \qquad \partial_\kappa f = \frac{1}{\kappa}(\frac{3}{2}f - 2f(\cdot - \kappa) + \frac{1}{2}f(\cdot - 2\kappa)).$$

The TR derivative is implicit: $u = \partial_\kappa f$ is defined as the only causal solution of

$$\frac{1}{\kappa}(f - f(\cdot - \kappa)) = \frac{1}{2}(u + u(\cdot - \kappa)).$$

The fully discrete CQ-BEM equations. In the Laplace domain, these equations are extremely simple. We just substitute s by s_κ in the two operators in (9.2), thus obtaining the problem (for all $s \in \mathbb{C}_+$):

$$\Lambda^h_\kappa(s) \in X_h, \tag{9.4a}$$

$$\mathrm{V}(s_\kappa)\Lambda^h_\kappa(s) - \mathrm{B}(s) \in X_h^\circ, \tag{9.4b}$$

$$\mathrm{U}^h_\kappa(s) = \mathrm{S}(s_\kappa)\Lambda^h_\kappa(s). \tag{9.4c}$$

Note how $s_\kappa \in \mathbb{C}_+$ for all $s \in \mathbb{C}_+$, which guarantees that equations (9.4) are actually meaningful.

Let us now have a look at the equations in the time domain. We need the sequences of operators generated by the expansions

$$\mathrm{V}(\tfrac{1}{\kappa}p(\zeta)) = \sum_{m=0}^\infty \zeta^m \mathrm{V}_m(\kappa), \qquad \mathrm{S}(\tfrac{1}{\kappa}p(\zeta)) = \sum_{m=0}^\infty \zeta^m \mathrm{S}_m(\kappa),$$

and the associated distributions

$$\mathcal{V}_\kappa = \sum_{m=0}^\infty \delta_{t_m} \otimes \mathrm{V}_m(\kappa), \qquad \mathcal{S}_\kappa = \sum_{m=0}^\infty \delta_{t_m} \otimes \mathrm{S}_m(\kappa).$$

Equations (9.4) are equivalent to finding a causal $H^{-1/2}(\Gamma)$-valued function (distribution) satisfying

$$\lambda^h_\kappa \in X_h, \tag{9.5a}$$

$$\mathcal{V}_\kappa * \lambda^h_\kappa - \beta \in X_h^\circ, \tag{9.5b}$$

$$u^h_\kappa = \mathcal{S}_\kappa * \lambda^h_\kappa. \tag{9.5c}$$

(Note that from this point of view u^h_κ is automatically a causal distribution.)

Further details. Equations (9.5) can be written as the search for a causal function

$$\lambda^h_\kappa : \mathbb{R} \to X_h, \tag{9.6a}$$

satisfying

$$\sum_{m=0}^\infty \langle \mu^h, \mathrm{V}_m(\kappa)\lambda^h_\kappa(t-t_m)\rangle_\Gamma = \langle \mu^h, \beta(t)\rangle_\Gamma \qquad \forall \mu^h \in X_h, \tag{9.6b}$$

followed by the potential formula

$$u^h_\kappa(t) = \sum_{m=0}^\infty \mathrm{S}_m(\kappa)\lambda^h_\kappa(t-t_m). \tag{9.6c}$$

Let us admit, for the moment being, that we have enough regularity on the data to ensure that all elements of these equations are causal functions. For a given t, the sums in (9.6) contain only a finite number of nonzero terms. As we have already mentioned in Chapter 4, we do not compute the complete functions λ^h_κ and u^h_κ. Instead, we compute

$$\lambda^h_\kappa[m] := \lambda^h_\kappa(t_m), \qquad u^h_\kappa[m] := u^h_\kappa(t_m), \qquad m \ge 0,$$

by progressively solving for $n \ge 0$

$$\lambda^h_\kappa[n] \in X_h, \qquad \sum_{m=0}^{n} \langle \mu^h, \mathrm{V}_m(\kappa)\lambda^h_\kappa[n-m]\rangle_\Gamma = \langle \mu^h, \beta(t_m)\rangle_\Gamma \qquad \forall \mu^h \in X_h,$$

and then computing

$$u^h_\kappa[n] = \sum_{m=0}^{n} \mathrm{S}_m(\kappa)\lambda^h_\kappa[n-m].$$

We also recall that the Laplace domain theory of CQ gives estimates for

$$\|\lambda^h(t) - \lambda^h_\kappa(t)\|_{-1/2,\Gamma} \quad \text{and} \quad \|u^h(t) - u^h_\kappa(t)\|_{1,\mathbb{R}^d}, \qquad \forall t \ge 0,$$

although we have only computed the fully discrete solution at discrete time steps (that is, in discrete time).

First time or first space. We might wonder if semidiscretization in time first, followed by Galerkin semidiscretization in space (using the same Galerkin scheme for all the time steps), would lead to the same system. The short answer is yes. A concise argument follows. Define $P_h : X_h \to H^{-1/2}(\Gamma)$ to be the canonical inclusion of X_h in $H^{-1/2}(\Gamma)$ and let $P_h' : H^{1/2}(\Gamma) \to X_h'$ be its adjoint operator, which can be understood as testing elements of $H^{1/2}(\Gamma)$ exclusively by elements of X_h (so that $\ker P_h' = X_h^\circ$). For the problem whose Laplace transform is

$$\mathrm{V}(s)\Lambda(s) = \mathrm{B}(s), \qquad \mathrm{U}(s) = \mathrm{S}(s)\Lambda(s),$$

the CQ scheme is the problem whose Laplace transform is

$$\mathrm{V}(s_\kappa)\Lambda_\kappa(s) = \mathrm{B}(s), \qquad \mathrm{U}_\kappa(s) = \mathrm{S}(s_\kappa)\Lambda_\kappa(s).$$

With the above notation, this corresponds to

$$\mathcal{V}_\kappa * \lambda_\kappa = \beta, \qquad u_\kappa = \mathcal{S}_\kappa * \lambda_k.$$

Galerkin discretization of this problem can be written as

$$P_h'\mathcal{V}_\kappa * (P_h\lambda^h_\kappa) = P_h'\beta, \qquad u^h_\kappa = \mathcal{S}_\kappa * (P_h\lambda^h_\kappa), \tag{9.7}$$

where we look for λ_κ^h with values in X_h. The Laplace transform of this problem is

$$P_h' \mathrm{V}(s_\kappa) P_h \Lambda_\kappa^h(s) = P_h' \mathrm{B}(s), \qquad \mathrm{U}_\kappa^h(s) = \mathrm{S}(s_\kappa)\Lambda_\kappa^h(s).$$

It is then easy to see that (9.7) can be written as

$$(P_h' \mathcal{V}_\kappa P_h) * \lambda_h^\kappa = P_h' \beta_0, \qquad u_\kappa^h = (\mathcal{S}_\kappa P_h) * \lambda_\kappa^h,$$

which shows that the processes of Galerkin discretization in space and CQ discretization in time can be applied in any order, leading to the same scheme.

The associated transmission problem. Problem (9.4) can be written as the search for $\mathrm{U}_\kappa^h(s)$ such that

$$s_\kappa^2 \mathrm{U}_\kappa^h(s) = \Delta \mathrm{U}_\kappa^h(s) \qquad \text{in } \mathbb{R}^d \setminus \Gamma, \tag{9.8a}$$

$$[\![\gamma \mathrm{U}_\kappa^h(s)]\!] = 0, \tag{9.8b}$$

$$\gamma \mathrm{U}_\kappa^h(s) - \mathrm{B}(s) \in X_h^\circ, \tag{9.8c}$$

$$[\![\partial_\nu \mathrm{U}_\kappa^h(s)]\!] \in X_h, \tag{9.8d}$$

followed by

$$\Lambda_\kappa^h(s) = [\![\partial_\nu \mathrm{U}_\kappa^h(s)]\!]. \tag{9.8e}$$

This is the Laplace domain expression for the problem looking for $u_\kappa^h \in \mathrm{TD}(H_\Delta^1(\mathbb{R}^d \setminus \Gamma))$ such that

$$\partial_\kappa^2 u_\kappa^h = \Delta u_\kappa^h, \tag{9.9a}$$

$$[\![\gamma u_\kappa^h]\!] = 0, \tag{9.9b}$$

$$\gamma u_\kappa^h - \beta \in X_h^\circ, \tag{9.9c}$$

$$[\![\partial_\nu u_\kappa^h]\!] \in X_h, \tag{9.9d}$$

followed by

$$\lambda_\kappa^h = [\![\partial_\nu u_\kappa^h]\!]. \tag{9.9e}$$

In turn, (9.9) is just a naive time domain (BE, BDF2, or TR) discretization of the equations:

$$\ddot{u}^h = \Delta u^h, \tag{9.10a}$$

$$[\![\gamma u^h]\!] = 0, \tag{9.10b}$$

$$\gamma u^h - \beta \in X_h^\circ, \tag{9.10c}$$

$$[\![\partial_\nu u^h]\!] \in X_h, \tag{9.10d}$$

followed by

$$\lambda^h = [\![\partial_\nu u^h]\!]. \tag{9.10e}$$

The error equations. In the above context, it is convenient to define the error and the consistency error as follows:

$$e_\kappa^h := u^h - u_\kappa^h, \qquad \theta_\kappa^h := \partial_\kappa^2 u^h - \ddot{u}^h.$$

Their respective Laplace transforms are

$$\mathrm{E}_\kappa^h(s) = \mathrm{U}^h(s) - \mathrm{U}_\kappa^h(s), \qquad \Theta_\kappa^h(s) = (s_\kappa^2 - s^2)\mathrm{U}^h(s).$$

Subtracting equations (9.9) from equations (9.10), we can focus on the problem of finding $e_\kappa^h \in \mathrm{TD}(H^1_\Delta(\mathbb{R}^d \setminus \Gamma))$ satisfying

$$\partial_\kappa^2 e_\kappa^h = \Delta e_\kappa^h + \theta_\kappa^h, \tag{9.11a}$$

$$[\![\gamma e_\kappa^h]\!] = 0, \tag{9.11b}$$

$$\gamma e_\kappa^h \in X_h^\circ, \tag{9.11c}$$

$$[\![\partial_\nu e_\kappa^h]\!] \in X_h, \tag{9.11d}$$

followed by the computation of the error for the density $[\![\partial_\nu e_\kappa^h]\!]$. It is important to notice that the error stays in the discrete domain of the Laplacian

$$D_h := \{u \in H^1(\mathbb{R}^d) \cap H^1_\Delta(\mathbb{R}^d \setminus \Gamma) \,:\, \gamma u \in X_h^\circ, \quad [\![\partial_\nu u]\!] \in X_h\},$$

and that we have the following integration by parts formula

$$(\Delta u, v)_{\mathbb{R}^d \setminus \Gamma} + (\nabla u, \nabla v)_{\mathbb{R}^d} = 0 \qquad \forall u, v \in D_h. \tag{9.12}$$

Proposition 9.1.1. *The solution of* (9.11) *satisfies:* $e_\kappa^h \in \mathrm{TD}(D_h)$ *and*

$$(\partial_\kappa^2 e_\kappa^h, v)_{\mathbb{R}^d} + (\nabla e_\kappa^h, \nabla v)_{\mathbb{R}^d} = (\theta_\kappa^h, v)_{\mathbb{R}^d} \qquad \forall v \in D_h. \tag{9.13}$$

Proof. Substitute (9.11a) in the integration by parts formula (9.12) to get the result, at least formally. Note that $\mathrm{E}_\kappa^h(s) \in D_h$ for all $s \in \mathbb{C}_+$, so there is no problem in considering e_κ^h as a D_h-valued distribution. To reach (9.13) we just have to notice that a similar result holds in the Laplace domain. □

For some future arguments, it will be convenient to consider the following quantity

$$f_\kappa^h = \partial_\kappa e_\kappa^h, \qquad \mathrm{F}_\kappa^h(s) = s_\kappa \mathrm{E}_\kappa^h(s).$$

Note then that (e_κ^h, f_κ^h) is in $\mathrm{TD}(D_h \times D_h)$ and satisfies

$$
\begin{aligned}
(\partial_\kappa \nabla e_\kappa^h, \nabla v)_{\mathbb{R}^d} - (\nabla f_\kappa^h, \nabla v)_{\mathbb{R}^d} &= 0 \qquad && \forall v \in H^1(\mathbb{R}^d), && (9.14a)\\
(\partial_\kappa f_\kappa^h, v)_{\mathbb{R}^d} + (\nabla e_\kappa^h, \nabla v)_{\mathbb{R}^d} &= (\theta_\kappa^h, v)_{\mathbb{R}^d} && \forall v \in D_h. && (9.14b)
\end{aligned}
$$

We will also use the norm:

$$
|||(u, v)|||^2 := \|\nabla u\|_{\mathbb{R}^d}^2 + \|v\|_{\mathbb{R}^d}^2.
$$

Bounds for all three methods will be given in terms of

$$
\max_{0 \le \tau \le t} \|(u^h)^{(k)}(\tau)\|_{\mathbb{R}^d} \le C H_3(\beta^{(k-1)}, t \,|\, H^{1/2}(\Gamma)), \tag{9.15}
$$

for some positive k. The last bound follows from the estimates for the Galerkin solver given in Theorem 7.5.3. It is at this point that the mapping properties of the Galerkin solver (which can be understood as stability estimates) are of key importance.

9.2 The Backward Euler method

Recall that now $s_\kappa = \frac{1}{\kappa}(1 - e^{-s\kappa})$, or, from the point of view of time

$$
\partial_\kappa f = \tfrac{1}{\kappa}(f - f(\cdot - \kappa)).
$$

The analysis follows a simple plan: for backward Euler discretization, everything boils down to an energy bound of a single step, where the error at the current time step is bounded by the error of the previous time step plus a consistency error term. We will see (Proposition 9.2.2) how to use this idea to prove an energy estimate, in the triple bar norm $|||\cdot|||$. The second group of estimates (one in L^2 and another one on the boundary) are proved by postprocessing the error bound of Proposition 9.2.2.

Proposition 9.2.1. *Assume that u^h is smooth enough so that e_κ^h is a continuous function of time. Then*

$$
(\partial_\kappa \nabla e_\kappa^h(t), \nabla e_\kappa^h(t))_{\mathbb{R}^d} + (\partial_\kappa f_\kappa^h(t), f_\kappa^h(t))_{\mathbb{R}^d} = (\theta_\kappa^h(t), f_\kappa^h(t))_{\mathbb{R}^d} \qquad \forall t.
$$

Proof. When e_κ^h is continuous, so is f_κ^h and then we can write (9.14) as

$$
(\partial_\kappa \nabla e_\kappa^h(t), \nabla v)_{\mathbb{R}^d} - (\nabla f_\kappa^h(t), \nabla v)_{\mathbb{R}^d} = 0 \qquad \forall v \in H^1(\mathbb{R}^d), \qquad \forall t, \tag{9.16a}
$$

$$
(\partial_\kappa f_\kappa^h(t), v)_{\mathbb{R}^d} + (\nabla e_\kappa^h(t), \nabla v)_{\mathbb{R}^d} = (\theta_\kappa^h(t), v)_{\mathbb{R}^d} \qquad \forall v \in D_h, \qquad \forall t. \tag{9.16b}
$$

We then test equation (9.16a) with $e_\kappa^h(t)$ and equation (9.16b) and $f_\kappa^h(t)$. Adding the results the proposition is proved. □

Proposition 9.2.2 (Basic energy norm estimate).

$$|||(e^h_\kappa(t), f^h_\kappa(t))||| \le \tfrac{5}{3}\kappa\, t \max_{0\le\tau\le t} \|(u^h)^{(3)}(\tau)\|_{\mathbb{R}^d} \qquad \forall t \ge 0.$$

Proof. A simple Taylor expansion argument shows that for a smooth enough function

$$\|\partial^2_\kappa g(t) - \ddot{g}(t)\| \le \tfrac{5}{3}\kappa \max_{t-2\kappa\le\tau\le t} \|g^{(3)}(\tau)\|.$$

If g is causal and smooth enough, we can bound

$$\begin{aligned}\sum_{j=0}^{\infty} \|\partial^2_\kappa g(t-t_j) - \ddot{g}(t-t_j)\| &\le \tfrac{5}{3}\kappa \sum_{j=0}^{\infty} \left(\max_{t-t_{j-2}\le\tau\le t-t_j} \|g^{(3)}(\tau)\| \right) \\ &\le \tfrac{5}{3}\kappa \max_{0\le\tau\le t} \|g^{(3)}(\tau)\| \#\{j \,:\, 0\le j < t/\kappa\} \\ &\le \tfrac{5}{3} t \max_{0\le\tau\le t} \|g^{(3)}(\tau)\|.\end{aligned}$$

This argument is valid independently of where the causal smooth function g takes values. In particular, we obtain

$$\sum_{j=0}^{\infty} \|\theta^h_\kappa(t-t_j)\|_{\mathbb{R}^d} \le \tfrac{5}{3} t \max_{0\le\tau\le t} \|(u^h)^{(3)}(\tau)\|_{\mathbb{R}^d}. \tag{9.17}$$

Proposition 9.2.1 can be rearranged to yield

$$|||(e^h_\kappa(t), f^h_\kappa(t))|||^2 = (\nabla e^h_\kappa(t), \nabla e^h_\kappa(t-\kappa))_{\mathbb{R}^d} + (f^h_\kappa(t), f^h_\kappa(t-\kappa))_{\mathbb{R}^d} + \kappa(\theta^h_\kappa(t), f^h_\kappa(t))_{\mathbb{R}^d}$$

and therefore

$$|||(e^h_\kappa(t), f^h_\kappa(t))||| \le |||(e^h_\kappa(t-\kappa), f^h_\kappa(t-\kappa))||| + \kappa\, \|\theta^h_\kappa(t)\|_{\mathbb{R}^d}.$$

Induction and (9.17) prove the result. □

Proposition 9.2.3 (Estimate in L^2).

$$\|e^h_\kappa(t)\|_{\mathbb{R}^d} \le \tfrac{5}{3}\kappa\, t^2 \max_{0\le\tau\le t} \|(u^h)^{(3)}(\tau)\|_{\mathbb{R}^d} \qquad \forall t \ge 0.$$

Proof. Note that $e^h_\kappa = e^h_\kappa(\cdot - \kappa) + \kappa f^h_\kappa$ and therefore

$$\|e^h_\kappa(t)\|_{\mathbb{R}^d} \le \|e^h_\kappa(t-\kappa)\|_{\mathbb{R}^d} + \kappa \|f^h_\kappa(t)\|_{\mathbb{R}^d}.$$

By induction,

$$\|e^h_\kappa(t)\|_{\mathbb{R}^d} \le \kappa \sum_{j=0}^{\infty} \|f^h_\kappa(t-t_j)\|_{\mathbb{R}^d} \le \tfrac{5}{3}\kappa t^2 \max_{0\le\tau\le t} \|(u^h)^{(3)}(\tau)\|_{\mathbb{R}^d},$$

where we have used Proposition 9.2.2 and the fact that the sum above involves only a finite number of terms, until $t_j \geq t$. □

Proposition 9.2.4 (Estimate on the boundary).

$$\|[\![\partial_\nu e_\kappa^h]\!](t)\|_{-1/2,\Gamma} \leq C\kappa\, t\big(\max_{0\leq\tau\leq t}\|(u^h)^{(3)}(\tau)\|_{\mathbb{R}^d} + \max_{0\leq\tau\leq t}\|(u^h)^{(4)}(\tau)\|_{\mathbb{R}^d}\big) \qquad \forall t \geq 0.$$

Proof. Note that $\Delta e_\kappa^h = \partial_\kappa f_\kappa^h - \theta_\kappa^h$ and therefore

$$\begin{aligned}
\|\Delta e_\kappa^h(t)\|_{\mathbb{R}^d\setminus\Gamma} &\leq \max_{t-\kappa\leq\tau\leq t}\|\dot f_\kappa^h(\tau)\|_{\mathbb{R}^d} + \|\theta_\kappa^h(t)\|_{\mathbb{R}^d}\\
&\leq \tfrac{5}{3}\kappa t \max_{0\leq\tau\leq t}\|(u^h)^{(4)}(\tau)\|_{\mathbb{R}^d} + \tfrac{5}{3}\kappa^2 \max_{0\leq\tau\leq t}\|(u^h)^{(3)}(\tau)\|_{\mathbb{R}^d}\\
&\leq \tfrac{10}{3}\kappa\, t \max_{0\leq\tau\leq t}\|(u^h)^{(4)}(\tau)\|_{\mathbb{R}^d},
\end{aligned}$$

where we have applied Proposition 9.2.2 and a bound for the consistency error seen in the proof of that result. □

Summary. For the BE discretization, Propositions 9.2.2–9.2.4 together with the bounds for the Galerkin solver (9.15) provide the following bounds

$$\begin{aligned}
\|\nabla u_\kappa^h(t) - \nabla u^h(t)\|_{\mathbb{R}^d} &\leq C\kappa t H_3(\ddot\beta, t\,|\,H^{1/2}(\Gamma)),\\
\|u_\kappa^h(t) - u^h(t)\|_{\mathbb{R}^d} &\leq C\kappa t^2 H_3(\ddot\beta, t\,|\,H^{1/2}(\Gamma)),\\
\|\lambda_\kappa^h(t) - \lambda^h(t)\|_{-1/2,\Gamma} &\leq C\kappa t H_4(\ddot\beta, t\,|\,H^{1/2}(\Gamma)),
\end{aligned}$$

where, as usual,

$$H_k(\psi, t\,|\,X) := \sum_{\ell=0}^{k}\int_0^t \|\psi(\tau)\|_X \mathrm{d}\tau$$

and therefore we can overestimate the upper bounds by simpler forms

$$H_k(\ddot\beta, t\,|\,H^{1/2}(\Gamma)) \leq \sum_{\ell=0}^{k+1}\int_0^t \|\beta^{(\ell)}(\tau)\|_{1/2,\Gamma}\mathrm{d}\tau.$$

9.3 The order two Backward Differentiation Formula

The analysis of BDF2 is slightly more complicated than the analysis of BE for a simple reason: There does not seem to be any simple way of choosing test functions in (9.14) leading to an energy estimate like Proposition 9.2.2 for BDF2. What is missing is a simple estimate of the current error in terms of past errors plus consistency error, which can be iterated to accumulate the effect of all past time steps. Instead, we will recur to a Laplace domain estimate to prove the energy estimate. Once the energy estimate is established, the L^2 and boundary estimates will follow with techniques similar to the ones of Section 9.2. We recall our way of denoting $\underline{\sigma} := \min\{1, \operatorname{Re} s\}$ for $s \in \mathbb{C}_+$ and the BDF2 definition of

$$s_\kappa = \frac{1}{\kappa}\left(\frac{3}{2} - 2e^{-s\kappa} + \frac{1}{2}e^{-2s\kappa}\right), \qquad \partial_\kappa f = \frac{1}{\kappa}\left(\frac{3}{2}f - 2f(\cdot - \kappa) + \frac{1}{2}f(\cdot - 2\kappa)\right).$$

Lemma 9.3.1. *Let* $\mathrm{V} \in D_h$ *satisfy*

$$s_\kappa^2 \mathrm{V} - \Delta \mathrm{V} = (s_\kappa^2 - s^2)\mathrm{G}, \qquad \mathrm{G} \in L^2(\mathbb{R}^d). \tag{9.18}$$

Then

$$|||(\mathrm{V}, s_\kappa \mathrm{V})||| \le C\kappa^2 \frac{1}{\underline{\sigma}} \|s^4 \mathrm{G}\|, \qquad \forall s \in \mathbb{C}_+, \qquad \forall \kappa.$$

Proof. Multiplying (9.18) by $\overline{\mathrm{V}}$, integrating over $\mathbb{R}^d$, and applying the integration by parts formula (9.12) it follows that

$$s_\kappa^2 \|\mathrm{V}\|_{\mathbb{R}^d}^2 + \|\nabla \mathrm{V}\|_{\mathbb{R}^d}^2 = ((s_\kappa^2 - s^2)\mathrm{G}, \overline{\mathrm{V}})_{\mathbb{R}^d},$$

and therefore

$$(\operatorname{Re} s_\kappa)|||(\mathrm{V}, s_\kappa \mathrm{V})||| \le \|(s_\kappa^2 - s^2)\mathrm{G}\|_{\mathbb{R}^d}. \tag{9.19}$$

Proposition 4.5.1 can be rephrased (see Exercise 2 in Chapter 4) as

$$|s_\kappa| \le C_1|s|, \qquad |s_\kappa - s| \le C_2\kappa^2|s|^3, \qquad \operatorname{Re} s_\kappa \ge C_3\underline{\sigma}, \qquad \forall s \in \mathbb{C}_+, \qquad \forall \kappa.$$

Therefore $|s_\kappa^2 - s^2| \le C_4\kappa^2|s|^4$ and the result follows from (9.19). □

Proposition 9.3.2 (Basic energy norm estimate).

$$|||(e_\kappa^h(t), f_\kappa^h(t))||| \le C\kappa^2 t^2 \max_{0 \le \tau \le t} \|(\mathcal{P}_2 u^h)^{(4)}(\tau)\|_{\mathbb{R}^d} \qquad \forall t \ge 0,$$

where $\mathcal{P}_2 g := g + 2\dot{g} + \ddot{g}$.

Proof. The error equations (9.11) can be rewritten in the Laplace domain as

$$\mathrm{E}_\kappa^h(s) \in D_h, \qquad s_\kappa^2 \mathrm{E}_\kappa^h(s) - \Delta \mathrm{E}_\kappa^h(s) = (s_\kappa^2 - s^2)\mathrm{U}^h(s).$$

Then, Lemma 9.3.1 proves that

$$|||(\mathrm{E}_\kappa^h(s), \mathrm{F}_\kappa^h(s))||| \le C\kappa^2 \frac{1}{\underline{\sigma}} \|s^4 \mathrm{U}^h\|_{\mathbb{R}^d}.$$

We can thus use Proposition 3.2.2 (estimates for convolution operators) with $\mu = 0$ to get

$$\|(e^h_\kappa(t), f^h_\kappa(t))\| \le C\kappa^2 \frac{t\max\{1,t\}}{1+t} \int_0^t \|(\mathcal{P}_2 u^h)^{(4)}(\tau)\|_{\mathbb{R}^d} \mathrm{d}\tau.$$

Some easy upper bounds applied to this inequality prove the result. □

Proposition 9.3.3 (Estimate in L^2).

$$\|e^h_\kappa(t)\|_{\mathbb{R}^d} \le C\kappa^2 t^3 \max_{0\le\tau\le t} \|(\mathcal{P}_2 u^h)^{(4)}(\tau)\|_{\mathbb{R}^d} \qquad \forall t \ge 0.$$

Proof. Solving the recurrence

$$\frac{3}{2} e^h_\kappa - 2e^h_\kappa(\cdot - \kappa) + \frac{1}{2} e^h_\kappa(\cdot - 2\kappa) = \kappa f^h_\kappa$$

(see exercises) we can write

$$e^h_\kappa = \kappa \sum_{j=0}^{\infty} (1 - 3^{-(j+1)}) f^h_\kappa(\cdot - t_j),$$

and therefore

$$\|e^h_\kappa(t)\|_{\mathbb{R}^d} \le \kappa \sum_{j=0}^{\infty} \|f^h_\kappa(t - t_j)\|_{\mathbb{R}^d} \le t \max_{0\le\tau\le t} \|f^h_\kappa(\tau)\|_{\mathbb{R}^d}.$$

The result is then a direct consequence of the energy estimate (Proposition 9.3.2). □

Proposition 9.3.4 (Estimate on the boundary).

$$\|[\![\partial_\nu e^h_\kappa]\!](t)\|_{-1/2,\Gamma} \le C\kappa^2 t^2 \sum_{\ell=4}^{7} \left(\max_{0\le\tau\le t} \|(u^h)^{(\ell)}(\tau)\|_{\mathbb{R}^d} \right) \qquad \forall t \ge 0.$$

Proof. By (9.11a) and the definition of $f^h_\kappa = \partial_\kappa e^h_\kappa$ we can write

$$\Delta e^h_\kappa = \partial_\kappa f^h_\kappa - (\partial^2_\kappa u^h - \ddot{u}^h),$$

and then bound

$$\begin{aligned}
\|\Delta e^h_\kappa(t)\|_{\mathbb{R}^d\setminus\Gamma} &\le \frac{3}{2} \left\| \frac{1}{\kappa}(f^h_\kappa(t) - f^h_\kappa(t-\kappa)) \right\|_{\mathbb{R}^d} + \frac{1}{2} \left\| \frac{1}{\kappa}(f^h_\kappa(t-\kappa) - f^h_\kappa(t-2\kappa)) \right\|_{\mathbb{R}^d} \\
&\quad + \|\partial^2_\kappa u^h(t) - \ddot{u}^h(t)\|_{\mathbb{R}^d} \\
&\le \frac{3}{2} \max_{t-2\kappa\le\tau\le t} \|\dot{f}(\tau)\|_{\mathbb{R}^d} + C\kappa^2 \max_{t-4\kappa\le\tau\le t} \|(u^h)^{(4)}(\tau)\|_{\mathbb{R}^d} \\
&\le C\kappa^2 t^2 \max_{0\le\tau\le t} \|(\mathcal{P}_2 u^h)^{(5)}(\tau)\|_{\mathbb{R}^d} + C\kappa^2 t^2 \max_{0\le\tau\le t} \|(u^h)^{(6)}(\tau)\|_{\mathbb{R}^d},
\end{aligned}$$

where we have applied simple estimates based on Taylor expansions and Proposition 9.3.2. The result then follows readily. □

Summary of results. Using Theorem 7.5.3, Propositions 9.3.2–9.3.4 yield the following bounds

$$\begin{aligned}\|\nabla u^h_\kappa(t) - \nabla u^h(t)\|_{\mathbb{R}^d} &\le C\kappa^2 t^2 H_5(\beta^{(3)}, t\,|\,H^{1/2}(\Gamma)) \le C\kappa^2 t^2 H_8(\beta, t\,|\,H^{1/2}(\Gamma)),\\ \|u^h_\kappa(t) - u^h(t)\|_{\mathbb{R}^d} &\le C\kappa^2 t^3 H_5(\beta^{(3)}, t\,|\,H^{1/2}(\Gamma)) \le C\kappa^2 t^3 H_8(\beta, t\,|\,H^{1/2}(\Gamma)),\\ \|\lambda^h_\kappa(t) - \lambda^h(t)\|_{-1/2,\Gamma} &\le C\kappa^2 t^2 H_5(\beta^{(4)}, t\,|\,H^{1/2}(\Gamma)) \le C\kappa^2 t^2 H_9(\beta, t\,|\,H^{1/2}(\Gamma)).\end{aligned}$$

9.4 The trapezoidal rule

While the trapezoidal rule is based on an implicit operator that approximates the derivative, the classical analysis of the method (and its implementation when it is not related to the CQ algorithm) is written in a completely different form. We need some new operators: first order differentiation and averaging

$$\hat\partial_\kappa g := \frac{1}{\kappa}(g - g(\cdot - \kappa)), \qquad \hat a_\kappa g := \frac{1}{2}(g + g(\cdot - \kappa))$$

and their squares

$$\hat\partial^2_\kappa g := \frac{1}{\kappa^2}(g - 2g(\cdot - \kappa) + g(\cdot - 2\kappa)), \qquad \hat a^2_\kappa g := \frac{1}{4}(g + 2g(\cdot - \kappa) + g(\cdot - 2\kappa)).$$

The error equations (9.11) are now written in the equivalent form

$$\hat\partial^2_\kappa e^h_\kappa = \Delta \hat a^2_\kappa e^h_\kappa + \hat\partial^2_\kappa u^h - \hat a^2_\kappa \ddot u^h =: \Delta \hat a^2_\kappa e^h_\kappa + \chi^h_\kappa. \tag{9.20}$$

The natural quantities to bound are

$$\hat e^h_\kappa := \hat a_\kappa e^h_\kappa, \qquad \hat f^h_\kappa := \hat\partial_\kappa e^h_\kappa. \tag{9.21}$$

Proposition 9.4.1 (Energy estimate).

$$|\!|\!|(\hat e^h_\kappa(t), \hat f^h_\kappa(t))|\!|\!| \le C\kappa^2 t \max_{0\le\tau\le t} \|(u^h)^{(4)}(\tau)\|_{\mathbb{R}^d} \qquad \forall t \ge 0.$$

Proof. In terms of the quantities defined in (9.21), the error equation (9.20) reads

$$\hat\partial_\kappa \hat f^h_\kappa = \Delta \hat a_\kappa \hat e^h_\kappa + \chi^h_\kappa.$$

Using the integration by parts formula (9.12), we can prove

$$(\hat{f}^h_\kappa(t)-\hat{f}^h_\kappa(t-\kappa), v)_{\mathbb{R}^d}+\frac{\kappa}{2}(\nabla\hat{e}^h_\kappa(t)+\nabla\hat{e}^h_\kappa(t-\kappa), \nabla v)_{\mathbb{R}^d} = \kappa(\chi^h_\kappa(t), v)_{\mathbb{R}^d} \qquad \forall v \in D_h.$$

We then choose

$$v := \frac{1}{\kappa}(e^h_\kappa(t) - e^h_\kappa(t-2\kappa)) = 2\hat{\partial}_\kappa \hat{e}^h_\kappa(t) = 2\hat{a}_\kappa \hat{f}^h_\kappa(t)$$

to obtain the energy estimate for the trapezoidal rule:

$$|||(\hat{e}^h_\kappa(t), \hat{f}^h_\kappa(t))|||^2 = |||(\hat{e}^h_\kappa(t-\kappa), \hat{f}^h_\kappa(t-\kappa))|||^2 + \kappa(\chi^h_\kappa(t), 2\hat{a}_\kappa \hat{f}^h_\kappa(t))_{\mathbb{R}^d}.$$

By induction

$$|||(\hat{e}^h_\kappa(t), \hat{f}^h_\kappa(t))|||^2 = \kappa \sum_{j=0}^{\infty}(\chi^h_\kappa(t-t_j), 2\hat{a}_\kappa \hat{f}^h_\kappa(t-t_j))_{\mathbb{R}^d}, \tag{9.22}$$

where, as usual, the sum involves only a finite number of terms, due to causality. Take now $t^\star \le t$ such that

$$|||(\hat{e}^h_\kappa(t^\star), \hat{f}^h_\kappa(t^\star))||| = \max_{0\le\tau\le t} |||(\hat{e}^h_\kappa(\tau), \hat{f}^h_\kappa(\tau))|||,$$

and note that

$$\|\hat{a}_\kappa \hat{f}^h_\kappa(\tau)\|_{\mathbb{R}^d} \le |||(\hat{e}^h_\kappa(\tau), \hat{f}^h_\kappa(\tau))||| \le |||(\hat{e}^h_\kappa(t^\star), \hat{f}^h_\kappa(t^\star))|||.$$

By (9.22)

$$\begin{aligned}|||(\hat{e}^h_\kappa(t^\star), \hat{f}^h_\kappa(t^\star))|||^2 &\le 2\kappa\,(\#\{j\,:\,t^* - t_j > 0\})|||(\hat{e}^h_\kappa(t^\star), \hat{f}^h_\kappa(t^\star))||| \max_{0\le\tau\le t^\star} \|\chi^h_\kappa(\tau)\|_{\mathbb{R}^d}\\ &\le 2t^\star |||(\hat{e}^h_\kappa(t^\star), \hat{f}^h_\kappa(t^\star))||| \max_{0\le\tau\le t} \|\chi^h_\kappa(\tau)\|_{\mathbb{R}^d}.\end{aligned}$$

Noticing that

$$\|\chi^h_\kappa(\tau)\|_{\mathbb{R}^d} \le C\kappa^2 \max_{\tau-2\kappa\le\rho\le\tau} \|(u^h)^{(4)}(\rho)\|_{\mathbb{R}^d},$$

the result follows readily. □

The proofs for the remaining results of this section are left as exercises. The reader is also invited to wrap the results up by writing all the estimates in terms of bounds for derivatives of the right-hand side.

Proposition 9.4.2 (L^2 **estimate).**

$$\|e^h_\kappa(t)\|_{\mathbb{R}^d} \le C\kappa^2 t^2 \max_{0\le\tau\le t} \|(u^h)^{(4)}(\tau)\|_{\mathbb{R}^d}.$$

Proposition 9.4.3 (Boundary estimate)**.**

$$\|\hat{a}_\kappa^2 [\![\partial_\nu e_\kappa^h]\!](t)\|_{-1/2,\Gamma} \le C\kappa^2 \left(\max_{0\le\tau\le t} \|(u^h)^{(4)}(\tau)\|_{\mathbb{R}^d} + t \max_{0\le\tau\le t} \|(u^h)^{(5)}(\tau)\|_{\mathbb{R}^d} \right).$$

9.5 Exercises

1. (Section 9.1) Prove (9.12).
2. (Section 9.3) Show that if a_n and b_n are causal sequences (vanishing for negative n) satisfying

$$\frac{3}{2}a_n - 2a_{n-1} + \frac{1}{2}a_{n-2} = b_n,$$

then

$$a_n = \sum_{j=0}^{n} \left(1 - 3^{-(j+1)}\right) b_{n-j}.$$

(**Hint.** Use the ζ transform of the recurrence.)
3. (Section 9.4) Prove Proposition 9.4.2. (Hint. e_κ^h can be easily written in terms of $\hat{f}_\kappa^h$.)
4. (Section 9.4) Prove Proposition 9.4.3.

Chapter 10
Patterns, extensions, and conclusions

In this short chapter we will go over the results of Chapters 7 and 8, trying to look for common arguments and patterns in the final estimates. It will be clear that problems are paired and that by looking at the type of transmission condition we can predict what kind of regularity the data needs to have. We will next give two simple extensions of this theory, to problems defined on screens and to problems on linear elasticity. The chapter will finish with an overview of part of the literature and mentioning some work in progress.

Notation in this chapter will be reduced to a bare minimum, trying to avoid long expressions, so that we can perceive common features of the results more easily. We will thus shorten

$$\mathcal{C}_0^k := \mathcal{C}_0^k([0,\infty);X), \quad \mathcal{W}_+^k := \mathcal{W}_+^k(\mathbb{R};X), \quad \mathcal{C}_+^k := \mathcal{C}_+^k(\mathbb{R};X), \quad H_k(\eta) := H_k(\eta, t\,|\,X)$$

letting the reader to place back the different incarnations of X. Notation will also be unified from the point of view of the solution of a transmission problem to be studied: in all cases this solution will be called u. In the cut-off setting, there will be a common space

$$H := L^2(B_T),$$

a bilinear form

$$a(u,v) := (\nabla u, \nabla v)_{B_T \setminus \Gamma} + (u,v)_{B_T},$$

F.-J. Sayas, *Retarded Potentials and Time Domain Boundary Integral Equations*, Springer Series in Computational Mathematics 50,
DOI 10.1007/978-3-319-26645-9_10

and a 'master energy space'

$$W := H^1_{\partial B_T}(B_T \setminus \Gamma),$$

which will contain the different choices of V. The operator A will be invariably the distributional Laplacian in $B_T \setminus \Gamma$.

10.1 A review of time domain estimates

10.1.1 The potentials

We gather here the two layer potentials

$$u = \mathcal{S} * \lambda \qquad \text{or} \qquad u = \mathcal{D} * \varphi.$$

The quantities of interest are u and

$$\left[\begin{array}{l} \{\gamma u\} = \mathcal{V} * \lambda, \\ \{\partial_\nu u\} = \mathcal{K}^t * \lambda, \end{array}\right. \qquad \text{or} \qquad \left[\begin{array}{l} \{\gamma u\} = \mathcal{K} * \varphi, \\ \{\partial_\nu u\} = -\mathcal{W} * \varphi. \end{array}\right.$$

The associated transmission conditions are

$$(\text{Nat}) \left[\begin{array}{l} [\![\gamma u]\!] = 0, \\ [\![\partial_\nu u]\!] = \lambda, \end{array}\right. \qquad \text{or} \qquad (\text{Ess}) \left[\begin{array}{l} [\![\gamma u]\!] + \varphi = 0, \\ [\![\partial_\nu u]\!] = 0. \end{array}\right.$$

We have tagged (Nat)/(Ess) to refer to natural/essential transmission conditions. Essential conditions are nonhomogeneous conditions related to the conditions that appear in the definition of the space V or, in other words, conditions related to the trace operator. Natural conditions are related to the normal derivative and are thus nonhomogeneous versions of the conditions appearing in the definition of $D(A)$. The cut-off setting uses the following spaces

$$\begin{aligned} V &:= H^1_0(B_T) = \{v \in W \,:\, [\![\gamma v]\!] = 0\}, \\ D(A) &:= H^1_0(B_T) \cap H^1_\Delta(B_T) = \{v \in V \,:\, \Delta v \in L^2(B_T \setminus \Gamma),\ [\![\partial_\nu v]\!] = 0\}. \end{aligned}$$

The lifting operator in the cut-off setting is either

$$(\text{Nat}) \left[\begin{array}{l} u_0 \in V, \\ a(u_0, v) = \langle \lambda, \gamma v \rangle_\Gamma \quad \forall v \in V, \end{array}\right. \qquad \text{or} \qquad (\text{Ess}) \left[\begin{array}{l} u_0 \in W, \quad [\![\gamma v]\!] + \varphi = 0, \\ a(u_0, v) = 0 \quad \forall v \in V. \end{array}\right.$$

Having to lift essential data always leads to higher smoothness requirements for data. The strong theory in the cut-off setting is only a first step to the final results, where requirements are relaxed using density and shifting arguments, and where the constant C_T in the abstract Poincaré inequality (the bound of the H norm by the V norm) disappears from the bounds. This is important because in many cases this constant is not easy to estimate as a function of T. Here are the requirements and the associated abstract theorem that are used:

$$\lambda \in \mathcal{C}_0^2 \quad \text{(Theorem 6.2.2)} \qquad \text{or} \qquad \varphi \in \mathcal{C}_0^3, \quad \text{(Theorem 6.2.3)}.$$

(There is also a chance of obtaining some estimates with $\varphi \in \mathcal{C}_0^2$ using Theorem 6.2.4, but these results are not used in the final shifting process.) With density and shifting arguments we can prove that for

$$\lambda \in \mathcal{W}_+^1 \qquad \text{or} \qquad \varphi \in \mathcal{W}_+^2$$

we have regularity

$$u \in \mathcal{C}_+^0(L^2(\mathbb{R}^d)) \cap \mathcal{C}_+^1(H^1(\mathbb{R}^d \setminus \Gamma)) \qquad \text{and therefore} \qquad \{\gamma u\} \in \mathcal{C}_+^0$$

and bounds

$$\|u\|_{1,\mathbb{R}^d\setminus\Gamma} \le C H_2(\partial^{-1}\lambda) \qquad \text{or} \qquad \|u\|_{1,\mathbb{R}^d\setminus\Gamma} \le C H_3(\partial^{-1}\varphi).$$

Using only density (and no shifting) yields that for

$$\lambda \in \mathcal{W}_+^2 \qquad \text{or} \qquad \varphi \in \mathcal{W}_+^3$$

we have

$$\{\partial_\nu u\} \in \mathcal{C}_+^0$$

and

$$\|\{\partial_\nu u\}\|_{-1/2,\Gamma} \le C H_2(\lambda) \qquad \text{or} \qquad \|\{\partial_\nu u\}\|_{-1/2,\Gamma} \le C H_3(\varphi).$$

10.1.2 The semidiscrete Dirichlet problem

We now focus on the Galerkin error operator and Galerkin solver corresponding to the integral representation

$$u = \mathcal{S} * \lambda, \qquad \text{with} \qquad \mathcal{V} * \lambda = \beta,$$

and a space X_h. We rename the potential so that it represents the solution of the problem that is actually studied. Therefore, we think of

$$\left[\begin{array}{l} u = \mathcal{S} * (\lambda^h - \lambda), \\ \lambda^h \in X_h, \\ \mathcal{V} * (\lambda^h - \lambda) \in X_h^\circ, \end{array}\right. \qquad \text{or} \qquad \left[\begin{array}{l} u = \mathcal{S} * \lambda^h, \\ \lambda^h \in X_h, \\ \mathcal{V} * \lambda^h - \beta \in X_h^\circ. \end{array}\right.$$

In addition to u, the quantities of interest are

$$\{\partial_\nu u\} = \lambda^h - \lambda \qquad \text{or} \qquad \{\partial_\nu u\} = \lambda^h.$$

The transmission conditions are

$$(\text{Nat}) \left[\begin{array}{l} [\![\gamma u]\!] = 0, \\ \{\gamma u\} \in X_h^\circ, \\ [\![\partial_\nu u]\!] + \lambda \in X_h, \end{array}\right. \qquad \text{or} \qquad (\text{Ess}) \left[\begin{array}{l} [\![\gamma u]\!] = 0, \\ \{\gamma u\} - \beta \in X_h^\circ, \\ [\![\partial_\nu u]\!] \in X_h. \end{array}\right.$$

Of the three conditions, one of them is never made nonhomogeneous. It reflects the fact that we are using a single layer potential representation. Because of the condition $[\![\gamma u]\!] = 0$, we can write $\gamma u = \{\gamma u\}$ in the second transmission condition. The spaces for the cut-off setting are

$$\begin{aligned} V &:= \{v \in W \,:\, [\![\gamma v]\!] = 0, \quad \gamma v \in X_h^\circ\} = \{v \in W_0 \,:\, \gamma v \in X_0\}, \\ D(A) &:= \{v \in V \,:\, \Delta v \in L^2(B_T \setminus \Gamma), \quad [\![\partial_\nu v]\!] \in X_h\}, \end{aligned}$$

where $W_0 := \{w \in W \,:\, [\![\gamma v]\!] = 0\}$. The lifting operators are

$$(\text{Nat}) \left[\begin{array}{l} u_0 \in V, \\ a(u_0, v) = \langle \lambda, \gamma v \rangle_\Gamma \quad \forall v \in V, \end{array}\right. \qquad \text{or} \qquad (\text{Ess}) \left[\begin{array}{l} u_0 \in W_0, \quad \gamma u_0 - \beta \in X_h^\circ \\ a(u_0, v) = 0 \quad \forall v \in V. \end{array}\right.$$

Depending on whether a natural or an essential condition has to be lifted we have different smoothness requirements (and different abstract evolution equation theorems used):

$$\lambda \in \mathcal{C}_0^2 \quad (\text{Theorem 6.2.2}) \qquad \text{or} \qquad \beta \in \mathcal{C}_0^3 \quad (\text{Theorem 6.2.3}).$$

With density and shifting we can prove that for

$$\lambda \in \mathcal{W}_+^1 \qquad \text{or} \qquad \beta \in \mathcal{W}_+^2,$$

we have regularity

$$u \in \mathcal{C}_+^0(L^2(\mathbb{R}^d)) \cap \mathcal{C}_+^1(H^1(\mathbb{R}^d \setminus \Gamma))$$

and bounds

$$\|u\|_{1,\mathbb{R}^d \setminus \Gamma} \le C H_2(\partial^{-1}\lambda) \qquad \text{or} \qquad \|u\|_{1,\mathbb{R}^d \setminus \Gamma} \le C H_3(\partial^{-1}\beta).$$

Using only density and no shifting we keep

$$\lambda \in \mathcal{W}^2_+ \qquad \text{or} \qquad \beta \in \mathcal{W}^3_+$$

to prove

$$\{\partial_\nu u\} \in \mathcal{C}^0_+$$

and bounds

$$\|\{\partial_\nu u\}\|_{-1/2,\Gamma} \le C H_2(\lambda) \qquad \text{or} \qquad \|\{\partial_\nu u\}\|_{-1/2,\Gamma} \le C H_3(\beta).$$

It is interesting to note that the choice $X_h = \{0\}$ leads to $\lambda^h = 0$ and therefore, for the Galerkin error operator we have $u = -\mathcal{S} * \lambda$. This means that the Galerkin error operator estimates include estimates of $\mathcal{S} * \lambda$, $\mathcal{V} * \lambda$, and $\mathcal{K}^t * \lambda$.

10.1.3 The semidiscrete Neumann problem

A double layer representation of the Neumann problem (we change the sign of the Neumann data to unify expressions with the preceding two sections)

$$u = \mathcal{D} * \varphi, \qquad \text{where} \qquad \mathcal{W} * \varphi = \alpha,$$

can be semidiscretized with a Galerkin scheme based on the space Y_h. The potential is now

$$\left[\begin{array}{l} u = \mathcal{D} * (\varphi^h - \varphi), \\ \varphi^h \in Y_h, \\ \mathcal{W} * (\varphi^h - \varphi) \in Y_h^\circ, \end{array}\right. \qquad \text{or} \qquad \left[\begin{array}{l} u = \mathcal{D} * \varphi^h, \\ \varphi^h \in Y_h, \\ \mathcal{W} * \varphi^h - \alpha \in Y_h^\circ. \end{array}\right.$$

In addition to u, the quantities of interest are

$$\{\gamma u\} = \varphi - \varphi^h \qquad \text{or} \qquad \{\gamma_\nu u\} = -\varphi^h.$$

The transmission conditions are

$$(\text{Ess}) \left[\begin{array}{l} [\![\gamma u]\!] + \varphi \in Y_h, \\ [\![\partial_\nu u]\!] = 0, \\ \{\partial_\nu u\} \in Y_h^\circ, \end{array}\right. \qquad \text{or} \qquad (\text{Nat}) \left[\begin{array}{l} [\![\gamma u]\!] \in Y_h, \\ [\![\partial_\nu u]\!] = 0, \\ \{\partial_\nu u\} - \alpha \in Y_h^\circ, \end{array}\right.$$

which means that natural/essential conditions are in reversed position for the error operator and solver of double layer representation of the Neumann problem in comparison with the single layer representation of the Dirichlet problem.

The common homogeneous transmission condition $[\![\partial_\nu u]\!] = 0$ allows to write $\{\!\{\partial_\nu u\}\!\} = \partial_\nu u$ in the third transmission condition of both problems. The spaces for the cut-off setting are

$$V := \{v \in W : [\![\gamma v]\!] \in Y_h\},$$
$$D(A) := \{v \in V : [\![\partial_\nu v]\!] = 0, \quad \partial_\nu v \in Y_h^\circ\},$$

with the novelty of the occurrence of a space of rigid motions

$$M := \operatorname{span}\{\chi_{\Omega_-}\},$$

which will be excited by the solver, producing the purely kinetic motion

$$m(t) := \rho(t)\,(\tfrac{1}{|\Omega_-|}\chi_{\Omega_-}) \qquad \text{where} \qquad \rho(t) := \int_0^t (t-\tau)\langle \alpha(t), 1\rangle_\Gamma \mathrm{d}\tau.$$

The lifting operators are

$$(\text{Ess})\begin{bmatrix} u_0 \in W, & [\![\gamma u_0]\!] + \varphi \in Y_h, \\ a(u_0, v) = 0 & \forall v \in V, \end{bmatrix} \quad \text{or} \quad (\text{Nat})\begin{bmatrix} u_0 \in V, \\ a(u_0, v) = \langle \alpha, [\![\gamma v]\!]\rangle_\Gamma \quad \forall v \in V. \end{bmatrix}$$

In the evolution equation setting, the needed regularity and associated theorem are

$$\varphi \in \mathcal{C}_0^3 \quad \text{(Theorem 6.2.3)} \qquad \text{or} \qquad \alpha \in \mathcal{C}_0^2 \quad \text{(Theorem 6.2.2)}.$$

With density and shifting we can prove that for

$$\varphi \in \mathcal{W}_+^2 \qquad \text{or} \qquad \alpha \in \mathcal{W}_+^1,$$

we have regularity

$$u \in \mathcal{C}_+^0(L^2(\mathbb{R}^d)) \cap \mathcal{C}_+^1(H^1(\mathbb{R}^d \setminus \Gamma)), \quad \text{and therefore} \quad \{\!\{\gamma u\}\!\} \in \mathcal{C}_+^0,$$

with bounds

$$\|u\|_{1,\mathbb{R}^d\setminus\Gamma} \le C\,H_3(\partial^{-1}\varphi) \qquad \text{or} \qquad \|u\|_{1,\mathbb{R}^d\setminus\Gamma} \le C\,(|\rho(t)| + H_2(\partial^{-1}\alpha)).$$

The normal derivative $\{\!\{\partial_\nu u\}\!\}$ is not a quantity of interest in principle. Nevertheless, the reader will be able to figure out the associated bounds and smoothness requirements, and is actually invited to do so. An interesting case to explore is the choice $Y_h = \{0\}$ in the study of the Galerkin error operator. This choice does not satisfy the condition (8.17) (constant functions are elements of Y_h), which is related to the occurrence of rigid motions, but is not relevant in the analysis of the Galerkin error operator.

10.2 Direct formulations

To see if the reader is able to foresee the kind of the results that can be proved for other situations, we invite them to try to guess the results for the following problems and then to go back to the rigorous techniques of Chapters 7 and 8 and fill the gaps.

Exterior Dirichlet problem. For the exterior Dirichlet problem we can use a direct formulation (see Section 1.6)

$$u = \mathcal{D} * \beta - \mathcal{V} * \lambda$$

where

$$\mathcal{V} * \lambda = -\tfrac{1}{2}\beta + \mathcal{K} * \beta.$$

Using a space X_h, Galerkin semidiscretization is equivalent to looking for

$$\lambda^h \in X_h \qquad \text{satisfying} \qquad \mathcal{V} * \lambda^h + \tfrac{1}{2}\beta - \mathcal{K} * \beta \in X_n^\circ.$$

It is clear that the error operator for this discrete problem is the same one as the error operator for the single layer representation of the same problem, since only the right-hand side of the equation changes. We can work in terms of the potential

$$u = \mathcal{D} * \beta - \mathcal{S} * \lambda^h,$$

and note that it satisfies the conditions

$$[\![\gamma u]\!] + \beta = 0, \qquad \gamma^- u \in X_h^\circ, \qquad [\![\partial_\nu u]\!] \in X_h.$$

These conditions seem to lead to a different functional setting than the one of Chapter 7. We can however change some formulas to fit in exactly the same frame. First of all, recall the spaces of Section 10.1.2

$$\begin{aligned} V &:= \{v \in W \,:\, [\![\gamma v]\!] = 0, \quad \gamma v \in X_h^\circ\}, \\ D(A) &:= \{v \in V \,:\, \Delta v \in L^2(B_T \setminus \Gamma), \quad [\![\partial_\nu v]\!] \in X_h\}, \end{aligned}$$

and write instead

$$V = \{v \in W \,:\, [\![\gamma v]\!] = 0, \quad \{\!\{\gamma v\}\!\} \in X_h^\circ\}.$$

Using this idea we rewrite the transmission conditions as

$$[\![\gamma u]\!] + \beta = 0, \qquad \{\!\{\gamma u\}\!\} - \tfrac{1}{2}\beta \in X_h^\circ, \qquad [\![\partial_\nu u]\!] \in X_h.$$

There are two nonhomogeneous essential boundary conditions, leading to the lifting

$$\left[\begin{array}{l} u_0 \in W \quad [\![\gamma u_0]\!] + \beta = 0, \quad \{\gamma u_0\} - \tfrac{1}{2}\beta \in X_h^\circ, \\ a(u_0, v) = 0 \quad \forall v \in V. \end{array}\right.$$

The quantities of interest are u and $[\![\partial_\nu u]\!]$. The reader is invited to take off from here, guess and then prove the bounds and mapping properties.

Exterior Neumann problem. Once this exercise is finished, the reader is also suggested to think of a Galerkin semidiscretization of a direct formulation for the Neumann problem: we represent

$$u = \mathcal{D} * \varphi - \mathcal{S} * \alpha$$

and impose

$$\mathcal{W} * \varphi = \tfrac{1}{2}\alpha - \mathcal{K}^t * \alpha.$$

This equation is semidiscretized using a space Y_h. The space $D(A)$ of Section 10.1.3 has to be expressed in a slightly different way

$$D(A) = \{v \in V \,:\, [\![\partial_\nu v]\!] = 0 \quad \{\partial_\nu v\} \in Y_h^\circ\},$$

but, other than that, the setting of that section is valid for this new problem. Nonhomogeneous transmission conditions are of the natural kind, but there is a space of rigid motions that might be excited by the lifting operator.

10.3 Screens

One unexpected outcome of the generality with which we have been working is the fact that some equations for screens have been analyzed in passing. We have already emphasized that we do not need the spaces X_h and Y_h to be finite dimensional, but only respective closed subspaces of $H^{-1/2}(\Gamma)$ and $H^{1/2}(\Gamma)$. Assume now that we have decomposed

$$\Gamma = \overline{\Gamma}_{\rm scr} \cup \overline{\Gamma}_{\rm fake}, \qquad \Gamma_{\rm scr} \cap \Gamma_{\rm fake} = \emptyset.$$

Here $\Gamma_{\rm scr}$ plays the role of an open screen and $\Gamma_{\rm fake}$ completes this screen to the closed boundary of a Lipschitz domain. In principle, there is no need for $\Gamma_{\rm scr}$ to have a Lipschitz boundary.

Dirichlet screens. We then consider the space

$$\widetilde{H}^{1/2}(\Gamma_{\rm fake}) = \{\varphi \in H^{1/2}(\Gamma) \,:\, \operatorname{supp}\varphi \subset \overline{\Gamma}_{\rm fake}\},$$

which can be seen as the polar set of

$$X_h := \{\lambda \in H^{-1/2}(\Gamma) \, : \, \langle \lambda, \varphi \rangle_\Gamma = 0 \quad \forall \varphi \in \widetilde{H}^{1/2}(\Gamma_{\text{fake}})\}.$$

We can also think of X_h as

$$X_h = \widetilde{H}^{-1/2}(\Gamma_{\text{scr}}) := (H^{1/2}(\Gamma_{\text{scr}}))',$$

where

$$H^{1/2}(\Gamma_{\text{scr}}) = \{\varphi|_{\Gamma_{\text{scr}}} \, : \, \varphi \in H^{1/2}(\Gamma)\}.$$

The problem

$$\lambda \in X_h, \qquad \mathcal{V} * \lambda - \beta \in X_h^\circ, \qquad u = \mathcal{S} * \lambda \tag{10.1}$$

is then a particular case of what we studied in Chapter 7. The condition $\mathcal{V} * \lambda - \beta \in X_h^\circ$ is a weak form of the equality $\mathcal{V} * \lambda = \beta$ on Γ_{scr}. Problem (10.1) corresponds to the scattering problem by a sound-soft screen Γ_{scr}. The same arguments are valid for any choice of a finite dimensional $X_h \subset \widetilde{H}^{-1/2}(\Gamma_{\text{scr}})$, thus providing an analysis of Galerkin semidiscretization of the screen problem.

Neumann screens. We can also choose

$$Y_h := \widetilde{H}^{1/2}(\Gamma_{\text{scr}})$$

and think of the problem

$$\varphi \in Y_h, \qquad \mathcal{W} * \varphi - \alpha \in Y_h^\circ, \qquad u = \mathcal{D} * \varphi,$$

corresponding to the scattering by a sound-hard screen Γ_{scr}. A choice of Y_h as a finite dimensional subspace of $\widetilde{H}^{1/2}(\Gamma_{\text{scr}})$ yields Galerkin semidiscretizations of the screen problem. The continuous and the discrete problem are particular instances of what we studied in Chapter 8, since they follow from particular choices of the space Y_h.

10.4 Linear elasticity

A nice surprise. Linear elastic waves are widely considered to be much more complicated than scalar acoustic waves. There are several reasons for that, but probably the main one is already seen in the form of plane waves for the system of linear elasticity, with two possible combinations of the direction of propagation and the direction of the field that is being propagated. This gives rise to pressure waves

(P-waves or primary waves) and shear waves (S-waves or secondary waves), which travel at different speeds. It is however a fact that the theory we have just presented for acoustic waves can be translated word by word to elastic waves. (The reader who is puzzled about this should also consider how the two- and three-dimensional acoustic waves are treated simultaneously by our theories, even when, physically speaking, they are quite different.)

A list of concepts. We will just expose the main concepts in the Laplace domain. The emphasis is, again, on the fact that each of the results in the Laplace domain can be translated word by word to the elasticity system. The modifications for the time domain results from the acoustic to the elastic case can be considered as an exercise in pure translation (or change of notation). The main unknown for the elasticity problem is a vector field $\mathbf{u} : \Omega \subset \mathbb{R}^d \to \mathbb{R}^d$ representing **displacement**. Strain is given by

$$\boldsymbol{\varepsilon}(\mathbf{u}) := \tfrac{1}{2}(\boldsymbol{D}\mathbf{u} + (\boldsymbol{D}\mathbf{u})^\top), \qquad \text{where} \quad (\boldsymbol{D}\mathbf{u})_{ij} = \frac{\partial u_i}{\partial x_j},$$

and **stress** is given by Hooke's Law

$$\boldsymbol{\sigma}(\mathbf{u}) := 2\mu\boldsymbol{\varepsilon}(\mathbf{u}) + \lambda(\nabla\cdot\mathbf{u})\boldsymbol{I}, \qquad \boldsymbol{I}_{ij} := \delta_{ij},$$

here written in terms of the Lamé constants λ, μ. The Laplace operator is substituted by Navier's operator

$$\Delta^*\mathbf{u} := \operatorname{div}\boldsymbol{\sigma}(\mathbf{u}) = \mu\Delta\mathbf{u} + (\lambda+\mu)\nabla(\nabla\cdot\mathbf{u}).$$

The role of the normal derivative is taken by the normal stress: in an open domain $\mathcal{O}$, this is defined by duality using Betti's formula (Green's formula for the elasticity system)

$$\begin{aligned}\langle \boldsymbol{t}(\mathbf{u}), \gamma\mathbf{v}\rangle_{\partial\mathcal{O}} &:= (\boldsymbol{\sigma}(\mathbf{u}), \boldsymbol{\varepsilon}(\mathbf{v}))_{\mathcal{O}} + (\operatorname{div}\boldsymbol{\sigma}(\mathbf{u}), \mathbf{v})_{\mathcal{O}} \\ &= 2\mu(\boldsymbol{\varepsilon}(\mathbf{u}), \boldsymbol{\varepsilon}(\mathbf{u}))_{\mathcal{O}} + \lambda(\nabla\cdot\mathbf{u}, \nabla\cdot\mathbf{v})_{\mathcal{O}} + (\Delta^*\mathbf{u}, \mathbf{v})_{\mathcal{O}}.\end{aligned}$$

The Laplace domain version of the elastic wave equation

$$\rho s^2\mathbf{u} = \Delta^*\mathbf{u}$$

has associated an energy norm

$$|\!|\!|\mathbf{u}|\!|\!|^2_{|s|,\mathcal{O}} := (\boldsymbol{\sigma}(\mathbf{u}), \boldsymbol{\varepsilon}(\overline{\mathbf{u}}))_{\mathcal{O}} + |s|^2\rho(\mathbf{u}, \overline{\mathbf{u}})_{\mathcal{O}}.$$

The parameter ρ is related to density and, even if it can be easily eliminated from the problem by rescaling s at the very end of the process, it is common to keep it in all the equations. Layer potentials and boundary integral operators are defined through jump conditions for the trace and the normal stress.

10.5 To conclude

This is as far as we are going to get with this program. We have given a detailed introduction to the layer potentials and integral operators for linear acoustics in any space dimension, developing two sets of tools:

- Laplace domain analysis (exploiting the properties of Laplace transforms of the operators and working on resolvent estimates).
- Time domain analysis (using estimates based on the classical theory of C_0-semigroups of operators).

We have been careful in laying a solid ground for the theory using serious (but not difficult) mathematical analysis. The theory for inversion of integral operators is developed at the same time for the continuous and Galerkin-semidiscrete problems. Finally, we have coupled the theory with multistep Convolution Quadrature tools.

Many interesting questions are left unexplored. Let us make here a small list:

- Integral equation methods based on direct formulations that are formally of the second kind. (Note that the theory for equations of the second kind does not follow from what we have studied, and there does not seem to be an easy way to do it.)
- Problems where sound-soft and sound-hard scatterers are combined. Much more challenging are problems where different parts of the boundary of a scatterer have different properties.
- Transmission problems, corresponding to penetrable scatterers with homogeneous properties, and to penetrable scatterers with variable coefficients. The first group can be represented using systems of boundary integral equations, while the second group can be described using boundary-field formulations (the continuous equivalent to BEM-FEM coupling methods).
- Scattering of electromagnetic waves. While the spaces for EM problems are really different, much of the current theory can be rewritten to deal with the layer potentials and integral operators for the Maxwell system.
- Runge-Kutta (or multistage) Convolution Quadrature techniques. (The theory is fully developed in the Laplace domain, but a complete understanding on behavior of constants with respect to time is still missing.)

Some of these problems have already been solved with some of the techniques explained in these notes (mainly the Laplace domain tools) and a few being studied with the time domain arguments.

Appendix A
A taste of Maxwell

In this short chapter we develop the main results associated with the Laplace domain analysis of the single layer potential for Maxwell's equations. These results were initially proved by Isabelle Terrasse in her doctoral thesis [84]. The language of this chapter is, however, closer to the one in [8]. We note that, while time domain techniques might provide better estimates than the ones obtained with Laplace domain bounds, with the exception of Bryan Rynne's article [77], time domain techniques do not seem to be fully developed in the literature of Maxwell retarded potentials. In particular, the adaptation of the techniques of Chapters 7 and 8 is not straightforward.

A.1 Informal introduction

Let $\Omega^- \subset \mathbb{R}^3$ be a bounded Lipschitz domain and let $\Omega^+ := \mathbb{R}^3 \setminus \overline{\Omega^-}$. The electric field on the exterior of a perfect conductor satisfies the curl-curl wave equation

$$c^{-2}\mathbf{u}_{tt} + \nabla \times \nabla \times \mathbf{u} = \mathbf{0} \qquad \text{in } \Omega^+ \times \mathbb{R}, \tag{A.1}$$

with a boundary condition on the surface of the conductor $\Gamma := \partial\Omega^-$

$$(\mathbf{u} + \mathbf{u}^{\mathrm{inc}}) \times \boldsymbol{\nu} = \mathbf{0} \qquad \text{on } \Gamma \times \mathbb{R}. \tag{A.2}$$

We assume that the incident wave has not reached the obstacle at time $t = 0$ and, therefore, we look for a causal vector field $\mathbf{u} : \Omega^+ \times \mathbb{R} \to \mathbb{R}$. One way to represent

F.-J. Sayas, *Retarded Potentials and Time Domain Boundary Integral Equations*, Springer Series in Computational Mathematics 50,
DOI 10.1007/978-3-319-26645-9

the electric field is by using a single layer electromagnetic potential. We are given a **causal and tangential** density $\boldsymbol{\lambda} : \Gamma \times \mathbb{R} \to \mathbb{R}$, that is,

$$\boldsymbol{\lambda}(\cdot, t) \equiv 0 \qquad t < 0, \qquad \boldsymbol{\lambda} \cdot \boldsymbol{\nu} \equiv 0.$$

This density is the input for the potential expression

$$\begin{aligned}\mathbf{u}(\mathbf{z}, t) := &- \int_\Gamma \frac{1}{4\pi|\mathbf{z}-\mathbf{y}|} \partial_t \boldsymbol{\lambda}(\mathbf{y}, t - c^{-1}|\mathbf{z}-\mathbf{y}|) \mathrm{d}\Gamma(\mathbf{y}) \\ &+ c^{-2}\nabla \int_\Gamma \frac{1}{4\pi|\mathbf{z}-\mathbf{y}|} \left(\int_0^{t-c^{-1}|\mathbf{z}-\mathbf{y}|} \nabla_\Gamma \cdot \boldsymbol{\lambda}(\mathbf{y}, \tau) \mathrm{d}\tau \right) \mathrm{d}\Gamma(\mathbf{y}), \qquad \mathbf{z} \notin \Gamma,\end{aligned}$$

where $\nabla_\Gamma \cdot \boldsymbol{\lambda}$ is the tangential divergence of $\boldsymbol{\lambda}$. At this moment, the reader can just assume that Γ is a polyhedral surface and we are taking the divergence of the smooth enough tangential vector field $\boldsymbol{\lambda}$ in tangential coordinates. This potential can be written in terms of the acoustic single layer potential in the following way. Let $\mathcal{S} * \boldsymbol{\xi}$ be the operator arising from applying convolution with the acoustic layer potential $\eta \mapsto \mathcal{S} * \eta$ to the components of the vector field $\boldsymbol{\xi}$. Then

$$\mathbf{u} = \mathcal{S}_{\mathrm{Mxw}} * \boldsymbol{\lambda} = -\mathcal{S} * \partial_t \boldsymbol{\lambda} + c^{-2} \nabla \mathcal{S} * \partial_t^{-1} (\nabla_\Gamma \cdot \boldsymbol{\lambda}).$$

The jump properties of this potential are:

$$\boldsymbol{\nu} \times (\mathbf{u}^- - \mathbf{u}^+) \times \boldsymbol{\nu} = \mathbf{0}, \qquad c(\nabla \times \mathbf{u}^- - \nabla \times \mathbf{u}^+) \times \boldsymbol{\nu} = -\partial_t \boldsymbol{\lambda}.$$

This means that the tangential component of the single layer potential is continuous across the boundary Γ, while its curl jumps proportionally to the time derivative of its density. The value of the single layer potential on the boundary can be formally given using the tangential gradient and the acoustic single layer operators

$$\mathcal{V}_{\mathrm{Mxw}} \times \boldsymbol{\lambda} := \boldsymbol{\nu} \times (\mathcal{S}_{\mathrm{Mxw}} * \boldsymbol{\lambda}) \times \boldsymbol{\nu} = -\boldsymbol{\nu} \times (\mathcal{V} * \partial_t \boldsymbol{\lambda}) \times \boldsymbol{\nu} + c^{-2} \nabla_\Gamma \mathcal{V} * \partial_t^{-1} (\nabla_\Gamma \cdot \boldsymbol{\lambda}).$$

The time domain electric field integral equation (time domain EFIE) looks for a causal tangential density $\boldsymbol{\lambda}$ such that

$$\mathcal{V}_{\mathrm{Mxw}} * \boldsymbol{\lambda} = -\boldsymbol{\nu} \times \mathbf{u}^{\mathrm{inc}} \times \boldsymbol{\nu}. \tag{A.3}$$

It is also common to work with the differentiated EFIE, arising from differentiating (A.3) with respect to, so that we have to solve

$$-\boldsymbol{\nu} \times (\mathcal{V} * \partial_t^2 \boldsymbol{\lambda}) \times \boldsymbol{\nu} + c^{-2} \nabla_\Gamma \mathcal{V} * (\nabla_\Gamma \cdot \boldsymbol{\lambda}) = -\boldsymbol{\nu} \times \partial_t \mathbf{u}^{\mathrm{inc}} \times \boldsymbol{\nu}.$$

A.2 Maxwell traces

A fully rigorous description of the traces for the Maxwell spaces on non-smooth boundaries was a missing link in the PDE literature until quite recently. The seminal papers of Analisa Buffa and Patrick Ciarlet Jr [25, 26] clarified the situation for polyhedral domains. The case of general Lipschitz boundaries followed shortly thereafter [27]. In this section, we will review the basic ingredients that are necessary to introduce the variational (Laplace domain) theory of the Maxwell single layer potential. We want to emphasize that a detailed introduction of all the spaces (even without proofs) is more complicated than one would suspect.

In this section we consider a bounded Lipschitz domain $\Omega \subset \mathbb{R}^3$ with boundary Γ. The space

$$\mathbf{H}(\mathrm{curl},\Omega) := \{\mathbf{u} \in \mathbf{L}^2(\Omega) := L^2(\Omega)^3 \,:\, \nabla\times\mathbf{u} \in \mathbf{L}^2(\Omega)\}$$

is a Hilbert space when endowed with the norm

$$\|\mathbf{u}\|^2_{\mathrm{curl},\Omega} := \|\mathbf{u}\|^2_\Omega + \|\nabla\times\mathbf{u}\|^2_\Omega.$$

For $\mathbf{u} \in \mathbf{H}^1(\Omega) := H^1(\Omega)^3$, we can define the tangential traces:

$$\pi_\tau\mathbf{u} := \boldsymbol{\nu}\times\gamma\mathbf{u}\times\boldsymbol{\nu} = \gamma\mathbf{u} - (\gamma\mathbf{u}\cdot\boldsymbol{\nu})\boldsymbol{\nu},$$
$$\gamma_\tau\mathbf{u} := \gamma\mathbf{u}\times\boldsymbol{\nu}.$$

These traces are rotated ninety degrees around the normal vector field with respect to each other. Both of them take values in the space

$$\mathbf{L}^2_\tau(\Gamma) := \{\boldsymbol{\xi} \in L^2(\Gamma)^3 \,:\, \boldsymbol{\xi}\cdot\boldsymbol{\nu} = 0\}.$$

A simple integration by parts argument shows that

$$(\mathbf{u},\nabla\times\mathbf{v})_\Omega - (\nabla\times\mathbf{u},\mathbf{v})_\Omega = \langle\gamma_\tau\mathbf{u},\pi_\tau\mathbf{v}\rangle_\tau \qquad \forall\mathbf{u},\mathbf{v}\in\mathbf{H}^1(\Omega), \tag{A.4}$$

where the bracket in the right-hand side of (A.4) is the inner product of $\mathbf{L}^2_\tau(\Gamma)$. We can then define the range of π_τ and γ_τ:

$$\mathbf{H}^{1/2}_\parallel(\Gamma) := \{\boldsymbol{\xi} \in \mathbf{L}^2_\tau(\Gamma) \,:\, \boldsymbol{\xi} = \pi_\tau\mathbf{u}, \quad \mathbf{u}\in\mathbf{H}^1(\Omega)\},$$
$$\mathbf{H}^{1/2}_\perp(\Gamma) := \{\boldsymbol{\xi} \in \mathbf{L}^2_\tau(\Gamma) \,:\, \boldsymbol{\xi} = \gamma_\tau\mathbf{u}, \quad \mathbf{u}\in\mathbf{H}^1(\Omega)\}.$$

These spaces are dense in $\mathbf{L}^2_\tau(\Gamma)$. Here is an important difficulty: they are different for non-smooth boundaries. These trace spaces are endowed with the image norms

$$\|\boldsymbol{\xi}\|_{1/2,\parallel,\Gamma} := \inf\{\|\mathbf{u}\|_{1,\Omega} \,:\, \pi_\tau\mathbf{u} = \boldsymbol{\xi}\}, \qquad \|\boldsymbol{\xi}\|_{1/2,\perp,\Gamma} := \inf\{\|\mathbf{u}\|_{1,\Omega} \,:\, \gamma_\tau\mathbf{u} = \boldsymbol{\xi}\}.$$

The next step is the introduction of the dual spaces

$$\mathbf{H}_{\|}^{-1/2}(\Gamma) := \mathbf{H}_{\|}^{1/2}(\Gamma)', \qquad \mathbf{H}_{\perp}^{-1/2}(\Gamma) := \mathbf{H}_{\perp}^{1/2}(\Gamma)',$$

or, more precisely, of the representation of the dual spaces of $\mathbf{H}_{\|}^{1/2}(\Gamma)$ and $\mathbf{H}_{\perp}^{1/2}(\Gamma)$ when we identify $\mathbf{L}^2_\tau(\Gamma) \equiv \mathbf{L}^2_\tau(\Gamma)'$ via the Riesz-Fréchet theorem. These new spaces are equipped with the respective dual norms

$$\|\boldsymbol{\eta}\|_{-1/2,\|,\Gamma} := \sup_{0\neq\boldsymbol{\xi}\in\mathbf{H}_{\|}^{1/2}(\Gamma)} \frac{|\langle\boldsymbol{\eta},\boldsymbol{\xi}\rangle_\tau|}{\|\boldsymbol{\xi}\|_{1/2,\|,\Gamma}}, \qquad \|\boldsymbol{\eta}\|_{-1/2,\perp,\Gamma} := \sup_{0\neq\boldsymbol{\xi}\in\mathbf{H}_{\perp}^{1/2}(\Gamma)} \frac{|\langle\boldsymbol{\eta},\boldsymbol{\xi}\rangle_\tau|}{\|\boldsymbol{\xi}\|_{1/2,\perp,\Gamma}},$$

where the bracket denotes the two possible extensions of the $\mathbf{L}^2_\tau(\Gamma)$ inner product to the duality pairings $\mathbf{H}_{\|}^{-1/2}(\Gamma) \times \mathbf{H}_{\|}^{1/2}(\Gamma)$ and $\mathbf{H}_{\perp}^{-1/2}(\Gamma) \times \mathbf{H}_{\perp}^{1/2}(\Gamma)$. Continuing on the boundary, we need the introduction of tangential differential operators. The key result in this direction is the possibility of extending the tangential divergence and (scalar) curl to weak tangential vector fields. If $H^{3/2}(\Gamma)$ is the range of all traces of elements of $H^2(\Omega)$ and $H^{-3/2}(\Gamma) := H^{3/2}(\Gamma)'$, then

$$\mathrm{div}_\Gamma : \mathbf{H}_{\|}^{-1/2}(\Gamma) \to H^{-3/2}(\Gamma) \qquad \mathrm{curl}_\Gamma : \mathbf{H}_{\perp}^{-1/2}(\Gamma) \to H^{-3/2}(\Gamma)$$

are bounded operators. These operators are actually defined by duality: we first define the tangential gradient and the rotated tangential gradient on $H^{3/2}(\Gamma)$ and then define the tangential divergence and curl by duality. Finally, the two trace spaces for $\mathbf{H}(\mathrm{curl}, \Omega)$ are defined as follows:

$$\mathbf{H}^{-1/2}(\mathrm{div}_\Gamma, \Gamma) := \{\boldsymbol{\eta} \in \mathbf{H}_{\|}^{-1/2}(\Gamma) \; : \; \mathrm{div}_\Gamma \boldsymbol{\eta} \in H^{-1/2}(\Gamma)\},$$
$$\mathbf{H}^{-1/2}(\mathrm{curl}_\Gamma, \Gamma) := \{\boldsymbol{\eta} \in \mathbf{H}_{\perp}^{-1/2}(\Gamma) \; : \; \mathrm{curl}_\Gamma \boldsymbol{\eta} \in H^{-1/2}(\Gamma)\}.$$

These spaces are endowed with the graph norms:

$$\|\boldsymbol{\eta}\|^2_{-1/2,\mathrm{div},\Gamma} := \|\boldsymbol{\eta}\|^2_{-1/2,\|,\Gamma} + \|\mathrm{div}_\Gamma \boldsymbol{\eta}\|^2_{-1/2,\Gamma},$$
$$\|\boldsymbol{\eta}\|^2_{-1/2,\mathrm{curl},\Gamma} := \|\boldsymbol{\eta}\|^2_{-1/2,\perp,\Gamma} + \|\mathrm{curl}_\Gamma \boldsymbol{\eta}\|^2_{-1/2,\Gamma}.$$

We can now state all the key results related to the theory of traces in $\mathbf{H}(\mathrm{curl}, \Omega)$. Their proofs are not obtained sequentially; in some way, all of them are more or less proved at the same time. The first key result is the surprising fact that the spaces $\mathbf{H}^{-1/2}(\mathrm{div}_\Gamma, \Gamma)$ and $\mathbf{H}^{-1/2}(\mathrm{curl}_\Gamma, \Gamma)$ form a dual pairing, namely, there exists a bounded nondegenerate bilinear form

$$\langle \cdot, \cdot \rangle_\tau : \mathbf{H}^{-1/2}(\mathrm{div}_\Gamma, \Gamma) \times \mathbf{H}^{-1/2}(\mathrm{curl}_\Gamma, \Gamma) \to \mathbb{R}$$

that extends the $\mathbf{L}^2_\tau(\Gamma)$ inner product and with respect to which the two spaces are dual, that is, there exist constants such that

$$C_1\|\boldsymbol{\eta}\|_{-1/2,\mathrm{div},\Gamma} \le \sup_{0\neq\boldsymbol{\mu}\in\mathbf{H}^{-1/2}(\mathrm{curl}_\Gamma,\Gamma)} \frac{|\langle\boldsymbol{\eta},\boldsymbol{\mu}\rangle_\tau|}{\|\boldsymbol{\mu}\|_{-1/2,\mathrm{curl},\Gamma}} \le C_2\|\boldsymbol{\eta}\|_{-1/2,\mathrm{div},\Gamma}$$

and

$$C_3\|\boldsymbol{\mu}\|_{-1/2,\mathrm{curl},\Gamma} \le \sup_{0\neq\boldsymbol{\eta}\in\mathbf{H}^{-1/2}(\mathrm{div}_\Gamma,\Gamma)} \frac{|\langle\boldsymbol{\eta},\boldsymbol{\mu}\rangle_\tau|}{\|\boldsymbol{\eta}\|_{-1/2,\mathrm{div},\Gamma}} \le C_4\|\boldsymbol{\mu}\|_{-1/2,\mathrm{curl},\Gamma}.$$

The second key result states that the spaces $\mathbf{H}^{-1/2}(\mathrm{div}_\Gamma,\Gamma)$ and $\mathbf{H}^{-1/2}(\mathrm{curl}_\Gamma,\Gamma)$ are precisely the ranges of the operators π_τ and γ_τ when these operators are extended to $\mathbf{H}(\mathrm{curl},\Omega)$, that is, π_τ and γ_τ can be extended to bounded linear surjective operators

$$\gamma_\tau : \mathbf{H}(\mathrm{curl},\Omega) \to \mathbf{H}^{-1/2}(\mathrm{div}_\Gamma,\Gamma) \qquad \text{and} \qquad \pi_\tau : \mathbf{H}(\mathrm{curl},\Omega) \to \mathbf{H}^{-1/2}(\mathrm{curl}_\Gamma,\Gamma).$$

The final result is the extension of the integration by parts formula (A.4) to $\mathbf{H}(\mathrm{curl},\Omega)$:

$$(\mathbf{u},\nabla\times\mathbf{v})_\Omega - (\nabla\times\mathbf{u},\mathbf{v})_\Omega = \langle\gamma_\tau\mathbf{u},\pi_\tau\mathbf{v}\rangle_\tau \qquad \forall\mathbf{u},\mathbf{v}\in\mathbf{H}(\mathrm{curl},\Omega). \tag{A.5}$$

A.3 Laplace domain estimates

Instead of working slowly from a motivation in the time domain to rigorous definitions in the Laplace domain of all our operators (potential, boundary integral operator, inverse of the boundary integral operator, semidiscrete inverse of the boundary integral operator, and associated Galerkin error operator), we are going to rush into a joint treatment of all these operators in the Laplace domain. When we move back to the time domain, we will separate the operators that are produced from this theory and we identify what they mean. We return to the setting of a closed Lipschitz surface Γ separating an interior domain Ω^- from its exterior Ω^+. We choose a closed subspace

$$X_h \subset \mathbf{H}^{-1/2}(\mathrm{div}_\Gamma,\Gamma)$$

and define its polar set

$$X_h^\circ := \{\boldsymbol{\mu}\in\mathbf{H}^{-1/2}(\mathrm{curl}_\Gamma,\Gamma) \,:\, \langle\boldsymbol{\xi}^h,\boldsymbol{\mu}\rangle_\tau = 0 \quad \forall\boldsymbol{\xi}^h\in X_h\}.$$

Given $\boldsymbol{\beta} \in \mathbf{H}^{-1/2}(\mathrm{curl}_\Gamma, \Gamma)$, $\boldsymbol{\xi} \in \mathbf{H}^{-1/2}(\mathrm{div}_\Gamma, \Gamma)$, and $s \in \mathbb{C}_+$, we look for a solution of

$$\mathbf{u} \in \mathbf{H}(\mathrm{curl}, \mathbb{R}^3) \cap \mathbf{H}(\mathrm{curl}^2, \mathbb{R}^3 \setminus \Gamma), \tag{A.6a}$$

$$\nabla \times \nabla \times \mathbf{u} + s^2 \mathbf{u} = 0 \qquad \text{in } \mathbb{R}^3 \setminus \Gamma, \tag{A.6b}$$

$$\pi_\tau \mathbf{u} - \boldsymbol{\beta} \in \boldsymbol{X}_h^\circ, \tag{A.6c}$$

$$[\![\gamma_\tau(\nabla \times \mathbf{u})]\!] + s\boldsymbol{\xi} \in \boldsymbol{X}_h. \tag{A.6d}$$

Here we have used the space

$$\mathbf{H}(\mathrm{curl}^2, \mathbb{R}^3 \setminus \Gamma) := \{\mathbf{u} \in \mathbf{H}(\mathrm{curl}, \mathbb{R}^3 \setminus \Gamma) \,:\, \nabla \times \nabla \times \mathbf{u} \in \mathbf{L}^2(\mathbb{R}^3 \setminus \Gamma)\}$$

and the fact that

$$\mathbf{H}(\mathrm{curl}, \mathbb{R}^3) = \{\mathbf{u} \in \mathbf{H}(\mathrm{curl}, \mathbb{R}^3 \setminus \Gamma) \,:\, [\![\pi_\tau \mathbf{u}]\!] = \mathbf{0}\}.$$

Introducing the space

$$\mathbf{U}_h := \{\mathbf{v} \in \mathbf{H}(\mathrm{curl}, \Omega) \,:\, \pi_\tau \mathbf{v} \in \boldsymbol{X}_h^\circ\},$$

and the bilinear form

$$a_s(\mathbf{u}, \mathbf{v}) := (\nabla \times \mathbf{u}, \nabla \times \mathbf{v})_{\mathbb{R}^3} + s^2(\mathbf{u}, \mathbf{v})_{\mathbb{R}^3},$$

we can easily give a variational formulation of (A.6):

$$\left[\begin{array}{ll} \mathbf{u} \in \mathbf{H}(\mathrm{curl}, \Omega), & \pi_\tau \mathbf{u} - \boldsymbol{\beta} \in \boldsymbol{X}_h^\circ, \\ a_s(\mathbf{u}, \mathbf{v}) = -s\langle \boldsymbol{\xi}, \pi_\tau \mathbf{v}\rangle_\tau & \forall \mathbf{v} \in \mathbf{U}_h. \end{array}\right. \tag{A.7}$$

This is step two of our usual four-step program in the Laplace domain (see Chapters 3 and 5). The main difference between the Maxwell and the acoustic cases lies in the fact that there is no optimal lifting operator for the tangential trace, that is, we cannot give a result like Proposition 2.5.1 for the lifting of π_τ from $\mathbf{H}^{-1/2}(\mathrm{curl}_\Gamma, \Gamma)$ to $\mathbf{H}(\mathrm{curl}, \mathbb{R}^3)$. Instead, we will use a lifting that is independent of $|s|$. Otherwise, the analysis of (A.7) is standard. We use the energy norm

$$|\!|\!|\mathbf{u}|\!|\!|_{|s|}^2 := \|\nabla \times \mathbf{u}\|_{\mathbb{R}^3}^2 + \|s\,\mathbf{u}\|_{\mathbb{R}^3}^2,$$

and notice that

$$|a_s(\mathbf{u}, \mathbf{v})| \le |\!|\!|\mathbf{u}|\!|\!|_{|s|} |\!|\!|\mathbf{v}|\!|\!|_{|s|}, \tag{A.8a}$$

$$\mathrm{Re}(\overline{s}\, a_s(\mathbf{u}, \overline{\mathbf{u}})) = \sigma |\!|\!|\mathbf{u}|\!|\!|_{|s|}^2, \tag{A.8b}$$

$$\underline{\sigma} \|\mathbf{u}\|_{\mathrm{curl}, \mathbb{R}^3} \le |\!|\!|\mathbf{u}|\!|\!|_{|s|} \le \frac{|s|}{\underline{\sigma}} \|\mathbf{u}\|_{\mathrm{curl}, \mathbb{R}^3} \tag{A.8c}$$

for all $\mathbf{u}, \mathbf{v} \in \mathbf{H}(\mathrm{curl}, \mathbb{R}^3)$ and $s \in \mathbb{C}_+$. From this moment on, we will use the symbol $\lesssim$ for inequalities when we want to hide a generic constant $C_\Gamma > 0$ which only depends on Γ (and is, therefore, independent of $s \in \mathbb{C}_+$ and the choice of $\boldsymbol{X}_h$ and, in the next section, of the time variable). Well-posedness of (A.7) and a first bound for the solution are given in the next proposition.

Proposition A.3.1. *Problems* (A.6) *and* (A.7) *are uniquely solvable for all* $s \in \mathbb{C}_+$. *Moreover,*

$$\|\mathbf{u}\|_{\mathrm{curl},\mathbb{R}^3} \lesssim \frac{|s|^2}{\sigma \underline{\sigma}^2} \left(\|\boldsymbol{\beta}\|_{-1/2,\mathrm{curl},\Gamma} + \|\boldsymbol{\xi}\|_{-1/2,\mathrm{div},\Gamma} \right).$$

Proof. Existence and uniqueness of solution follow from a coercivity argument that is implied in the next estimates. We decompose $\mathbf{u} = \mathbf{u}_\beta + \mathbf{u}_0$, where

$$\pi_\tau \mathbf{u}_\beta = \boldsymbol{\beta}, \qquad \|\mathbf{u}_\beta\|_{\mathrm{curl},\mathbb{R}^3} \lesssim \|\boldsymbol{\beta}\|_{-1/2,\mathrm{curl},\Gamma}$$

and $\mathbf{u}_0 \in \mathbf{U}_h$. We then test equations (A.7) with $\mathbf{v} = \overline{\mathbf{u}_0}$ and use (A.8) to bound

$$\begin{aligned}
\sigma |\!|\!| \mathbf{u}_0 |\!|\!|_{|s|}^2 &= \mathrm{Re}(\overline{s}\, a_s(\mathbf{u}_0, \overline{\mathbf{u}_0})) \\
&= \mathrm{Re}\left(-|s|^2 \langle \boldsymbol{\xi}, \pi_\tau \mathbf{u}_0 \rangle_\tau - \overline{s}\, a_s(\mathbf{u}_\beta, \overline{\mathbf{u}_0}) \right) \\
&\lesssim |s|^2 \|\boldsymbol{\xi}\|_{-1/2,\mathrm{div},\Gamma} \|\mathbf{u}_0\|_{\mathrm{curl},\mathbb{R}^3} + \frac{|s|^2}{\underline{\sigma}} \|\boldsymbol{\beta}\|_{-1/2,\mathrm{curl},\Gamma} |\!|\!| \mathbf{u}_0 |\!|\!|_{|s|} \\
&\lesssim \frac{|s|^2}{\underline{\sigma}} \left(\|\boldsymbol{\xi}\|_{-1/2,\mathrm{div},\Gamma} + \|\boldsymbol{\beta}\|_{-1/2,\mathrm{curl},\Gamma} \right) |\!|\!| \mathbf{u}_0 |\!|\!|_{|s|}.
\end{aligned}$$

Therefore

$$\|\mathbf{u}_0\|_{\mathrm{curl},\mathbb{R}^3} \lesssim \frac{|s|^2}{\sigma \underline{\sigma}^2} \left(\|\boldsymbol{\xi}\|_{-1/2,\mathrm{div},\Gamma} + \|\boldsymbol{\beta}\|_{-1/2,\mathrm{curl},\Gamma} \right),$$

from which the result follows readily. □

The jump of the tangential trace of the curl is estimated in the next result.

Proposition A.3.2. *If* $\mathbf{u}$ *is the solution of* (A.6)–(A.7), *then*

$$\|s^{-1} [\![\gamma_\tau (\nabla \times \mathbf{u})]\!] \|_{-1/2,\mathrm{div},\Gamma} \lesssim \frac{|s|^2}{\sigma \underline{\sigma}^3} \left(\|\boldsymbol{\beta}\|_{-1/2,\mathrm{curl},\Gamma} + \|\boldsymbol{\xi}\|_{-1/2,\mathrm{div},\Gamma} \right).$$

Proof. This result is a simple consequence of Proposition A.3.1 and of the fact that $\nabla \times \nabla \times \mathbf{u} = -s^2 \mathbf{u}$, as can be seen from the next chain of inequalities:

$$\begin{aligned}
\| [\![\gamma_\tau (\nabla \times \mathbf{u})]\!] \|_{-1/2,\mathrm{div},\Gamma} &\lesssim \|\nabla \times \mathbf{u}\|_{\mathbb{R}^3} + \|\nabla \times \nabla \times \mathbf{u}\|_{\mathbb{R}^3 \setminus \Gamma} \\
&\lesssim \max\{1, |s|\} |\!|\!| \mathbf{u} |\!|\!|_{|s|} \\
&\lesssim \frac{|s|^2}{\sigma \underline{\sigma}^2} \frac{|s|}{\underline{\sigma}} \left(\|\boldsymbol{\beta}\|_{-1/2,\mathrm{curl},\Gamma} + \|\boldsymbol{\xi}\|_{-1/2,\mathrm{div},\Gamma} \right).
\end{aligned}$$

To wrap up notation, we can consider two operators

$$\mathbf{G}_h^u(s) : \mathbf{H}^{-1/2}(\mathrm{curl}_\Gamma, \Gamma) \to \mathbf{H}(\mathrm{curl}, \mathbb{R}^3) \cap \mathbf{H}(\mathrm{curl}^2, \mathbb{R}^3 \setminus \Gamma),$$
$$\mathbf{E}_h^u(s) : \mathbf{H}^{-1/2}(\mathrm{div}_\Gamma, \Gamma) \to \mathbf{H}(\mathrm{curl}, \mathbb{R}^3) \cap \mathbf{H}(\mathrm{curl}^2, \mathbb{R}^3 \setminus \Gamma),$$

representing the solution operator to problem (A.6) in the form

$$\mathbf{u} = \mathbf{G}_h^u(s)\boldsymbol{\beta} + \mathbf{E}_h^u(s)\boldsymbol{\xi}.$$

We also consider the operators

$$s^{-1}[\![\gamma_\tau(\nabla \times \cdot)]\!]\mathbf{G}_h^u(s) \qquad s^{-1}[\![\gamma_\tau(\nabla \times \cdot)]\!]\mathbf{E}_h^u(s).$$

Propositions A.3.1 and A.3.2 are obviously bounds on the norms of these operators. In the next section we will give definitions of convolution operators associated with these Laplace domain transfer functions for different incarnations of the space $\boldsymbol{X}_h$.

A.4 Bounds in the time domain

For the sake of simplicity, we define the space

$$\mathbf{M} := \mathbf{H}(\mathrm{curl}, \mathbb{R}^3) \cap \mathbf{H}(\mathrm{curl}^2, \mathbb{R}^3 \setminus \Gamma),$$

where all the potentials defined in this section will take values. We are going to take $c = 1$ in all the expressions. A simple change of $s \mapsto s/c$ in the Laplace domain leads to the change of speed in the Maxwell equation to an arbitrary $c > 0$.

The electromagnetic single layer potential. Given $\boldsymbol{\xi} \in \mathrm{TD}(\mathbf{H}^{-1/2}(\mathrm{div}_\Gamma, \Gamma))$ we look for $\mathbf{u} \in \mathrm{TD}(\mathbf{M})$ such that

$$\ddot{\mathbf{u}} + \nabla \times \nabla \times \mathbf{u} = \mathbf{0}, \qquad [\![\gamma_\tau(\nabla \times \mathbf{u})]\!] + \dot{\boldsymbol{\xi}} = 0. \tag{A.9}$$

We can write

$$\mathbf{u} = \mathcal{S}_{\mathrm{Mxw}} * \boldsymbol{\xi},$$

where $\mathcal{L}\{\mathcal{S}_{\mathrm{Mxw}}\} = \mathbf{S}_{\mathrm{Mxw}}(s) := \mathbf{E}_h^u(s)$ for the particular choice $\boldsymbol{X}_h = \{0\}$. Proposition A.3.1 shows that the transient single layer potential is well defined and

$$\|\mathbf{S}_{\mathrm{Mxw}}(s)\|_{\mathbf{H}^{-1/2}(\mathrm{div}_\Gamma,\Gamma)\to\mathbf{H}(\mathrm{curl},\mathbb{R}^3)} \lesssim \frac{|s|^2}{\sigma\underline{\sigma}^2}.$$

The inversion theorem (Proposition 3.2.2) proves then the estimate

$$\|(\mathcal{S}_{\mathrm{Mxw}} * \boldsymbol{\xi})(t)\|_{\mathrm{curl},\mathbb{R}^3} \lesssim \frac{t^2}{1+t}\max\{1,t^2\}\int_0^t \|(\mathcal{P}_4\boldsymbol{\xi})(\tau)\|_{-1/2,\mathrm{div},\Gamma}\mathrm{d}\tau.$$

We can also define the single layer operator (the forward operator in the time domain EFIE) as the convolution operator with

$$\mathcal{V}_{\mathrm{Mxw}} := \pi_\tau \mathcal{S}_{\mathrm{Mxw}} = \mathcal{L}^{-1}\{\pi_\tau \mathbf{S}_{\mathrm{Mxw}}\},$$

and we can bound

$$\|(\mathcal{V}_{\mathrm{Mxw}} * \boldsymbol{\xi})(t)\|_{-1/2,\mathrm{curl},\Gamma} \lesssim \frac{t^2}{1+t}\max\{1,t^2\}\int_0^t \|(\mathcal{P}_4\boldsymbol{\xi})(\tau)\|_{-1/2,\mathrm{div},\Gamma}\mathrm{d}\tau.$$

The solution of the time domain EFIE. We are now given $\boldsymbol{\beta} \in \mathrm{TD}(\mathbf{H}^{-1/2}(\mathrm{curl}_\Gamma,\Gamma))$ and we look for a solution $\boldsymbol{\xi} \in \mathrm{TD}(\mathbf{H}^{-1/2}(\mathrm{div}_\Gamma,\Gamma))$ of

$$\mathcal{V}_{\mathrm{Mxw}} * \boldsymbol{\xi} = \boldsymbol{\beta}, \tag{A.10a}$$

which is then postprocessed to a potential

$$\mathbf{u} = \mathcal{S}_{\mathrm{Mxw}} * \boldsymbol{\xi}. \tag{A.10b}$$

We can look at (A.10) from the point of view of what $\mathbf{u}$ satisfies, namely, $\mathbf{u} \in \mathrm{TD}(\mathbf{M})$ is a solution of

$$\ddot{\mathbf{u}} + \nabla\times\nabla\times\mathbf{u} = 0, \qquad \pi_\tau\mathbf{u} - \boldsymbol{\beta} = \mathbf{0}, \tag{A.11a}$$

and

$$\boldsymbol{\xi} = -\partial^{-1}[\![\gamma_\tau(\nabla\times\mathbf{u})]\!]. \tag{A.11b}$$

Taking $\boldsymbol{X}_h = \mathbf{H}^{-1/2}(\mathrm{div}_\Gamma,\Gamma)$ in the previous section, we can understand the solution operator to (A.10) or (A.11) as the convolution operators

$$\mathbf{u} = \mathcal{G}_h^u * \boldsymbol{\beta}, \qquad \boldsymbol{\xi} = \partial^{-1}[\![\gamma_\tau(\nabla\cdot\times)]\!]\mathcal{G}_h^u * \boldsymbol{\beta},$$

where $\mathcal{L}\{\mathcal{G}_h^u\} = \mathbf{G}_h^u$, or, in a language closer to (A.10),

$$\mathbf{u} = \mathcal{S}_{\mathrm{Mxw}} * \mathcal{V}_{\mathrm{Mxw}}^{-1} * \boldsymbol{\beta}, \qquad \boldsymbol{\xi} = \mathcal{V}_{\mathrm{Mxw}}^{-1} * \boldsymbol{\beta}.$$

Using Proposition 3.2.2 and the estimates of Propositions A.3.1–A.3.2, we obtain the bounds

$$\|(\mathcal{V}_{\mathrm{Mxw}}^{-1} * \boldsymbol{\beta})(t)\|_{-1/2,\mathrm{div},\Gamma} \lesssim \frac{t^2}{1+t}\max\{1,t^3\}\int_0^t \|(\mathcal{P}_4\boldsymbol{\beta})(\tau)\|_{-1/2,\mathrm{curl},\Gamma}\mathrm{d}\tau$$

and

$$\|(\mathcal{S}_{\mathrm{Mxw}} * \mathcal{V}_{\mathrm{Mxw}}^{-1} * \boldsymbol{\beta})(t)\|_{\mathrm{curl},\mathbb{R}^3} \lesssim \frac{t^2}{1+t}\max\{1,t^2\}\int_0^t \|(\mathcal{P}_4\boldsymbol{\beta})(\tau)\|_{-1/2,\mathrm{curl},\Gamma}\mathrm{d}\tau.$$

A semidiscrete solution of the time domain EFIE. The input is again $\boldsymbol{\beta} \in \mathrm{TD}(\mathbf{H}^{-1/2}(\mathrm{curl}_\Gamma,\Gamma))$ and we look for $\boldsymbol{\xi}^h \in \mathrm{TD}(\mathbf{H}^{-1/2}(\mathrm{div}_\Gamma,\Gamma))$ satisfying the Galerkin semidiscrete equations

$$\boldsymbol{\xi}^h \in \boldsymbol{X}_h \qquad \mathcal{V}_{\mathrm{Mxw}} * \boldsymbol{\xi}^h - \boldsymbol{\beta} \in \boldsymbol{X}_h^\circ. \tag{A.12a}$$

We then postprocess the density to a potential

$$\mathbf{u}^h = \mathcal{S}_{\mathrm{Mxw}} * \boldsymbol{\xi}^h. \tag{A.12b}$$

In this case $\boldsymbol{X}_h$ is a finite element subspace of $\mathbf{H}^{-1/2}(\mathrm{div}_\Gamma,\Gamma)$. Once again, we look at $\mathbf{u}^h \in \mathrm{TD}(\mathbf{M})$ as the solution of

$$\ddot{\mathbf{u}}^h+\nabla\times\nabla\times\mathbf{u}^h = 0, \qquad \pi_\tau\mathbf{u}^h-\boldsymbol{\beta} \in \boldsymbol{X}_h^\circ, \qquad [\![\gamma_\tau(\nabla\times\mathbf{u}^h)]\!] \in \boldsymbol{X}_h, \tag{A.13a}$$

and

$$\boldsymbol{\xi}^h = -\partial^{-1}[\![\gamma_\tau(\nabla\times\mathbf{u}^h)]\!]. \tag{A.13b}$$

We can again understand the operator $\boldsymbol{\beta} \mapsto \mathbf{u}$ as the convolution with $\mathcal{G}_h^u = \mathcal{L}^{-1}\{\mathbf{G}_h^u\}$. The inversion of the Laplace transform yields the bounds

$$\|\boldsymbol{\xi}^h(t)\|_{-1/2,\mathrm{div},\Gamma} \lesssim \frac{t^2}{1+t}\max\{1,t^3\}\int_0^t \|(\mathcal{P}_4\boldsymbol{\beta})(\tau)\|_{-1/2,\mathrm{curl},\Gamma}\mathrm{d}\tau$$

and

$$\|\mathbf{u}^h(t)\|_{\mathrm{curl},\mathbb{R}^3} \lesssim \frac{t^2}{1+t}\max\{1,t^2\}\int_0^t \|(\mathcal{P}_4\boldsymbol{\beta})(\tau)\|_{-1/2,\mathrm{curl},\Gamma}\mathrm{d}\tau.$$

The semidiscretization error. We can finally compare the solutions of (A.10) and (A.12). The errors

$$\mathbf{e}^h := \mathbf{u}-\mathbf{u}^h, \qquad \boldsymbol{\varepsilon}^h := \boldsymbol{\xi}-\boldsymbol{\xi}^h = -\partial^{-1}[\![\gamma_\tau(\nabla\times\mathbf{e}^h)]\!],$$

can be interpreted as the solution $\mathbf{e}^h \in \mathrm{TD}(\mathbf{M})$ of the evolution equation

$$\ddot{\mathbf{e}}^h + \nabla \times \nabla \times \mathbf{e}^h, \qquad \pi_\tau \mathbf{e}^h \in \boldsymbol{X}_h^\circ, \qquad [\![\gamma_\tau(\nabla \times \mathbf{e}^h)]\!] + \dot{\boldsymbol{\xi}} \in \boldsymbol{X}_h, \tag{A.14a}$$

followed by the computation of

$$\boldsymbol{\varepsilon}^h = -\partial^{-1}[\![\gamma_\tau(\nabla \times \mathbf{e}^h)]\!]. \tag{A.14b}$$

This means that

$$\mathbf{e}^h = \mathcal{E}_h^u * \boldsymbol{\xi}, \qquad \boldsymbol{\xi}^h = -\partial^{-1}[\![\gamma_\tau(\nabla \times \cdot)]\!]\mathcal{E}_h^u * \boldsymbol{\xi},$$

where $\mathcal{L}\{\mathcal{E}_h^u\} = \mathbf{E}_h^u$. Therefore

$$\|\boldsymbol{\xi}(t) - \boldsymbol{\xi}^h(t)\|_{-1/2,\mathrm{div},\Gamma} \lesssim \frac{t^2}{1+t}\max\{1,t^3\}\int_0^t \|(\mathcal{P}_4\boldsymbol{\xi} - \boldsymbol{\Pi}_h\mathcal{P}_4\boldsymbol{\xi})(\tau)\|_{-1/2,\mathrm{div},\Gamma}\mathrm{d}\tau$$

and

$$\|\mathbf{u}(t) - \mathbf{u}^h(t)\|_{\mathrm{curl},\mathbb{R}^3} \lesssim \frac{t^2}{1+t}\max\{1,t^2\}\int_0^t \|(\mathcal{P}_4\boldsymbol{\xi} - \boldsymbol{\Pi}_h\mathcal{P}_4\boldsymbol{\xi})(\tau)\|_{-1/2,\mathrm{div},\Gamma}\mathrm{d}\tau,$$

where $\boldsymbol{\Pi}_h : \mathbf{H}^{-1/2}(\mathrm{div}_\Gamma, \Gamma) \to \boldsymbol{X}_h$ is the orthogonal projection onto the discrete space $\boldsymbol{X}_h$. Note that the introduction to this projection follows from an argument we have met before: by uniqueness of solution of (A.14), $\mathcal{E}_h^u * \boldsymbol{\Pi}_h\boldsymbol{\xi} = \mathbf{0}$.

Final remark. The bounds of this chapter are likely to be suboptimal. A time domain analysis along the lines of Chapters 7 and 8 might produce better estimates. At the time of finishing this monograph, this is still being researched.

Appendix B
Evolution equations by separation of variables

In this chapter we prove the main theorems of Chapter 6 using simple arguments based on an abstract view of the method of separation of variables. The main tools for the analysis are the Hilbert-Schmidt theorem and some basic results on continuity of functions defined by series. They will be reviewed in the last section of this chapter for ease of reference. These results are adapted and expanded from [43, Section 8] and [81, Section 3]. This chapter uses the notation of Chapter 6, which will not be repeated here.

B.1 Green's operator

Definition of the abstract Green operator. We return to the abstract setting of Section 6.1 and define the operator $G : H_0 \to H_0$ given by

$$Gf = u \qquad \text{where} \qquad \left[\begin{array}{l} u \in V_0, \\ [u, v] = (f, v)_H \quad \forall v \in V \quad \big(\Leftrightarrow \ \forall v \in V_0 \big). \end{array}\right.$$

This operator is well defined because it is just the Riesz-Fréchet representation in V_0 of the bounded linear functional $(f, \cdot)_H : V_0 \to \mathbb{R}$. Note that if $u \in V_0 \cap D(A)$ and $-Au = f$, then by the abstract Green Identity

$$0 = (Au, v)_H + [u, v] = -(f, v)_H + [u, v] \qquad \forall v \in V,$$

and therefore $u = Gf$. This means that G provides us a solution of the equation $-Au = f$. The following properties are simple consequences of the hypotheses for the spaces and the operators.

F.-J. Sayas, *Retarded Potentials and Time Domain Boundary Integral Equations*, Springer Series in Computational Mathematics 50,
DOI 10.1007/978-3-319-26645-9

Proposition B.1.1 (Functional properties of Green's operator). *Assuming all the hypotheses of Section 6.1, it follows that:*

(a) *With the constant* $C_\circ > 0$ *of inequality* (6.1)

$$\|Gf\|_V \leq C_\circ \|f\|_H \qquad \forall f \in H_0. \tag{B.1}$$

(b) *The operator* G *is compact.*
(c) *The operator* G *is self-adjoint and positive definite.*

Proof. To prove (a) we just need to notice that since $u = Gf \in V_0$, then

$$\|u\|_V^2 = [u, u] = (f, u)_H \leq \|f\|_H \|u\|_H \leq C_\circ \|f\|_H \|u\|_V.$$

Since the injection of V_0 in H_0 is compact, the bound (B.1) proves the compactness of $G : H_0 \to H_0$. The proof of (c) is a direct consequence of the definition:

$$(f, Gg)_H = [Gf, Gg] = [Gg, Gf] = (g, Gf)_H.$$

From this equality it also follows that G is positive semidefinite. Positive definiteness is then reduced to showing that G is injective: if $Gf = 0$ then

$$0 = [Gf, v] = (f, v)_H = 0 \qquad \forall v \in V,$$

but since V is dense in H it follows that $f = 0$. □

The associated spectral decomposition. Proposition B.1.1 leaves us in place to apply the well-known Hilbert-Schmidt decomposition theorem (its precise statement is given in Theorem B.5.1 below); we can find a Hilbert basis of H_0, $\{\phi_n\}$, and a sequence of positive nonincreasing eigenvalues $\lambda_n > 0$ converging to zero, and write

$$G = \sum_{n=1}^{\infty} \lambda_n (\,\cdot\,, \phi_n)_H \phi_n. \tag{B.2}$$

Every $u \in H$ can be expressed as an H-orthogonal sum.

$$u = Pu + \sum_{n=1}^{\infty} (u, \phi_n)_H \phi_n. \tag{B.3}$$

Note as a first observation that, since $G\phi_n = \lambda_n \phi_n$, then

$$[\phi_n, \phi_m] = \lambda_n^{-1} [G\phi_n, \phi_m] = \lambda_n^{-1} (\phi_n, \phi_m)_H = \lambda_n^{-1} \delta_{n,m} \qquad \forall n, m \tag{B.4}$$

and that therefore

$$\lambda_n = \frac{(\phi_n, \phi_n)_H}{[\phi_n, \phi_n]} = \frac{\|\phi_n\|_H^2}{\|\phi_n\|_V^2} \leq C_\circ^2, \tag{B.5}$$

following (6.1). We now move on to characterizing the spaces V and $D(A)$ and the operator A in terms of the elements (eigenvalues and eigenfunctions) that appear in the series representation (B.2) of Green's operator.

Proposition B.1.2 (Series characterization of the energy space). *Assuming the hypotheses of Section 6.1 and with the spectral decomposition of G given in* (B.2) *it follows that:*

(a) $\{\lambda_n^{1/2}\phi_n\}$ *is a Hilbert basis of* V_0.
(b) *The seminorm of* V *(the norm of* V_0*) can be represented in series form as*

$$[u, u] = \sum_{n=1}^{\infty} \lambda_n^{-1} |(u, \phi_n)_H|^2 \qquad \forall u \in V.$$

Therefore

$$\|u\|_V^2 = \sum_{n=1}^{\infty} \lambda_n^{-1} |(u, \phi_n)_H|^2 + \|Pu\|_H^2 \qquad \forall u \in V.$$

(c) $V_0 = R(G^{1/2})$, *where*

$$G^{1/2} = \sum_{n=1}^{\infty} \lambda_n^{1/2} (\,\cdot\,, \phi_n)_H \phi_n.$$

Consequently, $V = M \oplus R(G^{1/2})$.

Proof. The identities (B.4) prove that $\lambda_n^{1/2}\phi_n$ is an orthonormal set in V_0. Since

$$[u, \phi_n] = \lambda_n^{-1} (\phi_n, u)_H \qquad \forall n \quad \forall u \in V_0, \tag{B.6}$$

it follows that $\{\phi_n\}$ is complete in V_0 and (a) is thus proved. In turn, this shows that

$$[u, u] = \sum_{n=1}^{\infty} \left|[u, \lambda_n^{1/2}\phi_n]\right|^2 = \sum_{n=1}^{\infty} \lambda_n \left|[u, \phi_n]\right|^2 = \sum_{n=1}^{\infty} \lambda_n^{-1} |(u, \phi_n)_H|^2 \quad \forall u \in V_0$$

(we have applied (B.6) again), from which (b) follows readily. Finally, by Picard's criterion (whose statement the reader will find in Theorem B.5.2) applied to the compact self-adjoint positive definite operator $G^{1/2}$ (see Section B.5) it follows that

$$R(G^{1/2}) = \{u \in H_0 : \sum_{n=1}^{\infty} \lambda_n^{-1} |(u, \phi_n)_H|^2 < \infty\}.$$

Therefore $V_0 \subset R(G^{1/2})$ and $[u, u]^{1/2} = \|u\|_V = \|u\|_{R(G^{1/2})}$ for all $u \in V_0$. However, the linear combinations of the eigenfunctions $\{\phi_n\}$ form a dense subset of V_0 that is also dense in $R(G^{1/2})$, which implies the equality of both sets. □

Proposition B.1.3 (Series characterization of $D(A)$). *Assuming the hypotheses of Section 6.1 and with the spectral decomposition of G given in* (B.2) *it follows that*

$$R(G) = \{u \in H_0 : \sum_{n=1}^{\infty} \lambda_n^{-2} |(u, \phi_n)_H|^2 < \infty\} = D(A) \cap H_0 \tag{B.7}$$

and

$$\sum_{n=1}^{\infty} \lambda_n^{-2} |(u, \phi_n)_H|^2 + \|Pu\|_H^2 \leq \|u\|_{D(A)}^2$$

$$\leq (1 + C_\circ^2) \sum_{n=1}^{\infty} \lambda_n^{-2} |(u, \phi_n)_H|^2 + \|Pu\|_H^2 \quad \forall u \in D(A).$$

Proof. We first define $B : R(G) \to H_0$ ($B = -G^{-1}$) by

$$Bu := -\sum_{n=1}^{\infty} \lambda_n^{-1} (u, \phi_n)_H \phi_n.$$

If $u \in D(A) \cap H_0 \subset V_0$, then by Green's Identity (6.2)

$$[u, v] = -(Au, v)_H \qquad \forall v \in V,$$

and therefore, by the definition of G, it follows that $u = -GAu \in R(G)$. This also implies that

$$\begin{aligned}
(u, \phi_n)_H &= -(GAu, \phi_h)_H \\
&= -(Au, G\phi_n)_H \qquad (G \text{ is self-adjoint}) \\
&= -\lambda_n (Au, \phi_n)_H, \qquad (G\phi_n = \lambda_n \phi_n)
\end{aligned}$$

and therefore

$$
\begin{aligned}
Bu &= -\sum_{n=1}^{\infty} \lambda_n^{-1} (u, \phi_n)_H \phi_n \\
&= \sum_{n=1}^{\infty} (Au, \phi_n)_H \phi_n \\
&= Au. \qquad\qquad (Au \in H_0 \text{ by Proposition 6.1.1(b)})
\end{aligned}
$$

We have so far proved that $D(A) \cap H_0 \subset R(G)$ and that $Au = Bu$ for all $u \in D(A) \cap H_0$.

From Proposition 6.1.1(e) we already know that $A - I : D(A) \cap H_0 \to H_0$ is a bijection. Using the characterization of $R(G)$ given by Picard's criterion (Theorem B.5.2) we can easily show that $I - B : R(G) \to H_0$ is also a bijection. However, $(I - B)|_{D(A)\cap H_0} = I - A$, and therefore $D(A) \cap H_0 = R(G)$ and

$$
Au = -\sum_{n=1}^{\infty} \lambda_n^{-1} (u, \phi_n)_H \phi_n \qquad \forall u \in D(A). \tag{B.8}
$$

(Note that the expression (B.8) includes the fact that $Au = 0$ for $u \in M$.) Moreover, if $u \in D(A) \cap H_0 = R(G)$, then

$$
\|u\|_{D(A)}^2 = \sum_{n=1}^{\infty} \lambda_n^{-2} |(u, \phi_n)_H|^2 + [u, u] = \sum_{n=1}^{\infty} (\lambda_n^{-2} + \lambda_n^{-1}) |(u, \phi_n)_H|^2, \tag{B.9}
$$

by Proposition B.1.2(b), and therefore

$$
\|u\|_{D(A)}^2 = \sum_{n=1}^{\infty} (\lambda_n^{-2} + \lambda_n^{-1}) |(u, \phi_n)_H|^2 + \|Pu\|_H^2 \quad \forall u \in D(A). \tag{B.10}
$$

The final bound is a simple consequence of (B.10) and (B.5). □

Final remark. As a direct consequence of (B.8) it is clear that

$$
-A\phi_n = \lambda_n^{-1} \phi_n \qquad \forall n.
$$

Moreover, other than the reciprocals of the eigenvalues of the Green operator, the only possible eigenvalue of A is $\lambda = 0$ with the elements of the space of rigid motions M as eigenfunctions. Therefore, the eigenvalues of the (unbounded) operator $-A$ diverge to infinity.

B.2 The Cauchy problem

Abstract wave equations. We still assume all the hypotheses of Section 6.1 and consider the problem

$$\begin{bmatrix} \ddot{u}(t) = Au(t) & t \in [0,\infty), \\ u(0) = u_0, & \\ \dot{u}(0) = v_0, & \end{bmatrix} \tag{B.11}$$

for initial data $u_0 \in D(A)$ and $v_0 \in V$. Associated with the spectral series representation of the unbounded operator A,

$$Au = -\sum_{n=1}^{\infty} \lambda_n^{-1}(u,\phi_n)_H \phi_n,$$

we can define the natural frequencies (modes):

$$\xi_n := \lambda_n^{-1/2} > 0 \qquad \lim_{n\to\infty} \xi_n = \infty.$$

Proposition B.2.1 (Solution by separation of variables). *Let $u_0 \in D(A)$ and $v_0 \in V$. The function*

$$u(t) := Pu_0 + tPv_0 + \sum_{n=1}^{\infty} \cos(\xi_n t)(u_0,\phi_n)_H\phi_n + \sum_{n=1}^{\infty} \xi_n^{-1} \sin(\xi_n t)(v_0,\phi_n)_H\phi_n \tag{B.12}$$

is a strong solution of the Cauchy problem (B.11)

$$u \in C^2(\mathbb{R};H) \cap C^1(\mathbb{R};V) \cap C(\mathbb{R};D(A)). \tag{B.13}$$

Proof. Consider first the functions

$$c_n(t) := \cos(\xi_n t)(u_0,\phi_n)_H\phi_n, \qquad c_n : \mathbb{R} \to D(A).$$

By (B.10), it follows that

$$\|c_n(t)\|^2_{D(A)} = (\lambda_n^{-2} + \lambda_n^{-1})|\cos(\xi_n t)|^2|(u_0,\phi_n)_H|^2 \le (1 + C_\circ^2)\lambda_n^{-2}|(u_0,\phi_n)_H|^2$$

with

$$\sum_{n=1}^{\infty} \lambda_n^{-2}|(u_0,\phi_n)_H|^2 < \infty, \tag{B.14}$$

since $u_0 \in D(A)$ by Proposition B.1.3. The values $c_n(t)$ and $c_m(t)$ are $D(A)$-orthogonal for all t. We can then apply Lemma B.5.3 to show that

$$c(t) := Pu_0 + \sum_{n=1}^{\infty} \cos(\xi_n t)(u_0, \phi_n)_H \phi_n \tag{B.15}$$

defines a continuous function $c : \mathbb{R} \to D(A)$. Its value at $t = 0$ is $c(0) = u_0$ by (B.3). This function is also continuous from $\mathbb{R}$ to V. We next note that

$$\dot{c}_n(t) = -\xi_n \sin(\xi_n t)(u_0, \phi_n)_H \phi_n = -\lambda_n^{-1/2} \sin(\xi_n t)(u_0, \phi_n)_H \phi_n,$$

and that, by Proposition B.1.2(b),

$$\|\dot{c}_n(t)\|_V^2 = \lambda_n^{-1} \big|\lambda_n^{-1/2} \sin(\xi_n t)(u_0, \phi_n)_H\big|^2 \le \lambda_n^{-2} |(u_0, \phi_n)_H|^2.$$

Using Lemma B.5.4, we deduce that the function c defined in (B.15) is continuously differentiable with values in V (and therefore in H). Its derivative is

$$\dot{c}(t) = -\sum_{n=1}^{\infty} \lambda_n^{-1/2} \sin(\xi_n t)(u_0, \phi_n)_H \phi_n,$$

with convergence in V (and in H). Also $\dot{c}(0) = 0$. We can proceed similarly to show that $\dot{c} \in C^1(\mathbb{R}, H)$ and that

$$\ddot{c}(t) = -\sum_{n=1}^{\infty} \xi_n^2 \cos(\xi_n t)(u_0, \phi_n)_H \phi_n = -\sum_{n=1}^{\infty} \lambda_n^{-1} \cos(\xi_n t)(u_0, \phi_n)_H \phi_n = A\big(c(t)\big).$$

We can use exactly the same ideas to analyze

$$s(t) := t\,Pv_0 + \sum_{n=1}^{\infty} \xi_n^{-1} \sin(\xi_n t)(v_0, \phi_n)_H \phi_n = t\,Pv_0 + \sum_{n=1}^{\infty} s_n(t),$$

using now the bound

$$\sum_{n=1}^{\infty} \lambda_n^{-2} \big|\xi_n^{-1} (v_0, \phi_n)_H\big|^2 = \sum_{n=1}^{\infty} \lambda_n^{-1} |(v_0, \phi_n)_H|^2,$$

due to the fact that $v_0 \in V$ and using Proposition B.1.2. □

Going backwards in time. It is clear that the expression in (B.12) works also for negative values of t and that we have found a formula for the solution of (B.11) valid for all $t \in \mathbb{R}$.

Proposition B.2.2 (Energy conservation). *Let u be the solution of the Cauchy problem* (B.11) *given by* (B.12). *Then, the energy*

$$e(t) := \tfrac{1}{2}[u(t), u(t)] + \tfrac{1}{2}(\dot{u}(t), \dot{u}(t))_H$$

is a constant function of t. Finally, problem (B.11) *has a unique solution.*

Proof. Conservation of energy follows from the following facts (we use the same notation of the proof of Proposition B.2.1). First of all, we have the convergent V-orthogonal sum

$$u(t) = c(t) + s(t) = Pu_0 + tPv_0 + \sum_{n=1}^{\infty}(c_n(t) + s_n(t)),$$

and the convergent H-orthogonal sum

$$\dot{u}(t) = Pv_0 + \sum_{n=1}^{\infty}(\dot{c}_n(t) + \dot{s}_n(t)).$$

Next, using that $[\phi_n, \phi_n] = \lambda_n^{-1} = \xi_n^2$, we prove that

$$\begin{aligned}[c_n(t) + s_n(t), c_n(t) + s_n(t)] &= \xi_n^2 \cos^2(\xi_n t)|(u_0, \phi_n)_H|^2 + \sin^2(\xi_n t)|(v_0, \phi_n)_H|^2 \\ &\quad + 2\cos(\xi_n t)\xi_n \sin(\xi_n t)\,(u_0, \phi_n)_H(v_0, \phi_n)_H,\end{aligned}$$

while

$$\begin{aligned}(\dot{c}_n(t) + \dot{s}_n(t), \dot{c}_n(t) + \dot{s}_n(t))_H &= \xi_n^2 \sin^2(\xi_n t)|(u_0, \phi_n)_H|^2 + \cos^2(\xi_n t)|(v_0, \phi_n)_H|^2 \\ &\quad - 2\xi_n \sin(\xi_n t)\cos(\xi_n t)\,(u_0, \phi_n)_H(v_0, \phi_n)_H,\end{aligned}$$

Hence

$$\begin{aligned}2e(t) &= \sum_{n=1}^{\infty}[c_n(t) + s_n(t), c_n(t) + s_n(t)] + \|Pv_0\|_H^2 + \sum_{n=1}^{\infty}(\dot{c}_n(t) + \dot{s}_n(t), \dot{c}_n(t) + \dot{s}_n(t))_H \\ &= \|Pv_0\|_H^2 + \sum_{n=1}^{\infty}\xi_n^2|(u_0, \phi_n)_H|^2 + \sum_{n=1}^{\infty}|(v_0, \phi_n)_H|^2 \\ &= [u_0, u_0] + (v_0, v_0)_H.\end{aligned}$$

The energy conservation property can also be proved directly, assuming that u has the regularity (B.13) and showing that

$$\dot{e}(t) = (\ddot{u}(t), \dot{u}(t))_H + [u(t), \dot{u}(t)] = (Au(t), \dot{u}(t))_H + [u(t), \dot{u}(t)] = 0,$$

by the associated Green Identity. This also shows that if initial data are set to zero, then $\dot{u}(t) = 0$ for all t and therefore $u(t) \equiv u(0) = 0$. □

B.3 Strong solutions to nonhomogeneous problems

The problem. The context of this section is exactly the one of the preceding sections (hypotheses are exposed in Section 6.1). In this case we want to study the problem

$$\begin{bmatrix} \ddot{u}(t) = Au(t) + f(t) & t \in [0, \infty), \\ u(0) = 0, & \\ \dot{u}(0) = 0, & \end{bmatrix} \tag{B.16}$$

where $f : [0, \infty) \to V_0$ is a continuous function. The solution of this problem can be expressed with a causal convolution, which is nothing but the representation using the formula of variation of constants (a.k.a. Duhamel's principle). The more general case, when $f(t) \in V$ is a slight generalization, obtained by considering the solution corresponding to a right-hand side $Pf(t)$. (Note that this was done in Section 6.3.)

The following lemma is straightforward to prove.

Lemma B.3.1. *Let $g : [0, \infty) \to \mathbb{R}$ be a continuous function, $\omega > 0$ and*

$$\alpha(t) := \omega^{-1} \int_0^t \sin(\omega(t-\tau))g(\tau)\mathrm{d}\tau.$$

Then $\alpha \in \mathcal{C}^2([0, \infty))$ and

$$\alpha(0) = \dot{\alpha}(0) = 0 \qquad \ddot{\alpha} + \omega^2 \alpha = g \quad \textit{in } [0, \infty).$$

Proposition B.3.2 (Duhamel's principle in series form). *Let $f \in \mathcal{C}([0, \infty); V)$ and consider the functions*

$$u_n(t) := \Big(\int_0^t \xi_n^{-1} \sin(\xi_n(t-\tau))\, (f(\tau), \phi_n)_H \mathrm{d}\tau \Big) \phi_n \qquad n \ge 1. \tag{B.17}$$

Then

$$u(t) := \sum_{n=1}^{\infty} u_n(t)$$

is the unique solution of (B.16) *and*

$$u \in \mathcal{C}^2([0, \infty), H) \cap \mathcal{C}^1([0, \infty), V) \cap \mathcal{C}([0, \infty), D(A)).$$

Proof (Properties of the functions $u_n(t)$.). Applying Lemma B.3.1 to the functions $g := (f(\cdot), \phi_n)_H$, it follows that the functions u_n are in $\mathcal{C}^2([0, \infty), D(A))$. Also

$$\dot{u}_n(t) = \Big(\int_0^t \cos(\xi_n(t-\tau))\, (f(\tau), \phi_n)_H \mathrm{d}\tau \Big) \phi_n, \tag{B.18}$$

and

$$\begin{aligned}\ddot{u}_n(t) &= (f(t), \phi_n)_H \phi_n - \xi_n^2 u_n(t) && \text{(by Lemma B.3.1)}\\ &= (f(t), \phi_n)_H \phi_n + A u_n(t). && (A\phi_n = -\xi_n^2 \phi_n = -\lambda_n^{-1}\phi_n)\end{aligned} \tag{B.19}$$

It is also straightforward to see that, because of the orthogonality of the functions ϕ_n,

$$(u_n(t), u_m(\tau))_H = 0 \quad n \neq m, \quad t, \tau \in [0, \infty), \tag{B.20a}$$

$$[u_n(t), u_m(\tau)] = 0 \quad n \neq m, \quad t, \tau \in [0, \infty), \tag{B.20b}$$

$$(Au_n(t), Au_m(\tau))_H = 0 \quad n \neq m, \quad t, \tau \in [0, \infty). \tag{B.20c}$$

Continuity in $D(A)$ and first initial condition. Using (B.10), the series representation for the norm in $D(A)$, we can write for all $t \leq T$

$$\begin{aligned}\|u_n(t)\|_{D(A)}^2 &= (\lambda_n^{-2} + \lambda_n^{-1}) \left| \int_0^t \xi_n^{-1} \sin(\xi_n(t-\tau))(f(\tau), \phi_n)_H \mathrm{d}\tau \right|^2\\ &= (\lambda_n^{-1} + 1) \left| \int_0^t \sin(\xi_n(t-\tau))(f(\tau), \phi_n)_H \mathrm{d}\tau \right|^2 && (\xi_n^{-2}\lambda_n^{-1} = 1)\\ &\leq (\lambda_n^{-1} + 1) t \int_0^t |(f(\tau), \phi_n)_H|^2 \mathrm{d}\tau && \text{(Cauchy-Schwarz)}\\ &\leq T \int_0^T (\lambda_n^{-1} + 1) |(f(\tau), \phi_n)_H|^2 \mathrm{d}\tau =: M_n = M_n(T).\end{aligned}$$

On the other hand, by the Monotone Convergence Theorem

$$\sum_{n=1}^\infty M_n = T \int_0^T \sum_{n=1}^\infty (\lambda_n^{-1} + 1) |(f(\tau), \phi_n)|^2 \mathrm{d}\tau \leq T \int_0^T \left(\|f(\tau)\|_V^2 + \|f(\tau)\|_H^2 \right) \mathrm{d}\tau,$$

where we have applied that $\{\phi_n\}$ is orthonormal in H and Proposition B.1.2. Using Lemma B.5.3 in the space $D(A)$ (note the orthogonality given by (B.20a) and (B.20c)), it follows that $u \in \mathcal{C}([0, T], D(A))$ for all $T > 0$ and thus $u \in \mathcal{C}([0, \infty), D(A))$. Since convergence of the series defining u is valid for all t (it is uniform in compact intervals), it also follows that

$$u(0) = \sum_{n=1}^\infty u_n(0) = 0. \tag{B.21}$$

Also, using that $A : D(A) \to H$ is bounded, we can write

$$Au(t) = \sum_{n=1}^\infty Au_n(t) \qquad \text{in } H \text{, uniformly in } [0, T] \text{ for all } T. \tag{B.22}$$

Differentiability in V and second initial condition. By Proposition B.1.2 and the expression for $\dot{u}_n$ given in (B.18), we can write

$$\begin{aligned}\|u_n(t)\|_V^2 + \|\dot{u}_n(t)\|_V^2 &= \lambda_n^{-1}\Big|\int_0^t \xi_n^{-1}\sin(\xi_n(t-\tau))(f(\tau),\phi_n)_H \mathrm{d}\tau\Big|^2 \\ &\quad + \lambda_n^{-1}\Big|\int_0^t \cos(\xi_n(t-\tau))(f(\tau),\phi_n)_H \mathrm{d}\tau\Big|^2 \\ &\le (1+\lambda_n^{-1})t\int_0^t |(f(\tau),\phi_n)_H|^2 \mathrm{d}\tau \le M_n \qquad \forall t \le T.\end{aligned}$$

We then use Lemma B.5.4 in the space V to show that $u \in \mathcal{C}^1([0,T],V)$ for all T and thus $u \in \mathcal{C}^1([0,\infty),V)$. Since the series that defines u can be differentiated term by term (see Lemma B.5.4) it also follows that

$$\dot{u}(0) = \sum_{n=1}^{\infty} \dot{u}_n(0) = 0. \tag{B.23}$$

Second order differentiability in H and differential equation. We first note that

$$\begin{aligned}\sum_{n=1}^{\infty} \ddot{u}_n(t) &= \sum_{n=1}^{\infty} \big(Au_n(t) + (f(t),\phi_n)_H\phi_n\big) && \text{(by (B.19))} \\ &= Au(t) + f(t), && \text{(by (B.22) and Lemma B.5.5)}\end{aligned}$$

with convergence in H, uniformly for all $t \in [0,T]$ and for all $T > 0$. Since $\sum_{n=1}^{\infty} \dot{u}_n(t) = \dot{u}(t)$ in H, uniformly in $t \in [0,T]$, this implies that

$$\ddot{u} = Au + f \in \mathcal{C}([0,\infty),H). \tag{B.24}$$

Conclusion. Note that we have proved all desired continuity properties, the differential equation (B.24), and the initial conditions (B.21) and (B.23). Further, we have proved that the series $\sum_n u_n(t)$ converges in $D(A)$ uniformly on bounded intervals, that $\sum_n \dot{u}_n(t)$ converges in V uniformly on bounded intervals and $\sum_n \ddot{u}_n(t)$ converges in H uniformly on bounded intervals. □

Proposition B.3.3. *With the notation of the previous proposition, the following bounds hold for all $t \ge 0$:*

$$\|u(t)\|_{D(A)} \le \sqrt{1+C_\circ^2}\int_0^t \|f(\tau)\|_V \mathrm{d}\tau, \tag{B.25a}$$

$$\|Au(t)\|_H \le \int_0^t \|f(\tau)\|_V \mathrm{d}\tau, \tag{B.25b}$$

$$\|u(t)\|_V \leq \int_0^t \|f(\tau)\|_H \mathrm{d}\tau, \tag{B.25c}$$

$$\|\dot{u}(t)\|_V \leq \int_0^t \|f(\tau)\|_V \mathrm{d}\tau, \tag{B.25d}$$

$$\|\dot{u}(t)\|_H \leq \int_0^t \|f(\tau)\|_H \mathrm{d}\tau. \tag{B.25e}$$

Proof. Let us fix a value of t and consider the functions

$$[0,t] \ni \tau \longmapsto g_n(\tau;t) := \xi_n^{-1} \sin(\xi_n(t-\tau))(f(\tau),\phi_n)_H \phi_n,$$

as well as their sum

$$g(\tau;t) := \sum_{n=1}^{\infty} g_n(\tau;t), \tag{B.26}$$

assuming that this series converges. Since

$$\|g_n(\tau;t)\|_{D(A)}^2 \leq (\lambda_n^{-1}+1)|(f(\tau),\phi_n)_H|^2 \qquad \forall \tau \in [0,t] \quad \forall n, \tag{B.27}$$

it follows that $g(\tau;t)$ is well defined for all τ and that

$$\|g(\tau;t)\|_{D(A)}^2 = \sum_{n=1}^{\infty} \|g_n(\tau;t)\|_{D(A)}^2 \leq (1+C_\circ^2)\|f(\tau)\|_V^2 \leq C \quad \forall \tau \in [0,t].$$

This proves that $g(\,\cdot\,;t) \in L^2(0,t;D(A)) \subset L^1(0,t;D(A))$ and therefore

$$\left\| \int_0^t g(\tau;t)\mathrm{d}\tau \right\|_{D(A)} \leq \int_0^t \|g(\tau;t)\|_{D(A)}\mathrm{d}\tau \leq \sqrt{1+C_\circ^2} \int_0^t \|f(\tau)\|_V \mathrm{d}\tau.$$

Since the functions $g_n(\,\cdot\,;t)$ are orthogonal in $L^2(0,t;D(A))$, the bound (B.27) also shows that the series (B.26) converges in $L^2(0,t;D(A))$ and therefore in $L^1(0,t;D(A))$. Thus

$$u(t) = \sum_{n=1}^{\infty} u_n(t) = \sum_{n=1}^{\infty} \int_0^t g_n(\tau;t)\mathrm{d}\tau = \int_0^t \sum_{n=1}^{\infty} g_n(\tau;t)\mathrm{d}\tau = \int_0^t g(\tau;t)\mathrm{d}\tau$$

and the proof of (B.25a) is finished. The other bounds are very similar and left to the reader as an exercise. □

B.4 Weak solutions of nonhomogeneous equations

In this section we study problems of the form

$$\begin{bmatrix} \ddot{u}(t) = Au(t) + f(t) & t \in [0, \infty), \\ u(0) = 0, & \\ \dot{u}(0) = 0, & \end{bmatrix} \tag{B.28}$$

where $f : [0, \infty) \to H_0$ is a continuous function. We will look for some kind of weak solution to this equation. The series representation for A is still valid but convergence of the series

$$Au = -\sum_{n=1}^{\infty} \lambda_n^{-1}(u, \phi_n)_H \phi_n$$

is in V' (and not in H any more). A weak solution of (B.28) will thus be a solution of this problem with the differential equation valid in V' for all t, or equivalently, a solution of

$$\begin{bmatrix} \langle \ddot{u}(t), v\rangle_{V' \times V} + [u(t), v] = (f(t), v)_H \quad \forall v \in V & t \in [0, \infty), \\ u(0) = 0, & \\ \dot{u}(0) = 0, & \end{bmatrix} \tag{B.29}$$

Note that the solution given by Proposition B.3.2 is easily seen to be a weak solution.

Proposition B.4.1. *For any continuous $f : [0, \infty) \to H_0$, problem* (B.29) *has a unique solution*

$$u \in \mathcal{C}^2([0, \infty); V') \cap \mathcal{C}^1([0, \infty); H) \cap \mathcal{C}^0([0, \infty); V) \tag{B.30}$$

and the following bounds hold for all $t \geq 0$:

$$C_\circ^{-1} \|u(t)\|_H \leq \|u(t)\|_V \leq \int_0^t \|f(\tau)\|_H \mathrm{d}\tau,$$

$$\|\dot{u}(t)\|_H \leq \int_0^t \|f(\tau)\|_H \mathrm{d}\tau.$$

Moreover, the function

$$w(t) := \int_0^t u(\tau) \mathrm{d}\tau \tag{B.31}$$

is in $\mathcal{C}([0, \infty); D(A))$.

Proof. We first consider the following modified problem

$$\begin{bmatrix} \ddot{v}(t) = Av(t) + G^{1/2}f(t) & t \in [0, \infty), \\ v(0) = 0, & \\ \dot{v}(0) = 0. & \end{bmatrix} \tag{B.32}$$

Since $G^{1/2}f : [0, \infty) \to V_0$ is continuous, we can apply Propositions B.3.2 and B.3.3 and we obtain that problem (B.16) has a solution

$$v \in \mathcal{C}^2([0, \infty); H) \cap \mathcal{C}^1([0, \infty); V) \cap \mathcal{C}^0([0, \infty); D(A)). \tag{B.33}$$

From Proposition B.1.2 it follows that

$$\|Av(t)\|_H \le \int_0^t \|G^{1/2}f(\tau)\|_V \mathrm{d}\tau = \int_0^t \|f(\tau)\|_H \mathrm{d}\tau, \tag{B.34}$$

$$\|v(t)\|_V \le \int_0^t \|G^{1/2}f(t)\|_H \mathrm{d}\tau \le C_\circ \int_0^t \|f(\tau)\|_H \mathrm{d}\tau, \tag{B.35}$$

$$\|\dot{v}(t)\|_V \le \int_0^t \|G^{1/2}f(\tau)\|_V \mathrm{d}\tau = \int_0^t \|f(\tau)\|_H \mathrm{d}\tau. \tag{B.36}$$

Consider now the operator

$$G^{-1/2}w := \sum_{n=1}^{\infty} \lambda_n^{-1/2}(w, \phi_n)_H \phi_n = (-A)^{1/2} w.$$

(The notation is slightly misleading, because $G^{1/2}$ is invertible only when we eliminate rigid motions from the problem.) It is easy to prove that $G^{-1/2}$ is bounded $D(A) \to V$, $V \to H$ and therefore $H \to V'$. Note also that as operators $D(A) \to V'$

$$G^{-1/2}A = AG^{-1/2}, \tag{B.37}$$

which can first be proved on the span of the eigenfunctions ϕ_n and then by density.

We then define $u(t) := G^{-1/2}v(t)$ and note that the regularity (B.33) for $v(t)$ implies the regularity (B.30) for $u(t)$. Initial conditions for u are satisfied for obvious reasons. Finally, with equalities in V',

$$\begin{aligned} \ddot{u}(t) &= G^{-1/2}\ddot{v}(t) && (G^{-1/2} \text{ is bounded from } H \text{ to } V') \\ &= G^{-1/2}Av(t) + G^{-1/2}G^{1/2}f(t) && (v \text{ satisfies (B.32)}) \\ &= AG^{-1/2}v(t) + f(t) && (\text{(B.37) and } f \text{ takes values in } H_0) \\ &= Au(t) + f(t). \end{aligned}$$

In this sense, $u(t)$ is a solution of (B.28) by understanding the equation to hold in the space V'. Note also that $u(t) \in V_0$ for all t, which implies that $v(t) = G^{1/2}u(t)$, and therefore

$$\|Av(t)\|_H = \|AG^{1/2}u(t)\|_H = \|u(t)\|_V, \qquad \|v(t)\|_V = \|G^{1/2}u(t)\|_V = \|u(t)\|_H,$$

and

$$\|\dot{v}(t)\|_V = \|G^{1/2}\dot{u}(t)\|_V = \|\dot{u}(t)\|_H.$$

These equalities and (B.34)–(B.36) finish the bounds for u.

Let us finally consider the function w defined in (B.31). It is clear that $w \in \mathcal{C}^2([0,\infty); H) \cap \mathcal{C}^1([0,\infty); V)$ and that

$$Aw(t) = \ddot{w}(t) - \int_0^t f(\tau)\mathrm{d}\tau \qquad (\text{in } V') \qquad \forall t \geq 0. \tag{B.38}$$

However, the right-hand side of (B.35) is a continuous function with values in H and therefore, so is Aw. □

Proposition B.4.2. *If $f \in \mathcal{C}^1([0,\infty); H_0)$ satisfies $f(0) = 0$, then the problem*

$$\left[\begin{array}{ll} \ddot{u}(t) = Au(t) + f(t) & t \in [0,\infty), \\ u(0) = 0, & \\ \dot{u}(0) = 0, & \end{array}\right.$$

has a unique solution

$$u \in \mathcal{C}^2([0,\infty); H) \cap \mathcal{C}^1([0,\infty); V) \cap \mathcal{C}([0,\infty); D(A)), \tag{B.39}$$

satisfying the bounds of Proposition B.4.1 as well as for all $t \geq 0$

$$\|Au(t)\|_H \leq \|f(t)\|_H + \int_0^t \|\dot{f}(\tau)\|_H \mathrm{d}\tau \leq 2\int_0^t \|\dot{f}(\tau)\|_H \mathrm{d}\tau, \tag{B.40a}$$

$$\|\dot{u}(t)\|_V \leq \int_0^t \|\dot{f}(\tau)\|_H \mathrm{d}\tau. \tag{B.40b}$$

Proof. Let first $z : [0,\infty) \to V$ be the solution to

$$\left[\begin{array}{lll} \langle \ddot{z}(t), v\rangle_{V'\times V} + [z(t), v] = (\dot{f}(t), v)_H & \forall v \in V & t \in [0,\infty), \\ z(0) = 0, & & \\ \dot{z}(0) = 0, & & \end{array}\right.$$

and define

$$u(t) := \int_0^t z(\tau)\mathrm{d}\tau.$$

By Proposition B.4.1, it follows that u satisfies (B.39). (Note that we are integrating once in the time variable, which increases time regularity by one.) Integrating the weak differential equation satisfied by z and using that $f(0) = 0$, it follows that

$$(\ddot{u}(t), v)_H + [u(t), v] = \langle \ddot{u}(t), v\rangle_{V'\times V} + [u(t), v] = (f(t), v)_H \qquad \forall v \in V.$$

Therefore,

$$\ddot{u}(t) - f(t) = Au(t) \qquad \text{in } V' \quad t \geq 0,$$

but since both sides are in H, this is just the strong differential equation. Finally

$$\|Au(t)\|_H \leq \|\ddot{u}(t)\|_H + \|f(t)\|_H = \|\dot{z}(t)\|_H + \|f(t)\|_H$$

and (see Proposition B.4.1)

$$\|z(t)\|_V \leq \int_0^t \|\dot{f}(\tau)\|_H \mathrm{d}\tau, \qquad \|\dot{z}(t)\|_H \leq \int_0^t \|\dot{f}(\tau)\|_H \mathrm{d}\tau,$$

which proves (B.40). □

B.5 Background material

For ease of reference, we are including a section on some basic results in Functional Analysis, a small block of which is related to spectral decompositions of compact operators and a second part related to uniform convergence of series. Proofs of these results can be found in any basic text on Functional Analysis.

Hilbert-Schmidt theory

We start the section with a slightly restricted version of the Hilbert-Schmidt Theorem for compact self-adjoint operators in Hilbert space.

Theorem B.5.1 (Hilbert-Schmidt Decomposition). *Let H_0 be an infinite dimensional real Hilbert space. Let $G : H_0 \to H_0$ be a compact (therefore bounded) self-adjoint and positive definite (therefore injective) linear operator. Then there* *exists a nonincreasing sequence of positive numbers* $\lim_{n\to\infty} \lambda_n = 0$ *and a complete orthogonal set* $\{\phi_n\} \subset H_0$ *such that*

$$G = \sum_{n=1}^{\infty} \lambda_n(\,\cdot\,, \phi_n)\,\phi_n, \tag{B.41}$$

where the series converges in the operator norm.

Simple consequences plus some remarks.

(a) Uniform convergence of the series refers to convergence in operator norm. Therefore

$$\lim_{N\to\infty} \|G - \sum_{n=1}^{N} \lambda_n(\,\cdot\,, \phi_n)\phi_n\| = 0.$$

(b) It is obvious that

$$G\phi_n = \lambda_n \phi_n$$

and therefore the decomposition (B.41) singles out eigenvalues and eigenvectors of G.

(c) Since $\{\phi_n\}$ is a complete orthonormal set in H_0, we can write

$$u = \sum_{n=1}^{\infty} (u, \phi_n)\phi_n \qquad \forall u \in H_0. \tag{B.42}$$

In particular, the hypotheses of the Hilbert-Schmidt theorem, as given here, cannot hold unless H_0 is a separable Hilbert space. A more general result can be easily shown by assuming that G is only positive semidefinite. In this case, the decomposition (B.41) still holds, but (B.42) only holds in the orthogonal complement of the kernel of G, a subspace of H_0 that can be non-separable.

(d) It can be easily proved that the decomposition (B.41) shows all possible eigenvalues of G and that eigenfunctions are necessarily linear combinations of those given in the Hilbert basis.

(e) The converse statement to that of Theorem B.5.1 is easy to prove: if there exist sequences $\{\lambda_n\}$ and $\{\phi_n\}$ in the conditions above ($\lambda_n > 0$, $\lim_{n\to\infty} \lambda_n = 0$, $\{\phi_n\}$ complete orthonormal in H_0), then the operator G is a compact self-adjoint positive definite linear operator in H_0.

Picard's criterion goes one step further to characterize the range of an operator of the form (B.41) with the conditions above for the eigenvalues and eigenvectors.

Theorem B.5.2 (Picard's Criterion). *Let $G : H_0 \to H_0$ be an operator in the conditions of Proposition B.5.1 and consider its Hilbert-Schmidt decomposition* (B.41). *Then $f \in R(G)$ if and only if*

$$\sum_{n=1}^{\infty} \lambda_n^{-2} |(f, \phi_n)|^2 < \infty. \tag{B.43}$$

More remarks and observations.

(a) Picard's criterion is actually proved by finding the inverse

$$R(G) \ni f \qquad \longmapsto \qquad u := \sum_{n=1}^{\infty} \lambda_n^{-1} (f, \phi_n) \phi_n.$$

It is clear that this is an element of H_0 and that $Gu = f$.

(b) The condition (B.43) can actually be used to endow $R(G)$ with a norm:

$$\|f\|_{R(G)}^2 := \sum_{n=1}^{\infty} \lambda_n^{-2} |(f, \phi_n)|^2$$

This norm makes $R(G)$ a Hilbert space that is compact and densely embedded in H_0. Furthermore, $R : H_0 \to R(G)$ is an isometric isomorphism.

Some continuity tests

Lemma B.5.3. *Let X be a Hilbert space and I a closed interval in $\mathbb{R}$. Assume that $c_n : I \to X$ are continuous functions such that*

$$(c_n(t), c_m(t))_X = 0 \quad \forall n \neq m \qquad \forall t \in I$$

and

$$\|c_n(t)\|_X^2 \leq M_n \quad \forall t \in I \qquad \text{where} \qquad \sum_{n=1}^{\infty} M_n < \infty.$$

Then the series

$$\sum_{n=1}^{\infty} c_n(t)$$

converges uniformly to a function in $C(I; X)$.

Proof. Consider the partial sums

$$s_N := \sum_{n=1}^{N} c_n \in \mathcal{C}(I;X)$$

and note that for $M > N$

$$\|s_M(t) - s_N(t)\|_X^2 = \Big\| \sum_{n=N+1}^{M} c_n(t) \Big\|_X^2 = \sum_{n=N+1}^{M} \|c_n(t)\|_X^2 < \sum_{n=N+1}^{M} M_n \quad \forall t \in I.$$

This proves that s_N is Cauchy in $\mathcal{C}(I;X)$ and therefore, it is convergent in this space. □

A second variant of this result includes term-by-term differentiation. The proof is similar.

Lemma B.5.4. *Let X be a Hilbert space and let $I \subset \mathbb{R}$ be a closed interval. Assume that $c_n \in \mathcal{C}^1(I;X)$ satisfy*

$$(c_n(t), c_m(\tau))_X = 0 \quad \forall n \neq m \qquad \forall t, \tau \in I,$$

and

$$\|c_n(t)\|_X^2 + \|\dot{c}_n(t)\|_X^2 \le M_n \quad \forall t \in I \qquad \textit{where} \qquad \sum_{n=1}^{\infty} M_n < \infty.$$

Then the two series

$$\sum_{n=1}^{\infty} c_n(t) \qquad \textit{and} \qquad \sum_{n=1}^{\infty} \dot{c}_n(t)$$

converge in $\mathcal{C}(I;X)$ and the derivative of the first function is the second one.

Proof. The hypothesis of orthogonality for all values of t, τ implies that

$$(\dot{c}_n(t), \dot{c}_m(t))_X = 0 \qquad \forall n \neq m \qquad \forall t \in I.$$

Using the notation of the proof of Lemma B.5.3, we thus prove that for $M > N$

$$\|s_M(t) - s_N(t)\|_X^2 + \|\dot{s}_M(t) - \dot{s}_N(t)\|_X^2 \le \sum_{n=N+1}^{M} M_n \qquad \forall t \in I,$$

from where the result follows. □

Lemma B.5.5. *Let $I \subset \mathbb{R}$ be a closed bounded interval and $f : I \to X$ be continuous. Assume that X has a Hilbert basis $\{\phi_n\}$. Then, the series*

$$f(t) = \sum_{n=1}^{\infty} (f(t), \phi_n)_X \phi_n \tag{B.44}$$

converges in $C(I;X)$.

Proof. It is clear that the series (B.44) converges in X for all t and in particular

$$a_N(t) := \Big\| \sum_{n=1}^{N} (f(t), \phi_n)_X \phi_n \Big\|_X^2 = \sum_{n=1}^{N} |(f(t), \phi_n)_X|^2 \overset{N\to\infty}{\longrightarrow} \sum_{n=1}^{\infty} |(f(t), \phi_n)_X|^2 = \|f(t)\|_X^2$$

for all t. The functions $a_N : I \to \mathbb{R}$ are continuous and increase monotonically to their continuous limit $a(t) := \|f(t)\|_X^2$. By Dini's Theorem, $a_N \to a$ uniformly. Finally

$$\Big\| f(t) - \sum_{n=1}^{\infty} (f(t), \phi_n)_X \phi_n \Big\|_X^2 = a(t) - a_N(t) \qquad \forall t \in I, \quad \forall N,$$

which proves the result. □

B.6 Exercises

1. (Section B.1) Consider the spaces

$$H := L^2(0,L), \quad V := H_0^1(0,L), \quad D(A) := H_0^1(0,L) \cap H^2(0,L).$$

Describe the spaces (and their associated norms) using Fourier sine series. Note that the Fourier sine series is the series of eigenfunctions for the one-dimensional problem

$$-u'' = f \quad \text{in } (0,L), \qquad u(0) = u(L) = 0$$

Follow carefully where the constant of Poincaré's inequality appears.

2. (Section B.1) Describe the spaces (and their norms)

$$H := L^2(0,L), \quad V := H^1(0,L), \quad D(A) := \{u \in H^2(0,L) \,:\, u'(0) = u'(L) = 0\}$$

using Fourier cosine series. Note that the Fourier cosine series is the series of eigenfunctions for the one-dimensional problem

$$-u'' = f \quad \text{in } (0, L), \qquad u'(0) = u'(L) = 0$$

and that there is a one-dimensional space of rigid motions.

3. (Section B.1) Study the Green's operator associated with the problem with periodic boundary conditions

$$-u'' = f \quad \text{in } (0, L), \qquad u(0) = u(L), \qquad u'(0) = u'(L).$$

4. (Section B.1) **Associated Hilbert scales.** In the notation of Section B.1, for $s \geq 0$ we consider the set

$$H_s := \{u \in H_0 \, : \, \|u\|_s < \infty\}, \quad \text{where} \quad \|u\|_s^2 := \sum_{n=1}^{\infty} \lambda_n^{-s} |(u, \phi_n)_H|^2.$$

Note that $H_1 = V_0 = R(G^{1/2})$ and $H_2 = D(A) \cap H_0 = R(G)$. Note also that by Picard's Criterion $H_s = R(G^{s/2})$, where

$$G^{s/2} u := \sum_{n=1}^{\infty} \lambda_n^{s/2} (u, \phi_n)_H \phi_n.$$

(a) Show that H_s is a Hilbert space. (**Hint.** You can easily show H_s to be isomorphic to a space of sequences with a weighted ℓ^2 norm.)
(b) Find the best estimate for $C(r, s) > 0$ such that

$$\|u\|_s \leq C(r, s) \|u\|_r \qquad \forall u \in H_r \qquad r \geq s.$$

(c) Show that if $r > s$, then the inclusion $H_r \subset H_s$ is dense and compact.
(d) Show that $G : H_s \to H_{s+2}$ and $A : H_{s+2} \to H_s$ are bounded for all $s \geq 0$.

5. (Section B.2) **The evolution group.** Using notation of Section B.2, consider the matrix of operators

$$\mathbb{S}(t) := \begin{bmatrix} S_{11}(t) & S_{12}(t) \\ S_{21}(t) & S_{22}(t) \end{bmatrix},$$

where

$$S_{11}(t) u := \; Pu + \sum_{n=1}^{\infty} \cos(\xi_n t)(u, \phi_n)_H \phi_n,$$

$$S_{12}(t) v := t P v + \sum_{n=1}^{\infty} \xi_n^{-1} \sin(\xi_n t)(v, \phi_n)_H \phi_n,$$

$$S_{21}(t)u := \quad -\sum_{n=1}^{\infty} \xi_n \sin(\xi_n t)(u, \phi_n)_H \phi_n,$$

$$S_{22}(t)v := \; Pv + \sum_{n=1}^{\infty} \cos(\xi_n t)(v, \phi_n)_H \phi_n.$$

These operators are defined for all $t \in \mathbb{R}$. Note that, at least formally, the second row of $\mathbb{S}(t)$ is the derivative of the first one.

(a) Show that $\mathbb{S}(t)$ defines a bounded operator $V \times H \to V \times H$.
(b) Show that $\mathbb{S}(t)$ is an isometry in $V \times H$ for all $t \in \mathbb{R}$ and that $\mathbb{S}(0)$ is the identity operator.
(c) Show that $\mathbb{S}(t)\mathbb{S}(\tau) = \mathbb{S}(t+\tau)$ for all $t, \tau \in \mathbb{R}$. (**Hint.** This can be proved on a dense subspace.)
(d) Show that $\mathbb{S}(-t) = \mathbb{S}(t)^{-1}$. (**Hint.** This is an easy consequence of the above.)
(e) Show that $\mathbb{S}(t)$ is also bounded $D(A) \times V \to D(A) \times V$.

6. (Section B.3) Prove all the bounds of Proposition B.3.3.

Appendix C
Some algorithms for CQ

In this chapter we sketch some algorithms for the implementation of discrete convolutions and convolution equations with Convolution Quadrature. The exposition of this chapter follows closely part of the material in the notes written with Matthew Hassell [54], while most of its ideas can be found in [22].

C.1 CQ by time-stepping

The Discrete Fourier Transform (DFT) of a vector $\mathbf{x} := (x_0, \ldots, x_M) \in \mathbb{C}^{M+1}$ is the vector $\hat{\mathbf{x}} \in \mathbb{C}^{M+1}$ given by

$$\hat{x}_\ell := \sum_{n=0}^{M} x_n \zeta_{M+1}^{-\ell n}, \quad \ell = 0, \ldots, M+1, \qquad \text{where} \qquad \zeta_{M+1} := e^{\frac{2\pi\imath}{M+1}}. \tag{C.1}$$

The inverse transform (IDFT) is given

$$x_n := \frac{1}{M+1} \sum_{\ell=0}^{M} \hat{x}_\ell \zeta_{M+1}^{\ell n}, \qquad n = 0, \ldots, M. \tag{C.2}$$

We recall that the Fast Fourier Transform (FFT) is a fast implementation of the DFT (and of the IDFT, which has the same structure) when the length of the vector is a power of 2.

Let $\mathbf{y} = \hat{\mathbf{x}}$ be the DFT of a real vector $\mathbf{x}$. (By a real vector we mean a vector taking values of a Banach space equipped with a conjugation operator $x \mapsto \overline{x}$ and such that $\overline{x_n} = x_n$ for all n.) The vector $\mathbf{y}$ has the following symmetry property

$$y_{N+1-\ell} = \overline{y}_\ell \qquad \ell = 1, \ldots, N. \tag{C.3}$$

F.-J. Sayas, *Retarded Potentials and Time Domain Boundary Integral Equations*, Springer Series in Computational Mathematics 50,
DOI 10.1007/978-3-319-26645-9

We will say that an algorithm can be symmetrized when we know in advance that its output $\mathbf{y}$ satisfies (C.3). Then we will:

- Compute y_ℓ for $\ell = 0, \ldots, \lfloor \frac{N+1}{2} \rfloor$.
- Copy the missing components $y_{N+1-\ell} = \overline{y}_\ell$ for $\ell = 1, \ldots \lfloor \frac{N}{2} \rfloor$.

Approximate computation of CQ weights. We are now given a transfer function F and look for a way to compute the coefficients

$$\omega_n(\kappa) := \frac{1}{n!}\frac{\mathrm{d}^n}{\mathrm{d}\zeta^n}\left(\mathrm{F}\left(\frac{1}{\kappa}p(\zeta)\right)\right)\Big|_{\zeta=0}, \qquad n = 0, \ldots, N, \tag{C.4}$$

where $p(\zeta)$ is the discrete function determining the CQ method. It is clear that

$$\omega_0(\kappa) = \mathrm{F}(\tfrac{1}{\kappa}p(0)). \tag{C.5}$$

The other coefficients are computed using the Cauchy integral formula

$$\omega_n(\kappa) = \frac{1}{2\pi\imath}\oint_{C_R} \zeta^{-n-1}\mathrm{F}(\tfrac{1}{\kappa}p(\zeta))\mathrm{d}\zeta \qquad \text{where} \qquad C_R := \{\zeta \in \mathbb{C} \,:\, |\zeta| = R\}, \tag{C.6}$$

where

$$R = \epsilon^{\frac{1}{2(N+1)}},$$

$N+1$ is the number of coefficients we want to compute (to give N time steps), and ϵ is proportional to the machine epsilon. A justification for this choice of R is given in [22] (see also [66]). The integral (C.6) can be efficiently approximated with the trapezoidal rule, since, after parametrization, its integrand is analytic and periodic. We use as many points as coefficients we want to compute:

$$\begin{aligned}
\omega_n(\kappa) &= \frac{1}{2\pi\imath}\oint_{C_R} \zeta^{-n-1}\mathrm{F}(\tfrac{1}{\kappa}p(\zeta))\mathrm{d}\zeta && \text{(Cauchy formula)}\\
&= R^{-n}\int_0^1 e^{-2\pi\imath n\theta}\mathrm{F}(\tfrac{1}{\kappa}p(Re^{2\pi\imath\theta}))\mathrm{d}\zeta && \text{(parametrization } \zeta = Re^{2\pi\imath\theta})\\
&\approx \frac{R^{-n}}{N+1}\sum_{\ell=0}^{N} \zeta_{N+1}^{-n\ell}\mathrm{F}(\tfrac{1}{\kappa}p(R\zeta_{N+1}^{\ell})) && \text{(trapezoidal rule: } \zeta_{N+1}^{\ell} = e^{\frac{2\pi\imath\ell}{N+1}})\\
&= \frac{R^{-n}}{N+1}\sum_{\ell=0}^{N} \zeta_{N+1}^{n\ell}\mathrm{F}(\tfrac{1}{\kappa}p(R\zeta_{N+1}^{-\ell})). && \text{(reindexing)}
\end{aligned}$$

Assume now that:

- $\mathrm{F}(\overline{s}) = \overline{\mathrm{F}(s)}$, which is the case for all the integral operators and potentials associated with the Laplace domain acoustic equation;
- $p(\overline{\zeta}) = \overline{p(\zeta)}$, as happens for the BE, BDF2, and TR methods.

Then the vector of evaluations

$$\widehat{\mathrm{F}}_\ell := \mathrm{F}(\tfrac{1}{\kappa}p(R\zeta_{N+1}^{-\ell})), \qquad \ell = 0,\dots,N$$

satisfies the symmetry property (C.3) and, therefore, its computation can be reduced by symmetry. In algorithms to follow we will write **(Par+sym)** when the computation in that step can be parallelized and reduced by symmetry.

- **(Par+sym)** Evaluations in the discrete transformed domain

$$\widehat{\mathrm{F}}_\ell := \mathrm{F}(\tfrac{1}{\kappa}p(R\zeta_{N+1}^{-\ell})), \qquad \ell = 0,\dots,N.$$

- IDFT

$$\mathrm{F}_n := \frac{1}{N+1}\sum_{\ell=0}^{N}\widehat{\mathrm{F}}_\ell\zeta_{N+1}^{n\ell}, \qquad n = 0,\dots,N.$$

- Scaling

$$\omega_n(\kappa) = R^{-n}\mathrm{F}_n, \qquad n = 0,\dots,N.$$

  At this step it is also possible to substitute the approximation of $\omega_0(\kappa)$ by its exact value $\omega_0(\kappa) = \mathrm{F}(\frac{1}{\kappa}p(0))$.

Once the coefficients have been computed, the discrete convolutions

$$\sum_{n=0}^{m}\omega_n(\kappa)g_{m-n}, \qquad m = 0,\dots,N$$

can be computed with standard techniques based on the FFT.

C.2 All-steps-at-once CQ

We will now give a fast algorithm for the simultaneous computation of

$$\sum_{m=0}^{n}\omega_{n-m}(\kappa)g(t_m), \qquad n = 0,\dots,N, \tag{C.7}$$

assuming that we have approximated the CQ coefficients by

$$\omega_n(\kappa) \approx \frac{R^{-n}}{N+1}\sum_{\ell=0}^{N} \widehat{\mathrm{F}}_\ell \zeta_{N+1}^{\ell n}, \qquad \widehat{\mathrm{F}}_\ell := \mathrm{F}(\tfrac{1}{\kappa}p(R\zeta_{N+1}^{-\ell})).$$

Note that by Cauchy's Theorem

$$\omega_n(\kappa) = \oint_{C_R} \zeta^{-n+1}\mathrm{F}(\zeta)\mathrm{d}\zeta = 0, \qquad \forall n \leq -1,$$

which justifies the approximations

$$\tilde{\omega}_n(\kappa) := \frac{R^{-n}}{N+1}\sum_{\ell=0}^{N} \widehat{\mathrm{F}}_\ell \zeta_{N+1}^{\ell n} \qquad -N \leq n \leq N,$$

knowing that the approximated coefficients for $n \leq -1$ are almost zero. Then

$$\begin{aligned}
u_n &= \sum_{m=0}^{n} \omega_{n-m}(\kappa) g_m && \text{(exact convolution)}\\
&= \sum_{m=0}^{N} \omega_{n-m}(\kappa) g_m && (\omega_n(\kappa) = 0 \text{ for } n \leq -1)\\
&\approx \sum_{m=0}^{N} \left(\frac{R^{m-n}}{N+1} \sum_{\ell=0}^{N} \widehat{\mathrm{F}}_\ell \zeta_{N+1}^{\ell(n-m)} \right) g_m && (\omega_n(\kappa) \approx \tilde{\omega}_n(\kappa))\\
&= R^{-n} \left(\frac{1}{N+1} \sum_{\ell=0}^{N} \widehat{\mathrm{F}}_\ell \left(\sum_{m=0}^{N} R^m g_m \zeta_{N+1}^{-m\ell} \right) \zeta_{N+1}^{\ell n} \right).
\end{aligned}$$

Following the brackets of the latter expression, we can develop the algorithm for computation of the forward convolution (C.7).

- Sampling

$$g_n := g(t_n), \qquad n = 0, \ldots, N.$$

- Scaling

$$h_m := R^m g_m, \qquad m = 0, \ldots, N.$$

- DFT

$$\hat{h}_\ell := \sum_{m=0}^{N} h_m \zeta_{N+1}^{-m\ell}, \qquad \ell = 0, \ldots, N.$$

- **(Par+sym)** Application of discrete transfer functions

$$\hat{v}_\ell := \widehat{\mathrm{F}}_\ell \hat{h}_\ell, \qquad \widehat{\mathrm{F}}_\ell := \mathrm{F}(\tfrac{1}{\kappa} p(R\zeta_{N+1}^{-\ell})).$$

- IDFT

$$v_n := \frac{1}{N+1} \sum_{\ell=0}^{N} \hat{v}_\ell \zeta_{N+1}^{\ell n}, \qquad n = 0, \ldots, N.$$

- Scaling

$$u_n = R^{-n} v_n.$$

When we are dealing with convolution equations, our aim is the computation of the quantities g_m satisfying

$$\sum_{m=0}^{n} \omega_{n-m}(\kappa) g_m = h(t_n), \qquad n = 0, \ldots, N.$$

We then apply the previous algorithm to the transfer function $\mathrm{F}^{-1}(s)$. In Step (d), we would need to multiply by

$$\mathrm{F}^{-1}(\tfrac{1}{\kappa} p(R\zeta_{N+1}^{-\ell})),$$

which is equivalent to solving a linear problem associated with the operator

$$\widehat{\mathrm{F}}_\ell := \mathrm{F}(\tfrac{1}{\kappa} p(R\zeta_{N+1}^{-\ell})).$$

- Sampling

$$h_n := h(t_n), \qquad n = 0, \ldots, N.$$

- Scaling

$$v_m := R^m h_m, \qquad m = 0, \ldots, N.$$

- DFT: $\hat{v}_\ell$, $\ell = 0, \ldots, N$.
- **(Par+sym)** Solutions in the discrete transformed domain

$$\widehat{\mathrm{F}}_\ell \hat{w}_\ell := \hat{v}_\ell, \qquad \widehat{\mathrm{F}}_\ell := \mathrm{F}(\tfrac{1}{\kappa} p(R\zeta_{N+1}^{-\ell})).$$

- IDFT: w_ℓ, $\ell = 0, \ldots, N$.
- Scaling

$$g_n = R^{-n} w_n, \qquad n = 0, \ldots, N.$$

More algorithms to deal with CQ (including some recursive algorithms) can be found in the literature. See [22, 54] for more details.

Appendix D
Background material

In this appendix we include some precise definitions and results on Sobolev spaces that have been used throughout.

D.1 Lipschitz domains

In this section we are going to describe with precision what we understand by Lipschitz domains. We will introduce the concept of strongly Lipschitz domain. Most of the results on Sobolev spaces that we are going to use are also valid on the larger class of weakly Lipschitz domains.

Let then Ω^- be a bounded open domain in $\mathbb{R}^d$ with boundary Γ and exterior Ω^+. Assume now that for every point $\mathbf{x} \in \Gamma$, we can find:

(a) a vector $\mathbf{a} \in \mathbb{R}^d$ and an orthogonal matrix Q, determining a rigid transformation

$$\Phi(\mathbf{x}) = Q\mathbf{x} + \mathbf{a},$$

(b) a Lipschitz function $h : \mathbb{R}^{d-1} \to \mathbb{R}$, that is, a function for which there exists L such that

$$|h(\tilde{\mathbf{x}}) - h(\tilde{\mathbf{y}})| \le L|\tilde{\mathbf{x}} - \tilde{\mathbf{y}}| \qquad \forall \tilde{\mathbf{x}}, \tilde{\mathbf{y}} \in \mathbb{R}^{d-1},$$

(c) and a cylindrical reference domain

$$\{(\tilde{\mathbf{x}}, t) \in \mathbb{R}^{d-1} \times \mathbb{R} \,:\, |\tilde{\mathbf{x}}| \le \varepsilon, |t| \le \rho\},$$

F.-J. Sayas, *Retarded Potentials and Time Domain Boundary Integral Equations*, Springer Series in Computational Mathematics 50,
DOI 10.1007/978-3-319-26645-9

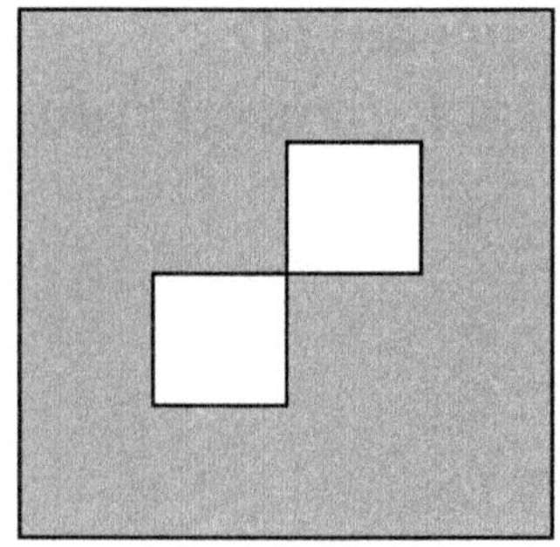

Fig. D.1 A non-Lipschitz planar polygon. The central point of the figure is troublesome, since the domain cannot be placed on one side of the boundary at this point.

so that

$$\begin{aligned} \Phi(\tilde{\mathbf{x}}, h(\tilde{\mathbf{x}}) + t) &\in \Omega^+, & |\tilde{\mathbf{x}}| &\le \varepsilon, & 0 < t \le \rho, \\ \Phi(\tilde{\mathbf{x}}, h(\tilde{\mathbf{x}})) &\in \Gamma, & |\tilde{\mathbf{x}}| &\le \varepsilon, \\ \Phi(\tilde{\mathbf{x}}, h(\tilde{\mathbf{x}}) - t) &\in \Omega^-, & |\tilde{\mathbf{x}}| &\le \varepsilon, & 0 < t \le \rho. \end{aligned}$$

(The elements of the above list –rigid motion, Lipschitz graph, and localization cylinder– depend on the point $\mathbf{x}$.) In this case we say that Ω^- is a (strongly) Lipschitz domain. We also say that Γ is locally a Lipschitz graph and Ω^- is locally a Lipschitz hypograph (it can be locally placed under a Lipschitz graph.) In some theoretical expositions it is common to flip the x_d axis so that the domain is a hypergraph. This obviously does not change the kind of domains.

Examples. Most friendly domains are Lipschitz domains. Circles, rectangles, simple polygons,... are planar Lipschitz domains. Domains with internal or boundary cracks are not Lipschitz domains. Domains with cuspidal points on the boundary are not Lipschitz domains. Similarly, spheres, tori, hexahedra,... are Lipschitz domains in the space, and once again, cracked domains are excluded (Figures D.1 and D.2).

The normal vector. Due to Rademacher's Theorem, the function h is differentiable almost everywhere in its domain of definition. This means that we can attach a tangent plane to almost every point in the graph $\Phi(\tilde{\mathbf{x}}, h(\tilde{\mathbf{x}}))$. The unit normal vector at the point $\mathbf{z} = (\tilde{\mathbf{x}}, h(\tilde{\mathbf{x}}))$ and pointing towards Ω^+ will be denoted $\boldsymbol{\nu}(\mathbf{z})$. It is not complicated to see that the normal vector at the point exists and coincides even if we change the point $\mathbf{x}$ around which we are localizing. We thus get an almost everywhere defined unit vector field $\boldsymbol{\nu} : \Gamma \to \mathbb{R}^d$. In the case of a polyhedral domain, the construction of the normal vector field is completely straightforward.

Integration on a Lipschitz boundary. Using the definition of Lipschitz domain, a compactness argument, and rescaling the local reference cylinder to a fixed cylinder, we can represent the boundary Γ with a collection of coordinate charts. Let us consider the reference cylinder and its lower half

$$\mho := B(\mathbf{0}; 1) \times (-1, 1) \subset \mathbb{R}^d, \qquad \mho_- := \{\mathbf{x} = (\tilde{\mathbf{x}}, t) \in \mho \, : \, t < 0\},$$

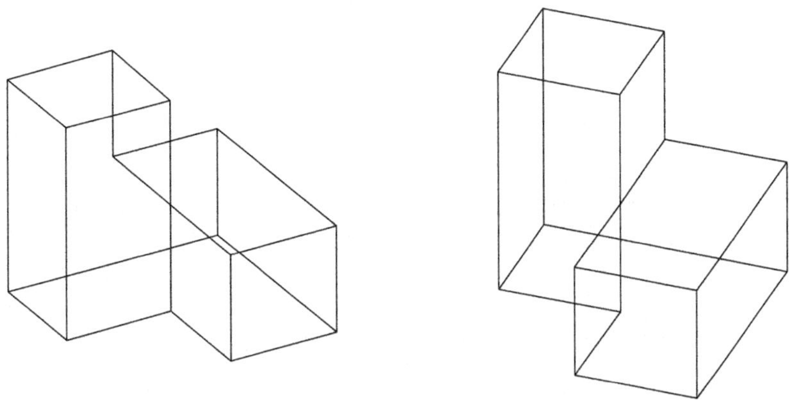

Fig. D.2 Two views of a popular non-Lipschitz polyhedron, made up of two crossing bricks. The crossing point is the only point where we cannot take a point of view allowing us to locally view the boundary as a graph. Note that 'topologically' the figure is quite simple, but this lack of visibility of the boundary is enough for the domain not to fit in the hypotheses. The domain is however an example of weakly Lipschitz domain.

as well as the disk $\partial_\circ\mho := B(\mathbf{0};1)\times\{0\}\equiv B(\mathbf{0};1)\subset\mathbb{R}^{d-1}$. We can then find a finite collection of open sets $\{\Omega_1,\ldots,\Omega_M\}$ covering Γ, and bi-Lipschitz (Lipschitz bijection with Lipschitz inverse) maps $\chi_j:\mho\to\Omega_j$ (for $j=1,\ldots,M$) such that $\chi_j(\mho_-)=\Omega_j\cap\Omega$ and $\chi_j(\partial_\circ\mho)=\Omega_j\cap\Gamma$. We can next construct a smooth partition of unity on a neighborhood of Γ associated with the covering of Γ by $\{\Omega_1,\ldots,\Omega_M\}$, i.e., we take functions $u_j\in\mathcal{C}^\infty(\mathbb{R}^d)$ such that

$$\operatorname{supp} u_j\subset\Omega_j\quad\forall j,\qquad \sum_{j=1}^M u_j\equiv 1\quad\text{on a neighborhood of }\Gamma.$$

The functions $\xi_j:=\chi_j|_{\partial_\circ\mho}:\partial_\circ\mho\equiv B(\mathbf{0};1)\to\Gamma\cap\Omega_j$ parametrize parts of Γ and can be used to properly define a concept of integral on Γ:

$$\int_\Gamma \phi(\mathbf{x})\mathrm{d}\Gamma(\mathbf{x})=\int_\Gamma\phi=\sum_{j=1}^M\int_{\Gamma\cap\Omega_j}\phi\,u_j=\int_{B(\mathbf{0};1)}(\phi\,u_j)\circ\chi_j\,|\det \mathrm{D}\xi_j|.$$

It can be proved that this definition is independent of the particular collection of charts and of the partition of unity (Figure D.3).

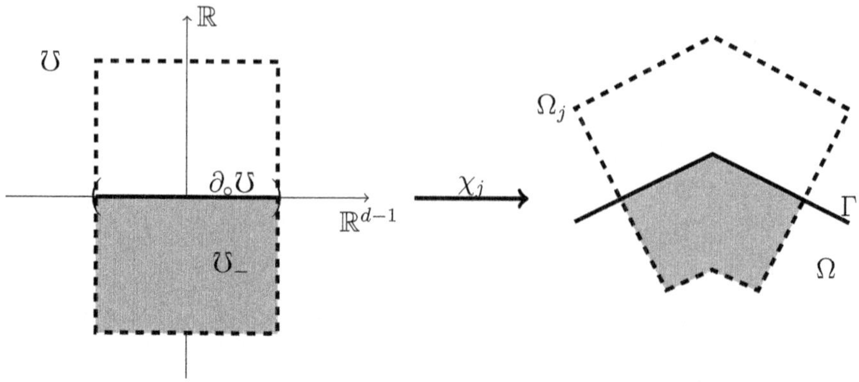

Fig. D.3 A pictorial representation of the reference cylinder in $\mathbb{R}^d$ being mapped to a portion of a cover of Γ.

D.2 Elementary results on Sobolev spaces

Let $\mathcal{O} \in \{\mathbb{R}^d, \mathbb{R}^d_+, \Omega^+, \Omega^-\}$, that is, let $\mathcal{O}$ be either the whole space, the half space, the interior Lipschitz domain, or its exterior. We consider the set

$$\mathcal{C}^\infty(\overline{\mathcal{O}}) := \{u|_{\mathcal{O}} \,:\, u \in \mathcal{C}^\infty(\mathbb{R}^d)\}.$$

Its completion with respect to the norm

$$\|u\|^2_{1,\mathcal{O}} := \|u\|^2_{\mathcal{O}} + \|\nabla u\|^2_{\mathcal{O}} = \int_{\mathcal{O}} (|u|^2 + |\nabla u|^2)$$

is the Sobolev space $H^1(\mathcal{O})$. The completion of $\mathcal{C}^\infty(\overline{\mathcal{O}})$ with respect to the norm

$$\|u\|^2_{\Delta} := \|u\|^2_{1,\mathcal{O}} + \|\Delta u\|^2_{\mathcal{O}}$$

is the space $H^1_\Delta(\mathcal{O})$. By definition, $H^1(\mathcal{O})$ and $H^1_\Delta(\mathcal{O})$ are Hilbert spaces. These two spaces can also be defined (but not by completion) for $\mathcal{O} = \Omega^- \cup \Omega^+ = \mathbb{R}^d \setminus \Gamma$ by attaching copies of the interior and exterior spaces. For example,

$$H^1(\mathbb{R}^d \setminus \Gamma) := \{u : \mathbb{R}^d \setminus \Gamma \,:\, u|_{\Omega^\pm} \in H^1(\Omega^\pm)\} \equiv H^1(\Omega^-) \times H^1(\Omega^+).$$

A Bi-Lipschitz map between the open sets Ω_1 and Ω_2 generates a natural isomorphism of $H^1(\Omega_1)$ and $H^1(\Omega_2)$.

Traces. Consider now the space $\mathcal{C}^\infty(\Gamma) := \{u|_\Gamma \,:\, u \in \mathcal{C}^\infty(\Gamma)\}$. Its closure with respect to the Sobolev-Slobodetskij norm

$$\|\phi\|_{1/2,\Gamma}^2 := \int_\Gamma |\phi(\mathbf{x})|^2 \mathrm{d}\Gamma(\mathbf{x}) + \int_\Gamma \int_\Gamma \frac{|\phi(\mathbf{x}) - \phi(\mathbf{y})|^2}{|\mathbf{x}-\mathbf{y}|^d} \mathrm{d}\Gamma(\mathbf{x})\mathrm{d}\Gamma(\mathbf{y}) \tag{D.1}$$

is denoted $H^{1/2}(\Gamma)$. This is again a Hilbert space, which is continuously and densely embedded into $L^2(\Gamma)$, the space of square integrable functions on Γ, endowed with its natural norm $\|\cdot\|_\Gamma$. There exist two bounded and surjective linear operators

$$\gamma^\pm : H^1(\Omega^\pm) \to H^{1/2}(\Gamma)$$

such that if $u \in \mathcal{C}^\infty(\overline{\Omega^\pm})$, then $\gamma^\pm u = u|_\Gamma$. The $H^{1/2}(\Gamma)$ norm is equivalent to the image norms associated with the interior and exterior trace operators

$$\inf\{\|u\|_{1,\Omega^\pm} \,:\, \gamma^\pm u = \phi\}.$$

Finally, if $u \in H^1(\mathbb{R}^d \setminus \Gamma)$, both traces $\gamma^\pm u$ are well defined and $\gamma^+ u = \gamma^- u$ if and only if $u \in H^1(\mathbb{R}^d)$.

Duality on Γ. Consider the inner product in $L^2(\Gamma)$:

$$\langle \mu, \phi \rangle_\Gamma := \int_\Gamma \phi(\mathbf{x})\mu(\mathbf{x})\mathrm{d}\Gamma(\mathbf{x}).$$

Since $H^{1/2}(\Gamma)$ is continuous and densely embedded in $L^2(\Gamma)$, the following norm

$$\|\mu\|_{-1/2,\Gamma} := \sup_{0 \neq \phi \in H^{1/2}(\Gamma)} \frac{|\langle \mu, \phi \rangle_\Gamma|}{\|\phi\|_{1/2,\Gamma}} \leq \|\mu\|_\Gamma$$

is well defined in $L^2(\Gamma)$. The space $H^{-1/2}(\Gamma)$ is the completion of $L^2(\Gamma)$ with respect to this norm and is isometrically isomorphic to the dual space of $H^{1/2}(\Gamma)$. It is common to refer to $H^{-1/2}(\Gamma)$ as the dual space of $H^{1/2}(\Gamma)$, although, properly speaking, $H^{-1/2}(\Gamma)$ is the representation of the dual of $H^{1/2}(\Gamma)$ obtained when $L^2(\Gamma)$ and its dual space are identified via the Riesz-Fréchet representation theorem. In this context, it can also be said that the $L^2(\Gamma)$ inner product can be extended to the duality product $H^{-1/2}(\Gamma) \times H^{1/2}(\Gamma)$.

Fourier representation of some Sobolev spaces. The Fourier transform of a smooth compactly supported function $u : \mathbb{R}^d \to \mathbb{C}$

$$\hat{u}(\omega) := \int_{\mathbb{R}^d} e^{-\imath \omega \cdot \mathbf{x}} u(\mathbf{x}) \mathrm{d}\mathbf{x}$$

satisfies the Plancherel identity

$$\|u\|_{\mathbb{R}^d}^2 = (2\pi)^d \|\hat{u}\|_{\mathbb{R}^d}^2,$$

which allows for an extension of the transform to a bounded linear isomorphism $L^2(\mathbb{R}^d) \to L^2(\mathbb{R}^d)$. The subspace of functions $u \in L^2(\mathbb{R}^d)$ such that

$$(2\pi)^d \int_{\mathbb{R}^d} (1 + |\omega|^2)|\hat{u}(\omega)|^2 \mathrm{d}\omega < \infty \tag{D.2}$$

equals the Sobolev space $H^1(\mathbb{R}^d)$ and the left-hand side of (D.2) equals the square of the $H^1(\mathbb{R}^d)$ norm. The subspace of functions $u \in L^2(\mathbb{R}^d)$ such that

$$(2\pi)^d \int_{\mathbb{R}^d} (1 + |\omega|^2)^{1/2}|\hat{u}(\omega)|^2 \mathrm{d}\omega < \infty \tag{D.3}$$

is called $H^{1/2}(\mathbb{R}^d)$ and coincides with the completion of the set of smooth compactly supported functions with respect to the norm (compare with (D.1))

$$\|u\|_{1/2,\mathbb{R}^d}^2 := \|u\|_{\mathbb{R}^d}^2 + \int_{\mathbb{R}^d}\int_{\mathbb{R}^d} \frac{|u(\mathbf{x}) - u(\mathbf{y})|^2}{|\mathbf{x}-\mathbf{y}|^{d+1}} \mathbf{dx}\, \mathbf{dy}.$$

Using the local transformations and the partition of unity introduced in Section D.1, localized pieces of $H^{1/2}(\Gamma)$ can be pulled back to the space $H^{1/2}(\mathbb{R}^{d-1})$.

Appendix E
Some numerical illustrations

In this appendix we include some numerical experiments with a piecewise polynomial Galerkin-CQ method for some simple two-dimensional scattering problems. Extensive experimentation with CQ-BEM schemes can be found in the literature and is not the goal of this little appendix.

E.1 Experiments

We start with a convergence study. The scatterer is the inverted L-shaped domain (Figure E.1)

$$\Omega^- := (-1,1)^2 \setminus ([0,1) \times [-1,0)),$$

and the incident wave is

$$u^{\mathrm{inc}}(\mathbf{x},t) := \psi(t - t_{\mathrm{lag}} - \mathbf{x}\cdot\mathbf{d}), \tag{E.1}$$

where $\mathbf{d} = \frac{1}{\sqrt{2}}(-1,1)$ (the wave is hitting the concave corner diagonally), $t_{\mathrm{lag}} = 1$ (thus preserving causality of the traces on Γ), and

$$\psi(t) := \sin(4\,t)\,H(t),$$

where $H \in \mathcal{C}^5(\mathbb{R})$ is a piecewise polynomial causal approximation of the Heaviside function. We solve the exterior Dirichlet problem (sound-soft scattering) with a single layer potential representation

$$\mathcal{V} * \lambda = -\gamma u^{\mathrm{inc}}, \qquad u = \mathcal{S} * \lambda,$$

F.-J. Sayas, *Retarded Potentials and Time Domain Boundary Integral Equations*, Springer Series in Computational Mathematics 50,
DOI 10.1007/978-3-319-26645-9

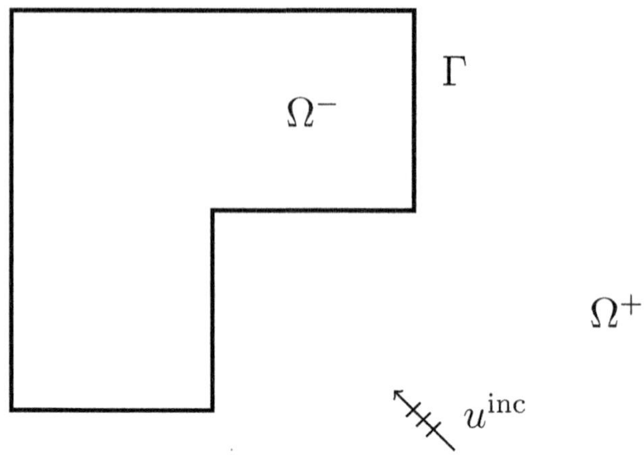

Fig. E.1 The domain used in the numerical illustrations.

and the exterior Neumann problem (sound-hard scattering) with a double layer potential representation

$$\mathcal{W} * \varphi = \partial_\nu u^{\text{inc}}, \qquad u = \mathcal{D} * \varphi.$$

Note that, by uniqueness of solution of the interior problem, in both cases $u = -u^{\text{inc}}$ in Ω^-. We will use this as a convergence test.

We first divide Γ using a piecewise uniform partition, where each edge of the polygon is divided into an equal number of equally sized intervals. Following traditional notation in the Finite Element community, we denote this partition $\mathcal{T}_h$. The total number of intervals will be denoted N_{elt}. We use the boundary element spaces

$$X_h := \{\lambda^h : \Gamma \to \mathbb{R} : \lambda^h|_T \in \mathcal{P}_0(T) \quad \forall T \in \mathcal{T}_h\} \subset H^{-1/2}(\Gamma),$$
$$Y_h := \{\varphi^h : \Gamma \to \mathbb{R} : \varphi^h \in \mathcal{C}(\Gamma), \varphi^h|_T \in \mathcal{P}_1(T) \quad \forall T \in \mathcal{T}_h\} \subset H^{1/2}(\Gamma).$$

Both spaces have dimension N_{elt}. We then use BDF2- and TR-based CQ discretizations in time to a final time $T = 5$ with M time steps. Finally, we compute the error

$$\max_{\mathbf{z} \in \text{Obs}} |u^h(\mathbf{z}, T) + u^{\text{inc}}(\mathbf{z}, T)|,$$

where

$$\text{Obs} := \{(-1/2, -1/2), (-1/2, 1/2), (1/2, 1/2)\} \subset \Omega^-.$$

Note that the regularity in space of the density λ is unknown, which makes the effect of space discretization difficult to estimate. Note also that the density is expected to have some kind of singularity in the corners of the domain, but we have not

Table E.1 Lowest order BEM with BDF2-CQ

N_{elt}	M	errorDirichlet	e.c.r.	errorNeumann	e.c.r.
12	40	1.4351e+00	-	7.2904e-02	-
24	80	4.1663e-01	1.7843	1.0310e-01	−0.5000
48	160	6.8267e-02	2.6095	1.6999e-02	2.6006
96	320	2.3990e-02	1.5087	1.1341e-02	0.5838
192	640	6.5110e-03	1.8815	3.7259e-03	1.6059

Table E.2 Lowest order BEM with TR-CQ

N_{elt}	M	errorDirichlet	e.c.r.	errorNeumann	e.c.r.
12	40	9.8232e-01	-	7.1641e-02	-
24	80	1.0604e-01	3.2116	5.8668e-02	2.8823
48	160	1.9271e-02	2.4600	9.3293e-03	2.6527
96	320	6.6482e-03	1.5354	1.6272e-03	2.5194
192	640	1.5678e-03	2.0842	6.5096e-04	1.3218

Table E.3 Second order BEM with BDF2-CQ

N_{elt}	M	errorDirichlet	e.c.r.	errorNeumann	e.c.r.
12	40	1.3587e+00	-	3.1516e-01	-
24	80	2.8198e-01	2.2685	8.2658e-02	1.9308
48	160	6.5535e-02	2.1053	2.1575e-02	1.9378
96	320	2.4346e-02	1.4286	1.3162e-02	0.7130
192	640	6.4955e-03	1.9062	4.0471e-03	1.7014

Table E.4 Second order BEM with TR-CQ

N_{elt}	M	errorDirichlet	e.c.r.	errorNeumann	e.c.r.
12	40	3.7334e-01	-	1.5065e-01	-
24	80	5.9190e-02	2.6570	2.4991e-02	2.5917
48	160	2.5243e-02	1.2295	1.4377e-02	0.7976
96	320	6.3915e-03	1.9817	4.0067e-03	1.8433
192	640	1.5443e-03	2.0492	9.6409e-04	2.0552

refined the grid there. (For a practical simulation, this would be a safe practice.) The results are reported in Tables E.1 and E.2, with estimated convergence rates (e.c.r.) computed with consecutive refinements.

We then change the discrete spaces to the $2N_{elt}$-dimensional spaces

$$\begin{aligned} X_h &:= \{\lambda^h : \Gamma \to \mathbb{R} \,:\, \lambda^h|_T \in \mathcal{P}_1(T) \quad \forall T \in \mathcal{T}_h\} \subset H^{-1/2}(\Gamma), \\ Y_h &:= \{\varphi^h : \Gamma \to \mathbb{R} \,:\, \varphi^h \in \mathcal{C}(\Gamma),\, \varphi^h|_T \in \mathcal{P}_2(T) \quad \forall T \in \mathcal{T}_h\} \subset H^{1/2}(\Gamma), \end{aligned}$$

and repeat the experiments (Tables E.3 and E.4).

Fig. E.2 Six views of the sound-soft scattering of an obstacle by a causal sinusoidal wave. The last image shows the situation once the (local) time-harmonic regime has been reached.

We next display some snapshots of an experiment showing the interaction of the plane wave (E.1) with $\mathbf{d} = (-1, 1)/\sqrt{2}$, $t_{\text{lag}} = 1$, and $\psi(t) = \sin(32t)$ with a sound-soft and a sound-hard obstacle. This is done in Figures E.2 and E.3.

Fig. E.3 Six views of the sound-hard scattering of an obstacle by a causal sinusoidal wave, showing the transition to time-harmonic regime.

E.2 A computational toy

For the reader interested in performing easy tests with time domain integral equations, we explain here the simplest uses of the `deltaBEM` MATLAB suite to obtain almost elementary discretizations of the boundary integral operators for two-dimensional smooth scatterers. We just present the discretization of the single layer potential formulation for the sound-soft scattering problem by a smooth scatterer in the plane. The discretization of all integral operator for acoustics on a finite collection of smooth scatterers can be found in [41], while the extension for the more involved operators of elastodynamic appears in [42]. The fully documented code to generate the fully discrete integral operators and potentials in the Laplace domain, ready to be used with CQ tools, can be found in

`http://www.math.udel.edu/~fjsayas/deltaBEM/`

The scatterer is represented by a smooth 1-periodic parametrization $\mathbf{x} : \mathbb{R} \to \Gamma$ satisfying

$$\mathbf{x}(1+t) = \mathbf{x}(t) \qquad |\mathbf{x}'(t)| > 0 \qquad \forall t \in \mathbb{R},$$

as well as $\mathbf{x}(\tau) \neq (t)$ if $t - \tau \notin \mathbb{Z}$ (Γ is a simple curve). We take a positive integer N, and sample the curve $3N$ times

$$\mathbf{m}_j := \mathbf{x}(j\,h), \qquad \mathbf{m}_j^{\pm} := \mathbf{x}((j \pm \tfrac{1}{6})h) \qquad j = 1, \ldots, N,$$

where $h := 1/N$. The Laplace domain single layer operator (needed for the Convolution Quadrature method) is discretized by

$$(\mathrm{S}_h(s)\boldsymbol{\lambda})(\mathbf{z}) = \frac{\imath}{4} \sum_{j=1}^{N} H_0^{(1)}(\imath s|\mathbf{z} - \mathbf{m}_j|)\, \lambda_j.$$

The density has been discretized by a vector $\boldsymbol{\lambda} \in \mathbb{C}^N$, which, in the time domain will become a causal function $\mathbb{R} \to \mathbb{R}^N$. The operator $\mathrm{V}(s)$ is discretized by a matrix-valued operator with entries

$$\mathrm{V}_h(s)_{ij} = \frac{1}{2}\left(\frac{\imath}{4} H_0^{(1)}(\imath s|\mathbf{m}_i^+ - \mathbf{m}_j|) + \frac{\imath}{4} H_0^{(1)}(\imath s|\mathbf{m}_i^- - \mathbf{m}_j|)\right), \qquad i,j = 1, \ldots, N.$$

This discretization can be understood as a Galerkin method with piecewise constant functions on a uniform mesh in parametric space, with reduced order quadrature. The incident wave is sampled in the points $\mathbf{m}_i^{\pm}$ to generate the right-hand side of the semidiscrete equation

$$-\frac{1}{2}\left(\partial_\nu u^{\mathrm{inc}}(\mathbf{m}_i^+, t) + \partial_\nu u^{\mathrm{inc}}(\mathbf{m}_i^-, t)\right), \qquad i = 1, \ldots, N.$$

The magical choice of the $\pm 1/6$ shift for the quadrature points is due to a formula on optimal sampling of periodic logarithmic integral operators discovered by Jukka Saranen and Liisa Schroderus [78].

Bibliography

1. T. Abboud, P. Joly, J. Rodríguez, I. Terrasse, Coupling discontinuous Galerkin methods and retarded potentials for transient wave propagation on unbounded domains. J. Comput. Phys. **230**(15), 5877–5907 (2011)
2. A. Aimi, M. Diligenti, A. Frangi, C. Guardasoni, A stable 3D energetic Galerkin BEM approach for wave propagation interior problems. Eng. Anal. Bound. Elem. **36**(12), 1756–1765 (2012)
3. A. Aimi, M. Diligenti, A. Frangi, C. Guardasoni, Neumann exterior wave propagation problems: computational aspects of 3D energetic Galerkin BEM. Comput. Mech. **51**(4), 475–493 (2013)
4. A. Aimi, M. Diligenti, A. Frangi, C. Guardasoni, Energetic BEM-FEM coupling for wave propagation in 3D multidomains. Int. J. Numer. Methods Eng. **97**(5), 377–394 (2014)
5. A. Aimi, M. Diligenti, C. Guardasoni, On the energetic Galerkin boundary element method applied to interior wave propagation problems. J. Comput. Appl. Math. **235**(7), 1746–1754 (2011)
6. F.P. Andriulli, K. Cools, F. Olyslager, E. Michielssen, Time domain Calderón identities and their application to the integral equation analysis of scattering by PEC objects. II. Stability. IEEE Trans. Antennas Propag. **57**(8), 2365–2375 (2009)
7. A. Bachelot, L. Bounhoure, A. Pujols, Couplage éléments finis–potentiels retardés pour la diffraction électromagnétique par un obstacle hétérogène. Numer. Math. **89**(2), 257–306 (2001)
8. J. Ballani, L. Banjai, S. Sauter, A. Veit, Numerical solution of exterior Maxwell problems by Galerkin BEM and Runge-Kutta convolution quadrature. Numer. Math. **123**(4), 643–670 (2013)
9. A. Bamberger, T. Ha Duong, Formulation variationnelle espace-temps pour le calcul par potentiel retardé de la diffraction d'une onde acoustique. I. Math. Methods Appl. Sci. **8**(3), 405–435 (1986)
10. A. Bamberger, T. Ha Duong, Formulation variationnelle pour le calcul de la diffraction d'une onde acoustique par une surface rigide. Math. Methods Appl. Sci. **8**(4), 598–608 (1986)
11. L. Banjai, M. Kachanovska, Sparsity of Runge-Kutta convolution weights for the three-dimensional wave equation. BIT **54**, 901–936 (2014)
12. L. Banjai, M. Kachanovska, Fast convolution quadrature for the wave equation in three dimensions. J. Comput. Phys. **279**, 103–126 (2014)
13. L. Banjai, A.R. Laliena, F.-J. Sayas, Fully discrete Kirchhoff formulas with CQ-BEM. IMA J. Numer. Anal. **35**(2), 859–884 (2015)

F.-J. Sayas, *Retarded Potentials and Time Domain Boundary Integral Equations*, Springer Series in Computational Mathematics 50,
DOI 10.1007/978-3-319-26645-9

14. L. Banjai, S. Sauter, Rapid solution of the wave equation in unbounded domains. SIAM J. Numer. Anal. **47**(1), 227–249 (2008/2009)
15. L. Banjai, Multistep and multistage convolution quadrature for the wave equation: algorithms and experiments. SIAM J. Sci. Comput. **32**(5), 2964–2994 (2010)
16. L. Banjai, Time-domain Dirichlet-to-Neumann map and its discretization. IMA J. Numer. Anal. **34**(3), 1136–1155 (2014)
17. L. Banjai, V. Gruhne, Efficient long-time computations of time-domain boundary integrals for 2D and dissipative wave equation. J. Comput. Appl. Math. **235**(14), 4207–4220 (2011)
18. L. Banjai, C. Lubich, An error analysis of Runge-Kutta convolution quadrature. BIT **51**(3), 83–496 (2011)
19. L. Banjai, C. Lubich, J.M. Melenk, Runge-Kutta convolution quadrature for operators arising in wave propagation. Numer. Math. **119**(1), 1–20 (2011)
20. L. Banjai, C. Lubich, F.-J. Sayas, Stable numerical coupling of exterior and interior problems for the wave equation. Numer. Math. **129**(4), 611–646 (2015)
21. L. Banjai, M. Messner, M. Schanz, Runge-Kutta convolution quadrature for the boundary element method. Comput. Methods Appl. Mech. Eng. **245/246**, 90–101 (2012)
22. L. Banjai, M. Schanz, Wave propagation problems treated with convolution quadrature and BEM, in *Fast Boundary Element Methods in Engineering and Industrial Applications*, ed. by U. Langer, M. Schanz, O. Steinbach, W.L. Wendland. Lecture Notes in Applied and Computational Mechanics (2012), pp. 145–187
23. E. Becache, T. Ha-Duong, A space-time variational formulation for the boundary integral equation in a 2D elastic crack problem. RAIRO Modél. Math. Anal. Numér. **28**(2), 141–176 (1994)
24. Y. Beghein, K. Cools, H. Bagcı, D. De Zutter, A space-time mixed Galerkin marching-on-in-time scheme for the time-domain combined field integral equation. IEEE Trans. Antennas Propag. **61**(3), 1228–1238 (2013)
25. A. Buffa, P. Ciarlet, Jr., On traces for functional spaces related to Maxwell's equations. I. An integration by parts formula in Lipschitz polyhedra. Math. Methods Appl. Sci. **24**(1), 9–30 (2001)
26. A. Buffa, P. Ciarlet, Jr., On traces for functional spaces related to Maxwell's equations. II. Hodge decompositions on the boundary of Lipschitz polyhedra and applications. Math. Methods Appl. Sci. **24**(1), 31–48 (2001)
27. A. Buffa, M. Costabel, D. Sheen. On traces for $\mathbf{H}(\mathbf{curl}, \Omega)$ in Lipschitz domains. J. Math. Anal. Appl. **276**(2), 845–867 (2002)
28. J.F.-C. Chan, P. Monk, Time dependent electromagnetic scattering by a penetrable obstacle. BIT **55**(1), 5–31 (2015)
29. D.J. Chappell, A convolution quadrature Galerkin boundary element method for the exterior Neumann problem of the wave equation. Math. Methods Appl. Sci. **32**(12), 1585–1608 (2009)
30. Q. Chen, P. Monk, X. Wang, D. Weile, Analysis of convolution quadrature applied to the time-domain electric field integral equation. Commun. Comput. Phys. **11**(2), 383–399 (2012)
31. Q. Chen, P. Monk, Discretization of the time domain CFIE for acoustic scattering problems using convolution quadrature. SIAM J. Math. Anal. **46**(5), 3107–3130 (2014)
32. K. Cools, F.P. Andriulli, F. Olyslager, E.Michielssen, Time domain Calderón identities and their application to the integral equation analysis of scattering by PEC objects. I. Preconditioning. IEEE Trans. Antennas Propag. **57**(8), 2352–2364 (2009)
33. M. Costabel, Time-dependent problems with the boundary integral equation method, in *Encyclopedia of Computational Mechanics*, Chapter 22, ed. by E. Stein, R. de Borst, T. Hughes (Wiley, Chichester, 2003)
34. M. Costabel, Boundary integral operators on Lipschitz domains: elementary results. SIAM J. Math. Anal. **19**(3), 613–626 (1988)
35. R. Dautray, J.-L. Lions, *Mathematical Analysis and Numerical Methods for Science and Technology*, vol. 5 (Springer, Berlin, 1992). Evolution problems. I, With the collaboration of Michel Artola, Michel Cessenat and Hélène Lanchon

36. P.J. Davies, Numerical stability and convergence of approximations of retarded potential integral equations. SIAM J. Numer. Anal. **31**(3), 856–875 (1994)
37. P.J. Davies, On the stability of time-marching schemes for the general surface electric-field integral equation. IEEE Trans. Antennas Propag. **44**(11), 1467–1473 (1996)
38. P.J. Davies, D.B. Duncan, Stability and convergence of collocation schemes for retarded potential integral equations. SIAM J. Numer. Anal. **42**(3), 1167–1188 (electronic, 2004)
39. P.J. Davies, D.B. Duncan, Convolution-in-time approximations of time domain boundary integral equations. SIAM J. Sci. Comput. **35**(1), B43–B61 (2013)
40. P.J. Davies, D.B. Duncan, Convolution spline approximations for time domain boundary integral equations. J. Integr. Equ. Appl. **26**(3), 369–410 (2014)
41. V. Domínguez, S.L. Lu, F.-J. Sayas, A Nyström flavored Calderón calculus of order three for two dimensional waves, time-harmonic and transient. Comput. Math. Appl. **67**(1), 217–236 (2014)
42. V. Domínguez, T. Sánchez-Vizuet, F.-J. Sayas, A fully discrete Calderón calculus for the two-dimensional elastic wave equation. Comput. Math. Appl. **69**(7), 620–635 (2015)
43. V. Domínguez, F.-J. Sayas, Some properties of layer potentials and boundary integral operators for the wave equation. J. Integr. Equ. Appl. **25**(2), 253–294 (2013)
44. K.-J. Engel, R. Nagel, *A Short Course on Operator Semigroups*. Universitext (Springer, New York, 2006)
45. C.L. Epstein, L. Greengard, T. Hagstrom, On the stability of time-domain integral equations for acoustic wave propagation. arXiv:10504.04047v1
46. S. Falletta, G. Monegato, An exact non reflecting boundary condition for 2D time-dependent wave equation problems. Wave Motion **51**(1), 168–192 (2014)
47. S. Falletta, G. Monegato, L. Scuderi, A space-time BIE method for nonhomogeneous exterior wave equation problems. The Dirichlet case. IMA J. Numer. Anal. **32**(1), 202–226 (2012)
48. A. Furukawa, T. Saitoh, S. Hirose, Convolution quadrature time-domain boundary element method for 2-D and 3-D elastodynamic analyses in general anisotropic elastic solids. Eng. Anal. Bound. Elem. **39**, 64–74 (2014)
49. T. Ha-Duong, B. Ludwig, I. Terrasse, A Galerkin BEM for transient acoustic scattering by an absorbing obstacle. Int. J. Numer. Methods Eng. **57**(13), 1845–1882 (2003)
50. T. Ha-Duong, On retarded potential boundary integral equations and their discretisation, in *Topics in Computational Wave Propagation*. Lect. Notes Comput. Sci. Eng., vol. 31 (Springer, Berlin, 2003), pp. 301–336
51. W. Hackbusch, W. Kress, S.A. Sauter, Sparse convolution quadrature for time domain boundary integral formulations of the wave equation. IMA J. Numer. Anal. **29**(1), 158–179 (2009)
52. W. Hackbusch, W. Kress, S.A. Sauter, Sparse convolution quadrature for time domain boundary integral formulations of the wave equation by cutoff and panel-clustering, in *Boundary Element Analysis*. Lect. Notes Appl. Comput. Mech., vol. 29 (Springer, Berlin, 2007), pp. 113–134
53. E. Hairer, G. Wanner, Stiff and differential-algebraic problems, in *Solving Ordinary Differential Equations. II*. Springer Series in Computational Mathematics, vol. 14, 2nd edn (Springer, Berlin, 1996)
54. M. Hassell, F.-J. Sayas, Convolution quadrature for wave simulations, in *Classnotes for the XVI Spanish-French School on Numerical Simulation in Physics and Engineering*. SEMA SIMAI Springer Series (to appear). arXiv:1407.0345
55. G.C. Hsiao, W.L. Wendland, *Boundary Integral Equations*. Applied Mathematical Sciences, vol. 164 (Springer, Berlin, 2008)
56. G.C. Hsiao, F.-J. Sayas, R.J. Weinacht, Time-dependent fluid-structure interaction. Math. Methods Appl. Sci. (to appear)
57. B. Kager, M. Schanz, Fast and data sparse time domain BEM for elastodynamics. Eng. Anal. Bound. Elem. **50**, 212–223 (2015)
58. S. Kesavan, *Topics in Functional Analysis and Applications* (Wiley, New York, 1989)

59. A.R. Laliena, F.-J. Sayas, A distributional version of Kirchhoff's formula. J. Math. Anal. Appl. **359**(1), 197–208 (2009)
60. A.R. Laliena, F.-J. Sayas, Theoretical aspects of the application of convolution quadrature to scattering of acoustic waves. Numer. Math. **112**(4), 637–678 (2009)
61. A.R. Laliena, Theoretical and algorithmic aspects of the convolution quadrature method applied to scattering of acoustic waves (Spanish). PhD thesis, Universidad de Zaragoza (2011)
62. A. Lechleiter, P. Monk, The time-domain Lippmann-Schwinger equation and convolution quadrature. Numer. Methods Partial Differ. Equ. **31**(2), 517–540 (2015)
63. P. Li, M. Schanz, Time domain boundary element formulation for partially saturated poroelasticity. Eng. Anal. Bound. Elem. **37**(11), 1483–1498 (2013)
64. C. Lubich, Convolution quadrature and discretized operational calculus. I. Numer. Math. **52**(2), 129–145 (1988)
65. C. Lubich, Convolution quadrature and discretized operational calculus. II. Numer. Math. **52**(4), 413–425 (1988)
66. C. Lubich. On the multistep time discretization of linear initial-boundary value problems and their boundary integral equations. Numer. Math. **67**(3), 365–389 (1994)
67. C. Lubich, A. Ostermann, Runge-Kutta methods for parabolic equations and convolution quadrature. Math. Comp. **60**(201), 105–131 (1993)
68. W. McLean, *Strongly Elliptic Systems and Boundary Integral Equations* (Cambridge University Press, Cambridge, 2000)
69. M. Messner, M. Schanz, A symmetric Galerkin boundary element method for 3d linear poroelasticity. Acta Mech. **223**(8), 1751–1768 (2012)
70. J.-C. Nédélec, J. Planchard, Une méthode variationnelle d'éléments finis pour la résolution numérique d'un problème extérieur dans R^3. Rev. Française Automat. Informat. Recherche Opérationnelle Sér. Rouge **7**(R-3), 105–129 (1973)
71. A. Pazy, *Semigroups of Linear Operators and Applications to Partial Differential Equations*. Applied Mathematical Sciences, vol. 44 (Springer, New York, 1983)
72. A.J. Pray, Y. Beghein, N.V. Nair, K. Cools, H. Bağci, B Shanker, *A Stable Higher Order Space-Time Galerkin Scheme for Time Domain Integral Equations* (2014). http://arxiv.org/abs/1401.2435
73. T. Qiu, F.-J. Sayas, The Costabel-Stephan system of boundary integral equations in the time domain. Math. Comp. (to appear)
74. A. Romero, P. Galvín, J. Domínguez, 3D non-linear time domain FEM-BEM approach to soil-structure interaction problems. Eng. Anal. Bound. Elem. **37**(3), 501–512 (2013)
75. T. Rüberg, M. Schanz, Coupling finite and boundary element methods for static and dynamic elastic problems with non-conforming interfaces. Comput. Methods Appl. Mech. Eng. **198** (3–4), 449–458 (2008)
76. T. Rüeberg, Non-conforming coupling of finite and boundary element methods in time domain. PhD thesis, Graz University of Technology (2007)
77. B.P. Rynne, The well-posedness of the electric field integral equation for transient scattering from a perfectly conducting body. Math. Methods Appl. Sci. **22**(7), 619–631 (1999)
78. J. Saranen, L. Schroderus, Quadrature methods for strongly elliptic equations of negative order on smooth closed curves. SIAM J. Numer. Anal. **30**(6), 1769–1795 (1993)
79. S. Sauter, A. Veit, A Galerkin method for retarded boundary integral equations with smooth and compactly supported temporal basis functions. Numer. Math. **123**(1), 145–176 (2013)
80. S.A. Sauter, C. Schwab, *Boundary Element Methods*. Springer Series in Computational Mathematics, vol. 39 (Springer, Berlin, 2011)
81. F.-J. Sayas, Energy estimates for Galerkin semidiscretizations of time domain boundary integral equations. Numer. Math. **124**(1), 121–149 (2013)
82. M. Schanz, *Wave Propagation in Viscoelastic and Poroelastic Continua: A Boundary Element Approach*. Lecture Notes in Applied and Computational Mechanics (Springer, Berlin, 2001)
83. L. Schwartz, *Méthodes mathématiques pour les sciences physiques*. Enseignement des Sciences (Hermann, Paris, 1961)

84. I. Terrasse, *Résolution mathématique et numérique des équations de Maxwell instationnaires par une méthode de potentiels retardés*. PhD thesis, École polytechnique (1993)
85. F. Trèves, *Topological Vector Spaces, Distributions and Kernels* (Academic, New York, 1967)
86. H.A. Ülkü, H. Bağcı, E. Michielssen, Marching on-in-time solution of the time domain magnetic field integral equation using a predictor-corrector scheme. IEEE Trans. Antennas Propag. **61**(8), 4120–4131 (2013)
87. F. Valdés, F.P. Andriulli, H. Bağci, E. Michielssen, Time-domain single-source integral equations for analyzing scattering from homogeneous penetrable objects. IEEE Trans. Antennas Propag. **61**(3), 1239–1254 (2013)
88. A. Veit, Numerical methods for time-domain boundary integral equations. PhD thesis, Universität Zürich (2012)
89. X. Wang, R.A. Wildman, D.S. Weile, P. Monk, A finite difference delay modeling approach to the discretization of the time domain integral equations of electromagnetics. IEEE Trans. Antennas Propag. **56**(8 Pt 1), 2442–2452 (2008)

Index

F.-J. Sayas, *Retarded Potentials and Time Domain Boundary Integral Equations*, Springer Series in Computational Mathematics 50,
DOI 10.1007/978-3-319-26645-9

CPSIA information can be obtained
at www.ICGtesting.com
Printed in the USA
BVOW07*0218260416
445625BV00001B/7/P

9 783319 266435